AF895859

Plant Domestication in the Middle Nile Basin

An Archaeoethnobotanical Case Study

Anwar Abdel-Magid

Cambridge Monographs in African Archaeology 35

BAR International Series 523

1989

B.A.R.

5, Centremead, Osney Mead, Oxford OX2 0DQ, England.

GENERAL EDITORS

A.R. Hands, B.Sc., M.A., D.Phil.
D.R. Walker, M.A.

BAR -S523, 1989: 'Plant Domestication in the Middle Nile Basin'

© Anwar A-Magid Osman, 1989

The author's moral rights under the 1988 UK Copyright, Designs and Patents Act are hereby expressly asserted.

All rights reserved. No part of this work may be copied, reproduced, stored, sold, distributed, scanned, saved in any form of digital format or transmitted in any form digitally, without the written permission of the Publisher.

ISBN 9780860546641 paperback
ISBN 9781407348162 e-book
DOI https://doi.org/10.30861/9780860546641
A catalogue record for this book is available from the British Library
This book is available at www.barpublishing.com

To my parents

Contents

Figures		ix
Tables		xi
Histograms and diagrams		xiii
Plates		xiii
Appendices		xiv
Preface		xv
CHAPTER I: INTRODUCTION		1
1.1	Objectives and Methodology	1
1.2	Hypothesis Formulation and Research Strategies	5
1.2.1	Formulation of hypothesis	5
1.2.2	Research strategy	6
1.2.2.1	Field work	6
1.2.2.1.1	Archaeological excavations	6
1.2.2.1.1.1	Selection of sites	6
1.2.2.1.1.2	Excavation of the sites	7
1.2.2.2	Ethnographic and ethnobotanic studies	14
1.2.2.3	Experimental studies	14
1.3	Perspectives on the Presentation of the Study	14
CHAPTER II: THE HISTORY AND THE STATE OF ARCHAEOETHNO-BOTANICAL RESEARCH IN THE CENTRAL SUDAN		17
CHAPTER III: THE GEOGRAPHICAL SETTING OF THE SITES		25
3.1	Introduction	25
3.2	Description of Aneibis/Abu-Darbein Area	27
3.2.1	Geographic location	27
3.2.1.1	The site of Aneibis	27
3.2.1.2	The site of Abu-Darbein	27
3.2.2	Geology	27
3.2.3	Topography	29
3.2.3.1	The site of Aneibis	29
3.2.3.2	The site of Abu-Darbein (Atbara)	30
3.2.3.3	Comparison between the topographic features of Aneibis and Abu-Darbein	31

3.2.4	The present climate	34
3.2.5	The present flora and fauna	37
3.2.5.1	Flora	37
3.2.5.1.1	Floristic composition	37
3.2.5.1.2	Discussion	37
3.2.5.2	Fauna	40
3.3	Description of the Shaqadud Area	41
3.3.1	Geographical location: The cave-site of the Shaqadud	41
3.3.2	Geology	43
3.3.3	Topography	43
3.3.4	Climate	44
3.3.5	The present flora and fauna	44
3.3.5.1	Flora	44
3.3.5.1.1	Floristic composition	44
3.3.5.1.2	Discussion	48
3.3.5.2	Fauna	48
3.4	Description of the Jebel Tomat/Rabak Area	49
3.4.1	Geographical location	49
3.4.1.1	The site of Jebel Tomat	49
3.4.1.2	The site of Rabak	49
3.4.2	Geology	51
3.4.3	Topography	51
3.4.4	Climate	55
3.4.5	The present flora and fauna	58
3.4.5.1	Flora	58
3.4.5.2	Fauna	62

CHAPTER IV: PLANT REMAINS: NATURE OF EVIDENCE, METHODS AND TECHNIQUES OF ITS RECOVERY 64

4.1	Background	64
4.2	Macrofossil Plant Remains	65
4.2.1	Nature of Evidence	65
4.2.2	Preservation	66
4.3	Methods and Techniques	67
4.3.1	Sampling in the field	67
4.3.2	Retrieval of macrofossil plant remains by froth flotation	68
4.3.3	The method of positive casts of plant impressions in pottery	71

4.4	The Identification Procedure	74
4.4.1	Comparison of the archaeological specimens with a reference collection of extant plants	75
4.4.1.1	The reference collection	75
4.4.1.1.1	Experiments and results of carbonization	75
4.4.1.1.2	Experiments on and results from impressions in pottery	78
4.4.1.1.3	Results of identifying the archaeological specimens	81
4.4.1.1.3.1	The sites in the Atbara area	96
4.4.1.1.3.2	The area of West Butana. The Shaqadud cave-site	99
4.4.1.1.3.3	The sites in the Khartoum area	102
4.4.1.1.3.3.1	The sites on the east bank of the Nile	102
4.4.1.1.3.3.2	The sites on the west bank	104
4.4.1.1.3.3.3	The sites in the southern areas	105

CHAPTER V: ARCHAEOLOGICAL ARTEFACTS ASSOCIATED WITH THE EXPLOITATION OF FOOD-PLANTS 108

5.1	Introduction	108
5.2	Pottery	108
5.2.1	Pottery of the Early Khartoum tradition	110
5.2.2	Pottery of the Shaheinab tradition	113
5.2.3	The pottery of the post-Shaheinab traditions	122
5.2.4	Association of pottery with exploitation of food-plants	123
5.3	Lithic Artefacts	129
5.3.1	Lunates	135
5.3.2	Knife (?sickle)-blades	142
5.3.3	Grinders	144
5.3.4	Rubbers	154
5.3.5	Sandstone-rings	154
5.3.6	Gouges	157
5.4	Summary	177

CHAPTER VI: ETHNOGRAPHIC AND ETHNOBOTANIC STUDY ON GATHERING OF FOOD-PLANTS: CASE STUDY W. SUDAN 178

6.1	Gathering Wild Food-Plants: "General Remarks"	178
6.2	Gathering Food-Plants in West Sudan (Darfur area)	180

CHAPTER VII: TRADITIONAL LAND-USE AND CULTIVATION IN THE CENTRAL SUDAN 201
7.1 Introduction 201
7.2 Traditional Land-Use 202
7.2.1 Rainfed land 202
7.2.2 Riverain lands 204
7.3 Traditional Cultivation 206
7.3.1 Rainfed cultivation 207
7.3.1.1 Shifting cultivation 209
7.3.1.2 The *hariq* cultivation 210
7.3.1.3 Intensive cultivation 211
7.3.2 Irrigated cultivation 211
7.3.2.1 Natural flooding 212
7.3.2.2 Basin irrigation 215
7.3.2.3 The counter-weighted lever irrigation 215
7.3.2.4 The water-wheel 216
7.4 Summary 221

CHAPTER VIII: FROM GATHERING TO CULTIVATION OF WILD FOOD-PLANTS TO THEIR DOMESTICATION 223
SECTION A: FROM GATHERING TO CULTIVATION OF WILD FOOD-PLANTS 223
8.1 Introduction 223
8.2 Gathering of Food-Plants during Early Khartoum Tradition 224
8.3 Cultivation of Wild Food-Plants. Cultivation: Definition and Criteria 228
8.4 The Transition from Gathering to Cultivation of Wild Food-Plants 229
8.5 The Central Sudan Case: The Shaheinab Tradition and the Post-Shaheinab and Meroitic Period 231
8.6 Summary 235

SECTION B: THE DOMESTICATION OF CEREALS IN THE CENTRAL SUDAN: WITH EMPHASIS ON SORGHUM 238
8.7 Introduction 238
8.8 The Process of Domestication: Definition and Criteria 238
8.9 Domestication of Cereal Crops 239
8.10 The Role of Harvesting and Sowing in the Domestication of Cereals 247

8.10.1	Essence of harvesting	247
8.10.2	Essence of harvesting tools: Re-evaluation	249
8.10.3	Essence of sowing	255

CHAPTER IX 264

SECTION A:	SOME OBSTACLES IN THE PROCESS OF THE DOMESTICATION OF CEREALS IN THE CENTRAL SUDAN: WITH EMPHASIS ON SORGHUM	264
9.1	Irregular Cultivation Practices	264
9.2	Hybridization, Backcrossing and Cereal Domestication	265
9.2.1	*Sorghum bicolor* (L.) Moench. A case study	266
SECTION B:	AN ALTERNATIVE EXPLANATION FOR THE LATE DATE OF CEREAL DOMESTICATION THE CENTRAL SUDAN	272
9.3	Isolation: A Key to the Domestication of Cereals	273
9.4	Forms of Isolation	274
9.4.1	Isolation within the natural habitat	274
9.4.2	Isolation outside the natural habitat	278
9.5	An Alternative Explanation for the Late Date of Domestication of African Cereals: The Case of the Central Sudan	279
9.6	Implications of the Alternative Explanation	287

Figures:

1.	The Central Sudan, the excavated sites and other names mentioned in the text	2
2.	Distribution of wild sorghum in Africa and suggested zone of its domestication	3
3.	Distribution of known sites of wild pearl millet	3
4.	The excavated sites correlated with culture history and economy in the Central Sudan	8
5.	Contour map and sites in the Khartoum area	12
6.	Abu-Darbein: Contour map and excavated area	28
7.	Types of soils in the Sudan	32
8.	The Central Sudan: past and present vegetation	33
9.	The complex sites of Shaqadud: Contour map	42
10.	Average annual rainfall in the Sudan (1930-1960)	45

11. Past and present extent of the course of the White Nile 50
12. Rabak site contour map and excavated area 52
13. Geology and types of soil in the area of Jebel Tomat/Rabak and further north and east 53
14. Upland and lowland clays and drainage in the White Nile area 54
15. Salinity and alkalinity of soil in the are of Jebel Tomat/Rabak 56
16. Depressions seasonally filled with rain/or flood waters 57
17. Different parts of the flotation apparatus 69
18. Plots showing changes in dimensions of grains after carbonization 77
19. Plots showing changes in dimensions of grain impressions 80
20. Some of the present regions, towns and tribes in the Sudan cited in the text 109
21. Wavy -line pottery(Early Khartoum). 111
22. Some of the decoration patterns of the pottery of Shaheinab type. 114
23 Some illustrations of the pottery of Jebel Moya 124
24 Examples for the variability of shapes of the Jebel Moya pottery 125
25 Post-Shaheinab type of pottery from the Shaqadud cave-site(seriation) 126
26 Huge clay-jars for storage of grain-foods (Darfur,W.Sudan) 128
27 Lunates from Shaheinab type of sites 136
28 Suggested method of hafting of lunate-sickle 136
29 Archaeological evidence of lunate-arrow tips 141
30 Lower grinder of the Early Khartoum type 145
31 Lower grinder of the Shaheinab type 145
32 Fine-grained lower grinder (Post Shaheinab:Shaqadud cave-site) 146
33 Coarse-grained lower grinder (Post Shaheinab:Shaqadud cave-site) 147
34 Sandstone rubbers 155
35 Sandstone rings 156
36. Sketch illustrating different parts of a gouge 158
37. Gouges type I 160
38. Gouges type II 160
39. Gouges type III 161
40. Gouges type IV 161
41. Gouges type V 162
42. A gouge with missing distal and proximal ends 166
43. Illustration of the suggested methods of hafting and use of gouges 169
44 Cultivation wooden tools: hoes and sowing tools 172
45. Cultivation wooden tools: weeding and threshing tools 173
46A. Cultivation tools: rakes 174
46B. Earth-levelling-board 174

47.	Resharpening and size reduction of gouges	176
48.	Simplified illustration of the traditional land use in the Central Sudan	203
49.	Subdivisions of riverain land	205
50.	The actual limits and the desertification hazard-limits of Rainfed Cultivation in the Sudan	208
51.	The *saqia* in operation	217
52.	Sketch of five basic spikelet types of domesticated sorghum	244
53.	Simplified classification of domesticated sorghum and spontaneous taxa	257
54A.	Distribution of the basic and intermediate taxa of Sorghum bicolor (L.) Moench (Guinea and caudatum)	258
54B.	Distribution of the basic and intermediate taxa of Sorghum bicolor (L.) Moench (durra and kafir)	259
55.	Some "incipient" forms of man-induced isolation of the cultivated population	275
56.	Optimal conditions for animal keeping during the dry and wet seasons in the prehistoric Central Sudan	282
57.	Climate shifts in Sudan from 9000 to present	285

Tables:

1.	An example for the plant species exploited and their uses in the Aneibis Atbara area	38
2.	An example for the plant species exploited in the Shaqadud area	47
3.	An example for the plant species exploited and their uses Jebel Tomat/Rabak area	60
4.	A list of plant species at Jebel Tomat	61
5.	Carbonized and desiccated plant remains: Abu-Darbein site	82
6.	Plant impressions in pottery: Abu-Darbein site	83
7.	Plant impressions in pottery: The Shaqadud cave-site	84
8.	Carbonized and desiccated plant remains: Shaqadud cave-site	85
9.	Plant impressions in pottery: Um Direwia site	87
10.	Plant impressions in pottery: Kadero site	88
11.	Plant impressions in pottery: Zakiab site	89
12.	Carbonized and desiccated plant remains: The Nofalab site	90
13.	Plant impressions in pottery: Shaheinab site	91
14.	Plant impressions in pottery: Jebel Tomat site	92
15.	Plant impressions in pottery: Rabak site	93
16.	Distribution of wavy line and dotted wavy line (Early Khartoum) pottery and other types at Islang site	112

17. Decoration patterns of pottery of Shaheinab type excavated from 2 trenches (8 m^2) at Kadero site 115

18. Distribution of potsherds of Shaheinab type according to their decoration patterns at the Nofalab site 116

19. Decoration patterns of pottery of Shaheinab type excavated from square 107x/100y at Rabak site 117

20. Diameters and wall thickness of pottery of the Shaheinab type 119

21. Diameters of pottery of the Shaheinab type as estimated from measurements of rimsherds recovered from square 100x/100y at Um Direwia site 120

22. Shaheinab type of potsherds with holes from Zakiab site 121

23. Cross tabulation of lithic tools and raw-materials from Abu-Darbein (Early Khartoum type of sites) 130

24. Cross tabulation of lithic tools and raw-materials from Islang site (Late Early Khartoum - Early Shaheinab type of sites) 131

25. Cross tabulation of lithic tools and raw-materials from Nofalab (Shaheinab type of sites) 132

26. Cross tabulation of lithic tools and raw-materials from Kadero (Shaheinab type of sites) 133

27. Cross tabulation of lithic tools and raw-materials from Rabak (Shaheinab type of sites) 134

28. Raw-material and measurements of lunates from Islang (Late Early Khartoum - early Shaheinab type of sites) 137

29. Raw-material and measurements of lunates from Nofalab (Shaheinab type of sites) 138

30. Ground tools from Islang (Late Early Khartoum - early Shaheinab type of sites 150

31. Ground tools from Um Direwia site (Shaheinab type of sites) 151

32. Ground tools from Zakiab (Shaheinab type of sites) 152

33-37. Darfur (West Sudan): Some of the wild plant species which are common both as food-plants and for other purposes 183

38. An example of tools used in gathering food-plants 195

39. List of traditional means of irrigation and the main food-plants irrigated by each 213

40. A classic example of cultivated plants and their cycle in the saqia-land: from both north and Central Sudan 220

41. Culture history and economy as regards plants exploitation in the Central Sudan (c. 9000-1700) 225

42. Archaeological evidence from the Central Sudan and possible comparable ethnographic evidence of some basic cultivation activities 232

43. Some basic diagnostic differences between wild and domesticated cereal: Example of *Sorghum & Pennisetum* 243

44. Adaptation syndromes resulting from automatic selection due to planting harvested seed 248

45. Some of the most commonly used domesticated sorghums in the Sudan at present 262

Histograms and Diagrams:

1. Histogram of rainfall trends in Central Sudan: From three stations between 1941-1984 35
2. Histogram of temperature, evaporation and rainfall in the Atbara area 36
3. Scatter diagram showing tendency of decrease in size and number of lunates over time 139
4. Histogram: Comparison of length of the five gouge-types 163
5. Histogram: Comparison of breadth of the proximal and distal ends of the five gouge-types 164
6. Histogram: Comparison of thickness of the proximal and distal ends of the five gouge-types 165
7. Histogram: Comparison of average length and of the average thickness and breadth at the proximal and distal ends of the five gouge-types 167
8. Diagram of differences between wild and domesticated cereals: spikes 241
9. Diagram: Sketches of iron shears from different views 254
10. Scatter diagram indicating the range of morphological variability of wild varieties of S. bicolor, and some segregating populations of hybrids among them. 267

Plates:

1. Representation of sorghum on Meroitic rock engraving from Jebel Qeili 20
2. Desiccated seeds of *Celtis integrifolia* 98
3. Desiccated seeds of *Setaria* sp. 98
4. Carbonized seed of *Solanum dubium* 101
5. Positive cast of,an impression of a seed of *Pennisetum* sp. 101
6. Positive casts of impressions of seeds of wild *Sorghum* sp. 103
7. Positive cast of an impression of a seed of *Sesamum* sp. 103
8. An example for wear pattern on lunates 141
9. An example for wear pattern on knives 143
10. Querns 148
11. An example for wear pattern on gouges 171
12. Iron-sickle: late X-Group period 253
13. *Sorghum bicolor* (L.) Moench: spontaneous taxa 260
14A-B. *Sorghum bicolor* (L.) Moench: domesticated taxa, illustration of spikelets 260

Appendices:

1. Radiocarbon Dates 289
2. Preparation of samples for SEM examination 298
3. Experiments on carbonization of seeds and their impressions in pottery 301
4. Distribution, habit, and habitat of the Macrofossil plant species recovered from C. Sudan 316
5. Results of measurements of one hundred gouges 319
6. Pollination of *Pennisetum* spp. and *Sorghum* spp. 341

Bibliography: 344

Preface.

This work is a study of macrofossil plant remains (mainly seeds and grains), their impressions in pottery and the items of material culture that can be associated with them which were recovered from eleven archaeological sites in the Central Sudan. The research aims at studying the utilization of food-plants between c. 9000 and 1700 B.P. and at explaining the late date of their domestication.

Ethnographic and ethnobotanic materials as well as experimental studies are used in the interpretation of the archaeological data. On this basis it is possible to draw the following conclusions:

1) During the period under consideration the Central Sudan was characterized by three main economic activities which roughly correspond with three major periods in its culture history. During the period of the Early Khartoum tradition c. 9000-6000 B.P., the earliest of these periods, gathering food plants, particularly for immediate consumption, and perhaps also grain was the main activity. The second phase of economic activity consisted in the cultivation of wild food-plants, mainly cereal crops and had its beginning during the period of the Shaheinab tradition c. 6000-5000 B.P. The last phase is the cultivation of domesticated food-plants for which the earliest direct botanical evidence is dated to c. 1700 B.P., corresponding to the post-Shaheinab period c. 5000-1700 B.P. which includes the Shaqadud cave-site and the Jebel Moya tradition. It should, however, be emphasized that it is very difficult to define especially for the transition from one type of activity to another. Further, since the cultivation of wild food-plants started c. 6000 B.P., these three economic activities have co-existed until the present day.

2) It is observed that the degree of reliance upon the exploitation of food-plants and the plant species exploited differ from one area to another. This is probably largely conditioned by microenvironmental, topographical and geomorphological differences.

3) The evidence recovered for domesticated sorghum, *S. bicolor* (L.) Moench, from an area outside the natural zone of its wild growing relatives as well as the absence of harvesting tools and other cultivation tools appropriate for this crop strongly indicates that the necessary conditions for the domestication of cereals are as follows:
    a) regular sowing of the previous year's harvest of inflorescences of cereals,
    b) isolation of the cultivated population from the free-breeding gene pool in order to eliminate cross-breeding and to promote the selection pressures exerted by man and hence to initiate the domestication of the cultivated wild species.

4) The floral and faunal evidence recovered from the Central Sudan indicates that it was only after the main emphasis of the economy had shifted to livestock (mainly cattle) and in consequence the population of the Central Sudan had started to move with their herds including areas outside the natural belt of wild sorghum that its domestication was initiated

CHAPTER I

Introduction

1.1 Objectives and Methodology.

The theme of this book is the utilization of food-plant resources and its developments. The main focus is on the questions of the manner and effects of such utilization in the process of plant domestication with an emphasis on the two main African cereal crops, *Sorghum* sp. and *Pennisetum* sp. The geographical area of the study is the Central Sudan (c. 12° to 18°N, 32° to 36°E) in the period between c. 9000 -1700 B.P. (cf. fig. no.1). This area is chosen because it forms the eastern part of the natural zone of the wild stands of *Sorghum* sp. and *Pennisetum* sp. (figs. no.2 and 3) and where the initial stages of their domestication are thought to have taken place (Harlan, J.R., and A.B.L. Stemler, 1976: 473). This also includes areas marginal to the natural habitat of *Sorghum* sp. and *Pennisetum* sp. (Harlan, J.R., 1977: 375; Harlan, J.R., and A.B.L. Stemler, 1976: 473). These marginal areas seem to be where the process of domestication started. The reason for studying the utilization of plant resources in the period between c. 9000-1700 B.P. is mainly because it is within this time span the major developments in the utilization of the food-plants took place.

The problems on the prehistoric utilization of food-plants and the effect of these activities on plant domestication have been studied either on the basis of archaeological approach with little biological analysis or on the basis of biological approach with little application of archaeological analysis. I will, in this book, attack these problems on the basis of a new interdisciplinary field of research in the Sudan, which combines these two and is known as archaeoethnobotany.

Archaeoethnobotany can generally be described as an interdisciplinary field of archaeological and botanical research which integrates the study of plant remains (macrofossils plant remains, carbonized/desiccated and their impression in pottery and microfossil plant remains, e.g. pollen analysis) and cultural features recovered from archaeological sites which can be associated with the exploitation of the recovered plant remains. The basic objectives of such an integrated study is to identify the type and form (i.e.wild or domesticated) of the exploited plants, to reconstruct the strategies of exploitation and to explain how and why cultivation of wild plants and their domestication has taken place. The recovered plant remains may also include species

Fig. no.1 The Central Sudan, the excavated archaeological sites and other names mentioned in the text.

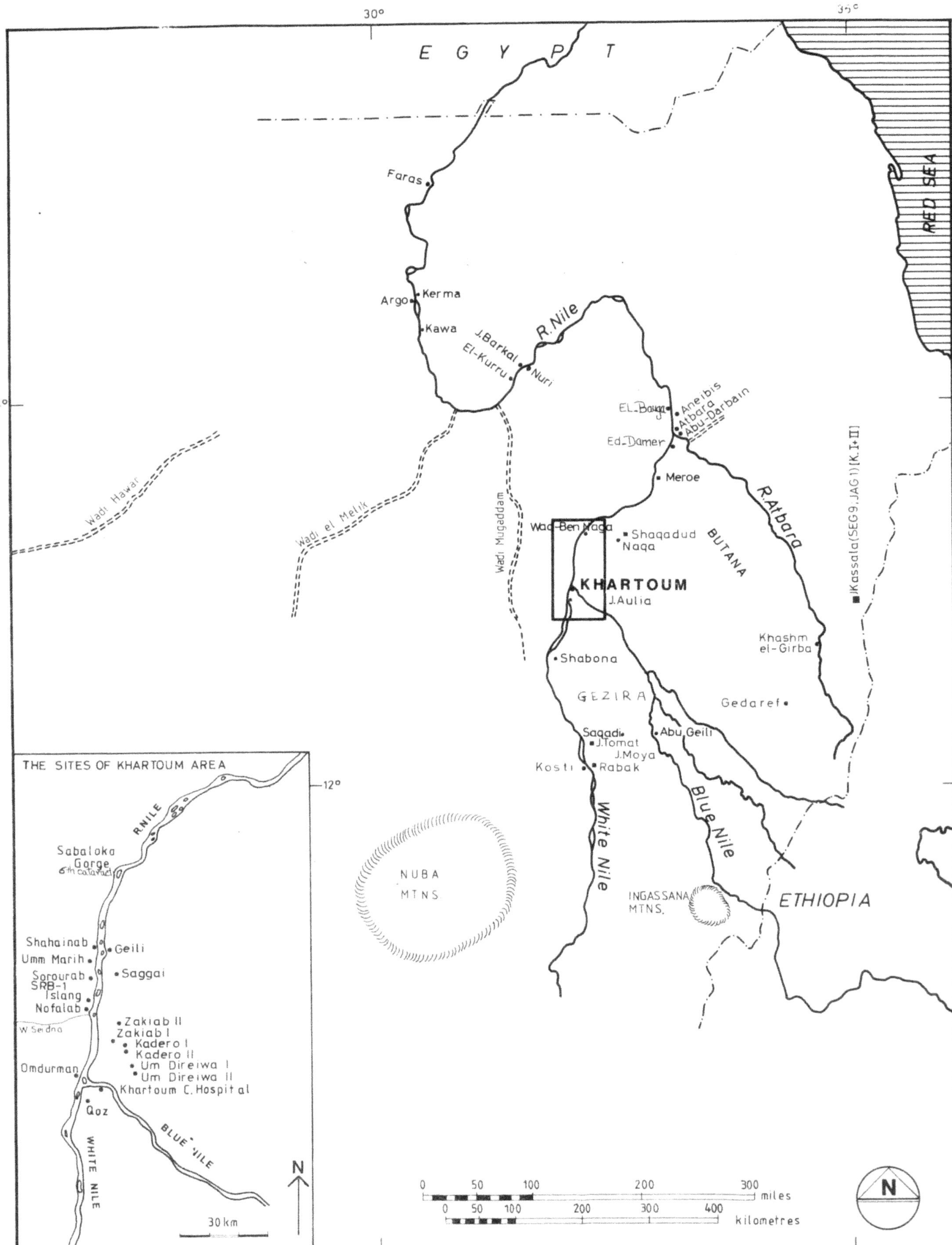

Fig.no.2 Distribution of wild races of sorghum in Africa generally and the Sudan in particular (dotted lines indicate approximate limit of the tropical forest; shaded area indicates suggested zone of sorghum domestication in the Sudan).

all of spp.arundinaceum:-

- o race arundinaceum
- ◉ race aethiophicum
- • race verticilliflorum
- ◉ race virgatum

After Harlan, J.R., et al, 1976: .454 and Harlan, J.R., 1977:374 with some modification.

Fig.no.3 Distribution of known sites of wild pearl millet. (dark shading indicates the northern pearl millet belt; light shading indicates the zone in which pearl millet is grown but sorghum is the dominant crop. Areas of culture in eastern and southern Africa are more diffuse and are not shown).

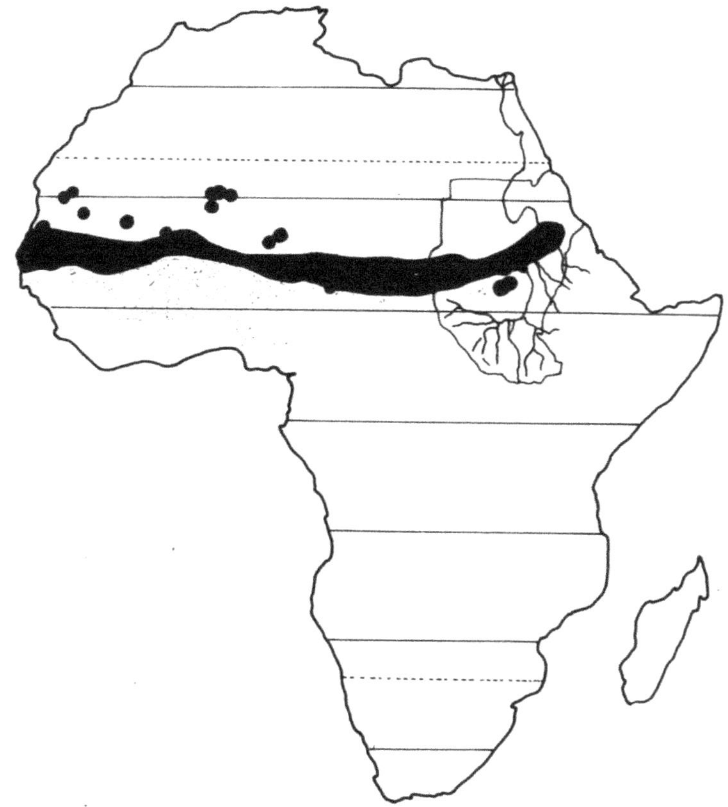

After Harlan, J.R., 1977: 376

which could have been used for purposes other than food and some of these can be used as climatic indicators, e.g. rainfall.

The data on which the study is based consist of macrofossil plant remains, both carbonized/desiccated remains, mainly seeds/grains and their impressions in pottery and the cultural artefacts, e.g. tools, pottery, etc. by which the food-plants were utilized. The geographical locations of the archaeological sites are also seen as important data. Archaeoethnobotanical evidence which has previously been recovered from this area and elsewhere will also be used as will relevant ethnographic evidence that show how comparative gatherers and traditional peasants use simple technology in the exploitation of food-plants.

The theory that plant domestication is a process of genetic manipulation through human activity (Harlan, J.R., et al, 1976: 6) will be used for the interpretation of the data and the definition of cultivation both as regards domesticated and wild food-plants. In other words, I shall on the basis of my data discuss the interpretation that when people start to sow seeds from the previous year's harvest strong selection pressures which are different from the natural ones, will be exerted on the cultivated gene pool and hence the process of its domestication would be started (ibid). Another theory is that the principal technique and the tools of harvesting are equally significant to regular sowing and harvesting in the process of domestication. All of these theoretical and methodological issues of plant domestication shall be used and debated while discussing the problems of plant domestication in the Central Sudan.

The culture history of the Central Sudan during the period between c. 9000-1700 B.P. include three main traditions. The oldest of these is the Early Khartoum tradition between 9000-6000 B.P., characterized by its unburnished wavy and dotted wavy line pottery, the bone harpoons and backed tools. The economy is mainly based on exploitations of aquatic resources, wild game and gathering of wild food-plants, (Arkell, A.J., 1949; Clark, J.D., 1974: 54-65; 1984a: 115-118; Haaland, R., 1987a: 16, 200-201, etc.). This tradition is followed by the Shaheinab one between c.6000-5000 B.P. It was characterized by the gouge (Arkell, A.J., 1953: 31) but this diagnostic tool of the culture was found to be restricted in its geographical distribution to a limited area, namely the Khartoum area (Haaland, R., 1987a: 208). Nevertheless, the tradition is also characterized by its burnished pottery and the, mixed, food gathering and food-producing economies (Clark, J.D., 1984a: 118-122). Cultivation of wild food-plants, e.g. cereal crops was believed to have been initiated during this period (Haaland, R., 1987a: 204-205; Krzyzaniak, L., 1978: 160-166; Magid, A., 1982: 101-106). The post-Shaheinab traditions c. 5000-1700 B.P. is divided into two traditions, the Shaqadud cave-site and the Jebel Moya tradition. Information on the characteristics of the material culture recovered from these is discussed in chapter IV but as regards their economies, the former is based on animal husbandry, gathering and cultivation of wild food-plants. The latter is characterized by its mixed economy of animal husbandry, cultivation of domesticated cereals, e.g. *Sorghum* sp., and hunting and fishing as supplementary forms of food (Clark, J.D., 1984a: 122-124).

These perspectives were the background for my research, both as regards the collection of the primary data on the prehistoric utilization of food plants and the effect of this in their domestication as well as the collection of ethnographic and ethnobotanic materials which are relevant to the interpretation of the prehistoric data.

1.2 Hypothesis Formulation and Research Strategies.

1.2.1 <u>Formulation of hypothesis</u>.

The study of plant domestication in the Central Sudan and perhaps Sub-Saharan Africa is confronted with problems of retrieval of evidence and more seriously with problems of theory which enable us to ask the right questions as regards the important dimensions in the process and results of domestication. In connection to the last point, my standing point is as follows:

1) Some of the basic characteristic features of domesticated plants are their morphological changes brought about as a consequence of selection processes resulting from man's activity utilizing these plants, e.g. *Sorghum* sp. Such morphological changes are in fact about the only basic criteria used to distinguish between these from the wild forms.
2) Domestication of plants was an unintentional consequence resulting from the effects of human activities of exploiting food-plants. The main objective of these activities was the provision of adequate supplies of food, however, the pursuit of these activities have unintentionally led to the domestication of plants. Thus, one of the indirect effects of some of the human activities (e.g., regular sowing and harvesting) with the intention of exploiting food-plants, is the domestication of these (Haaland, R., 1987a: 202-203).
3) In order for the selection pressure to result through human activity, in genetic modification and morphological changes, they must operate in a fairly uniform way over the entire free-breeding plant population. This is particularly important if these pressures are operating within the natural habitat of the wild plant species. But if these selection pressures are operating only on a fragment of a total free-breeding population, it is difficult to see how human activities and the resulting selection pressures could produce changes in the gene pool of the plant population. In this connection, it is peculiar that several thousands of years have lapsed between the

beginning of cultivation of wild food-plants and their domestication in the Central Sudan.

4) Isolation of the exploited plant population, e.g. sorghum, is the most likely essentialcondition under which the selection pressures (caused by man through regular activities of harvesting and resowing) could produce changes in the exploited genepool; that is to say to initiate their domestication. Therefore, I will argue that thedomestication of cereals, e.g. sorghum, in the Central Sudan was started outside thenatural habitat of the wild growing species. In other words, stock of grain-heads wasbrought and was regularly cultivated outside the area of its free-breeding population,most likely, by pastoral-nomadic people who were already dependent on theutilization of wild plants before moving outside their natural habitat.

In short, selection pressures operating and exerted through regular cultivation activities of a gene pool in an isolated habitat would lead to genetic and morphological changes and hence the domestication of the cultivated species.

On the basis of the foregoing, I will look for the following:

a) evidence which may show or may give an indication of the human activities which set the selection pressures at work on the plant population, and

c) evidence which indicates that there was an isolation of the plant population which was subject to both human utilization and the resulting selection pressures from this utilization.

1.2.2 Research strategy.

1.2.2.1 Field work.

1.2.2.1.1 Archaeological excavations.

1.2.2.1.1.1 Selection of sites.

The area of the study, Central Sudan, is divided into three regions on the basis of the past and present vegetational zones (Andrews, F.W., 1948: 32-61; Lebon, J.H.G., 1965: 19-42; Wickens, G.E., 1982: 23-50) and the past natural habitats of the wild *Sorghum* sp. and *Pennisetum* sp. (Harlan, J.R., 1977: 375-376; Harlan, J.R., and A.B.L. Stemler, 1976: 473). These are: the northern, the middle and the southern regions. The latitudinal limits of the middle region generally correspond to those of the natural habitat and initial zone within which wild *Sorghum* sp. and *Pennisetum* sp. were growing and thought to have been domesticated. The northern and southern regions are situated outside these natural habitats.

The selection and the excavation of the sites in these regions were based on the objectives of the research. The field work strategy is to excavate, in each of the three regions, as many sites dating to between c. 9000-1700 B.P. as possible in order to obtain evidence indicative of different activities of exploitation of plants (cf. fig. no.4) and hence to test the above mentioned hypothesis.

1.2.2.1.1.2 Excavation of the sites.

(1) Sites in the northern region: 17° 00'N and 18° 00'N.

According to Harlan's and Stemler's early belt of domesticated sorghum (Harlan, J.R., and A.B.L. Stemler, 1976: 473), the northern region is situated outside the northern limit of that belt. According to my hypothesis, this region could have been a marginal area for the growth of wild sorghum and an optimal one for its isolation from cross breeding and hence its domestication. Furthermore I assumed that the northern region was the ideal area to throw light on the utilization of food-plants for several reasons. Among these: the main physical features of this area are similar to those of Khartoum area in the middle region (cf. chapter III) where the earliest evidence of cultivation of sorghum and millet was found, as I shall discuss later. These features are the rivers (Atbara and the Nile) alluvial plains, the gravel-ridges, the desert plateau. Thus I expect that the adaptive strategies would most likely be the same but marked by more seasonal variability due to the sharp changes from one season to another as they prevail today.

An archaeological survey was conducted along both the eastern and western banks of the Atbara and the Nile rivers and in the hinterlands. The surveyed area covered 1600 km^2 in total; 40 kilometres along these rivers and 40 kilometres away from their banks. Only two sites were found in a state of preservation that made it worthwhile to excavate them. These sites are the Abu-Darbein and the Aneibis sites (cf. fig. no.1).

a) The site of Abu-Darbein.

The site is situated on the eastern bank of the Atbara river; the western part of it is badly eroded by the seasonal flood and the eastern part is disturbed due to recent human activities. The recovered cultural remains are related to the Early Khartoum tradition,c. 9000-6000 B.P., e.g wavy-line pottery, bone harpoons, etc. As regards the plant remains, a large number of mummified seeds of *Celtis integrifolia* Lam., and a few stones of *Ziziphus spina-christi* (L.) Willd. Lam. were recovered but no evidence is found of carbonized/ desiccated grains of cereals or their impressions in pottery except for two desiccated seeds of *Setaria* sp. Thus, both the recovered material culture and plant remains indicate that the occupants of the site were gatherers of food-plants, mainly readily eaten fruits, but there is no indication of any cultivation practices. In other words, the evidence recovered from this site indicates adaptive strategies which were based only on gathering of food-plants but no evidence that would have enable to test my hypothesis

Fig. no. 4. The excavated sites correlated with culture history and economy in the Central Sudan.

Culture history and chronology	The regions and excavated sites Economic strategy	Excavated sites in Northern region	Middle region	Southern Region
Early Khartoum c. 9000-6000 B.P.	Gathering of food-plants	Abu-Darbein Aneibis	Islang (bottom levels)	-
The Shaheinab c. 6000-5000 B.P.	Cultivation of wild food-plants	-	Kadero Um Direwia Zakiab Nofalab Shaheinab Islang (top levels)	Jebel Tomat (bottom levels) Rabak (mainly levels 14 to 4)
Post-Shaheinab c. 5000-1700 B.P.	Cultivation of wild and domesticated food-plants	-	Shaqadud cave site	Jebel Tomat (top levels) Rabak (mainly levels 4 to 1)

on plant domestication, was recovered. Therefore, it was needed to check this further at the site of Aneibis.

b) The site of Aneibis.

The site of Aneibis is situated in an area where conditions both for irrigated and rainfed cultivation were optimal; the flood plains west of the site and the rainfed cultivable plains to the east. Information about the size of the site, and the excavated area are given in chapter III.

In spite of these favourable conditions for cultivation practices, no evidence was found in the material culture, e.g. cultivation tools, or in the plant remains, e.g. grains of cereals or their impression in pottery that indicate any practices of cultivation. Not only that but also no evidence was found of any carbonized/desiccated plant remains or their impressions in pottery. However, the recovered remains of material culture, e.g., wavy-line pottery, bone and lithic tools are related to the Early Khartoum tradition and this was further supported by several radiocarbon dates among which 7730±110 B.P. was the oldest and 7290±150 B.P. was the youngest. Thus, this site was occupied during a period when cultivation of wild cereals had not yet been started in the Central Sudan. That is to say that although the site is relatively younger than that of Abu-Darbein, yet, it is of an Early Khartoum type of site. The botanical evidence recovered from all the sites which are associated with the Early Khartoum tradition indicates an adaptive strategy based on gathering of wild food-plants. But the absence of such evidence from the Aneibis site breaks this pattern of adaptive strategy in these types of sites and needs to be investigated further.

(2) Sites in the middle region: c. 15° 00'N to 16° 30'N.

This region, as previously mentioned, is assumed to have been part of both the natural habitat of the wild-growing cereals and the initial zone where these were domesticated (Harlan, J.R., and A.B.L. Stemler, 1976: 473). The excavated sites in this region are divided into two groups according to their geographical locations. These are: the sites in the Shaqadud area and those in the Khartoum area.

(a) The cave-site (S1A) in Shaqadud.

This site is situated about c. 140 kilometres north-east of Khartoum and c. 50 kilometres east of the Nile at Wad Ben-Naqa at the west fringe of the Butana plains (cf. fig. no.1). The oldest radiocarbon date from this site is 4123±86 B.P. and the youngest is 3165±88 B.P. (Marks, A.E., et al, 1984: 32). Unfortunately there was no other site in this area contemporary with or having material culture similar to this site. However, this site is younger than those of the Shaheinab tradition and, hence, is expected to bear evidence which would give some indications on the practices of the utilization of food-plants and the type of the exploited food-plants. Furthermore, since the site is situated in an area away from the Nile environment and has different ecology, topography and soils, etc., I would expect to recover evidence which indicates a different adaptive strategy as regards exploitation of plants.

The examination of the recovered plant remains attested for the presence of a variety of plant species, but no direct evidence for any domesticated species was found among these. It is apparent from the available evidence (chapter IV), that the occupants have practised both gathering of wild plants and cultivation of crops. But some of the gathered plant species were not present in the Khartoum area; the evidence of these indicate a drier and ecologically different habitat. This was further confirmed by the retrieval of carbonized grains of millet and their impression on pottery. Millet, as we know from its present habitat, grows well in sandy soils and it is more resistant to drought as compared to sorghum (Bacon, G.H., 1948: 318).

The lithic industry of the Shaqadud cave-site is generally poor and the recovered lithic tools did not include any tool that can be associated with cultivation activities except quite a small number of grinders found. Furthermore, no evidence of gouges existed at this site. As I previously mentioned, the distribution of gouges was restricted to the Khartoum Nile environment, between the sixth cataract in the north and the Dam of Jebel Aulia in the south (fig. no.1). My hypothesis that the domestication of cereal crops was initiated outside the belt of their natural growth is also supported by the results obtained, so far, from the study of the Shaqadud area: although the area of Shaqadud is situated within the past natural habitat of wild *Sorghum* sp. and *Pennisetum* sp. (where they were thought to have been domesticated), no evidence of domesticated species was found in this area.

(b) The sites of the Khartoum area.

These include sites on both the east and west banks of the Nile river north of Khartoum and comprise Kadero, Um Direwia and Zakiab on the east bank (Haaland, R., 1981; El-Mahi, A.T., 1982) and the Nofalab, the Islang and the Shaheinab sites on the west bank (Arkell, A.J., 1953; Haaland, R., 1981; Magid, A., 1982). The location of each of these sites (cf. fig. no.1) and the excavated area within each of them are discussed by Haaland,R.,1981 and Magid,A.,1982.All of these sites are situated within the same environment and they all are more or less contemporary as the radiocarbon dates indicate (cf. appendix no.1). The material culture from all these (except those recovered from the bottom levels of the site of Islang) are similar to those associated with the Shaheinab tradition. Artefacts from the lower levels of the site of Islang such as wavy line pottery, are similar to those of the Early Khartoum tradition. The material culture from the top levels of the same site is similar to those of the Shaheinab tradition[1], e.g.

---

[1] A few number of gouges were found during a study of the surface collection of this site (Mahir, H.: pers.comm.).

burnished pottery. There are, however, microenvironmental and topographical differences between the two banks. Accordingly, I assumed there were differences in adaptive strategies related to the exploitation of plants. The wide alluvial plains on the east bank (cf. fig. no.5) made it possible for the occupants to practise cultivation beside the activities of gathering. On the other hand the topography on the west bank (cf. fig. no.5) is characterized by a very narrow flood plain limited by gravel ridges. Thus exploitation of plants was restricted to activities of gathering of wild food-plants. The excavated remains, both material culture, e.g. grinders and plant remains, turned out to support this hypothesis (El-Mahi, A.T., 1982; Haaland, R., 1981; Magid, A., 1982). These assumptions are confirmed by the recovery of several grain impressions of wild cereals, namely wild sorghum *S. verticilliflorum* (Steud.) Stapf. from all the sites situated on the east bank and only one impression of these was recovered from the site of the Shaheinab on the west bank (cf. chapter IV) but no evidence of any domesticated plant species was found at any of these sites. Despite the recovery of large numbers of seeds of *Celtis integrifolia* Lam. from all the sites on the east and west banks of the Nile, only a few fragments of these were recovered from the site of Islang. The economy in the latter seemed to be mainly oriented towards exploitation of aquatic resources.

In general, the archaeoethnobotanical evidence indicates that both gathering and cultivation of wild plants were practised in the Khartoum area between c. 6000-5000 B.P. This period was characterized by the Shaheinab cultural tradition. Thus, the earliest evidence of cultivation of wild plants from the area of Khartoum and the Central Sudan at large is most likely associated with this cultural tradition, however, no evidence of tools which can be related directly to these activities, such as sickles, was found. Although such activities of cultivation of wild plants were apparently practised for over a thousand of years in this area, the domestication of these has not been initiated. One of the most likely reasons for this is that since this area was situated within the natural habitat of the wild cereals, e.g. *Sorghum* sp., the cross-breeding between the cultivated population and its free-breeding gene pool hindered the process of the domestication.

(3) <u>Sites in the southern region</u>: c. 12° 00'N to 13° 30'N.

This region provides the basis to test my hypothesis in that it was situated outside the natural belt of the wild growing cereals and that the earliest evidence of domesticated cereals, e.g. *Sorghum bicolor* (L.) Moench from the Central Sudan was found in this region (Clark, J.D., and A.B.L. Stemler, 1975: 588-590).

An archaeological survey was conducted (1983) in the area along the east bank of the White Nile and about 40 kilometres east of this river and as far south as the city of Kosti (cf. fig. no.1). The sites of Jebel Tomat and Rabak (cf. fig. no.1) were the only sites excavated in this area due to the bad preservation.

Fig.no5 Contour map and sites north of Khartoum.

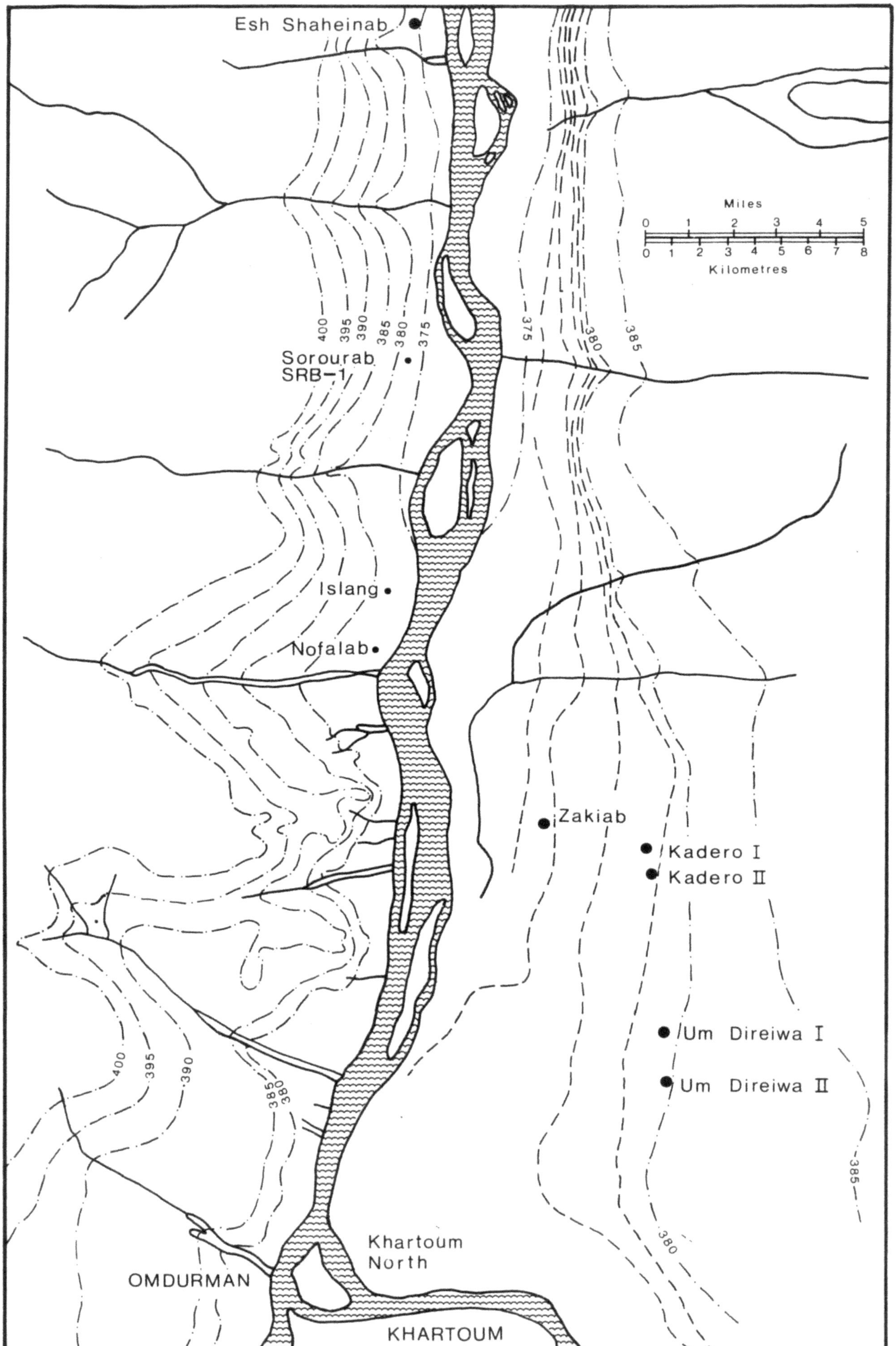

(a) The site of Jebel Tomat.

The earliest evidence of domesticated sorghum, *S. bicolor* (L.) Moench from the Central Sudan dated to 245±60 B.P. was recovered from this site (Clark, J.D., and A.B.L. Stemler, 1975: 588-590). This site is situated about 50 kilometres north of the site of Rabak (cf. fig. no.1). The radiocarbon dates indicate that the earliest occupation of the site was 4140±90 B.P. and the youngest was 1600±80 B.P.

The examination of plant impressions in pottery revealed the presence of evidence of wild sorghum as well as evidence of other wild edible food-plants (cf. chapter IV). One of the immediate indications of the correlation of evidence for wild and domesticated sorghum from this site, is that the occupants of the site (disregarding the question of their being one or different groups) have kept regular cultivation of wild sorghum in this "marginal" area until it was domesticated. Since no evidence was found of any artefact that can be associated with cultivation activities such as harvesting tools, it becomes rather uncertain if harvesting with cutting tools is as crucial a technical factor as it was stated in the process of domestication of e.g. sorghum. I therefore suggested an alternative explanation, i.e. that domestication of sorghum was achieved by isolating the species so that cross hybridization was eliminated to a point that genetic manipulation worked faster towards a breed which is highly dependent on being protected by man. Of course there are some basic factors that have to be observed, e.g. regular sowing of selected heads of grain from the previous year's harvest.

As regards the pottery recovered from this site it was typologically related to two cultural periods. The older of these (found at the bottom levels) is related to the Shaheinab tradition. The younger one found at the top levels is related to the Jebel Moya tradition (cf. chapter IV).

(b) The site of the Rabak.

Several radiocarbon dates were obtained from this site, the oldest being 6050±100 B.P. and the youngest is 4490±100 B.P.

The pottery recovered from levels 14 to 4 bear similar surface treatment and decorative patterns to that excavated from the sites in the Khartoum area which are associated mainly with the Shaheinab tradition. The upper 3 levels have a few pottery types similar to that found in levels 14-4 and pottery with decorative patterns and rims similar to that found first at the site of Jebel Moya (Haaland, R., 1984: 43). No lithic artefacts which can be associated with cultivation activities are found but there is evidence for those associated with food preparation, e.g. grinders.

No remains of carbonized/desiccated plants were retrieved but many impressions of plant materials in pottery were found. The majority of these impressions were identified as wild *Sorghum* sp., however, no evidence of domesticated sorghum or any other species was recovered. Thus, it is most likely that an adaptive strategy based on the cultivation of wild sorghum was practised. The extended plains east of the White Nile constitute a fertile clayey soil for the cultivation of this cereal crop.

1.2.2.2 Ethnographic and ethnobotanic studies.

Ethnographic and ethnobotanic studies relevant to the objectives of the present research are conducted on gathering and traditional cultivation activities. The results of the analysis of the data will be used, when relevant, in the interpretation of the archaeological remains.

1.2.2.3 Experimental studies.

Since the field of archaeoethnobotany is newly introduced in archaeological research in the Sudan (cf. chapter II), it is still in the stage of early development. This stage includes the building up of reference collections of comparative data of extant plant species. The present work includes some experimental studies on carbonization and impressions of plants in pottery, using in these experiments extant plant materials, mainly seeds/grains. Measurements of the data, e.g. breadth and thickness, was done before and after the experiments. The results of these measurements are used together with those of fresh comparative data in the procedure of identification of the recovered archaeological remains (cf. chapter IV). It is also expected that these experiments and the obtained results will be a base on which one can build up a variable reference collection, e.g. immature seeds/grains, leaves, etc., beside the fresh reference collection. Detailed information on the results of the experiments is given in chapter IV.

1.3 Perspectives on the Presentation of the Study.

The foregoing account is meant to furnish the reader with a general outline of the objectives and the main problems in this dissertation and the theoretical and methodological approaches which were used in studying them. It was not unexpected that new questions would rise but also the results I obtained in this study might change with the recovery of new evidence from this area.

The following chapters in this thesis will deal with detailed study and analysis of the data recovered and how the results of this archaeoethnobotanical research from the Central Sudan were obtained. But since this interdisciplinary field of research was newly introduced in the Sudan, I shall first outline its history and present state with the main emphasis on the Central Sudan. Then I shall present an account on the physical environment and the geographical setting within which the studied archaeological sites are situated. The geology, topography, the prevailing climate and the present flora, fauna, etc. will be included in this chapter. This description is important in providing a background in our attempt to understand the traditional ways of life, how far back these can be traced and if these have been subject to change. Records of this type will also assist in enriching our knowledge and scope while trying to reconstruct past human cultures.

Since the archaeoethnobotanical approach is the core of the present study, chapter four shall be devoted to a detailed presentation of the recovered plant remains from the excavated archaeological sites in the Central Sudan. Other issues which will be discussed are sampling methods in the field, methods of retrieval of botanical evidence and the identification procedure including experimental studies. The following chapter will be on the presentation and discussion of the recovered archaeological evidence which is assumed to be functionally related to the different activities of utilization of food-plants, e.g. gathering, cultivation, food-preparation, storage, etc. Comparative evidence of traditional tools as used in different parts of the Sudan shall be used when interpreting the archaeological data.

Gathering of food-plants was an essential stage before the domestication of these. It is also a practice with different types of activities that had co-existed with cultivation of wild and domesticated food-plants. But these will not be shown or indicated in the archaeological remains. The ethnographic studies and observations among contemporary communities of gatherers would certainly broaden our scope as regards the interpretation of these activities and the probable adaptive strategies at different conditions and circumstances, e.g. during the different seasons, at times of food shortages, etc. Therefore, I conducted an ethnographic and ethnobotanic study on gathering of food-plants in Darfur in western Sudan. The results of this study are presented in the sixth chapter of this book. In connection with the use of ethnographic information, the seventh chapter of this study will deal mainly with a general survey of the traditional land-use and traditional rainfed and irrigated cultivation in the Sudan. The main emphasis of the study is on the means of irrigation, the cultivated food-plants and how far back these can be traced and to what extent they can be incorporated in the interpretation of the archaeoethnobotanical data. Moreover, this study is meant to draw attention to more detailed and elaborate studies of this nature in the future in order to build up ethnographic collections of material culture and descriptions which would assist in future archaeoethno-botanical and ethnoarchaeological research.

Since gathering and the cultivation of the wild and domesticated food-plants are central in this book, chapter eight will be devoted to a detailed study of the development of these activities and the factors that led to these developments in the prehistoric Central Sudan. The recovered evidence from the excavated sites in this area, will be the main basis of the interpretation of these activities. But ethnographic materials will also be used particularly when the archaeological evidence is lacking or is controversial.

The last chapter of the book will focus on the interpretation of the biological and cultural evidence we obtained as regards the questions of the process of cereals domestication in the Central Sudan and the explanation of the late date of its initiation. The implications of these in the general culture history of the area and a summary of the results of this work will be the closing issues of this chapter.

CHAPTER II

The History and the State of Archaeoethnobotanical Research in the Central Sudan

During the past five decades, increasing attention has been given to palaeoenvironmental and palaeoeconomic studies in the Central Sudan on the basis of archaeological research. In the recent publications of their works, A. Mohammed-Ali and R. Haaland have summed up the nature of the past archaeological research in general (Haaland, R., 1987a: 16-25; 1981: 19-29; Mohammed-Ali, A., 1982: 10-21). However, since the theme of the present study is on the development of the past economies, from gathering to domestication of food-plants on the basis of an archaeoethnobotanical research, it is appropriate to give a brief summary of the history and state of this type of research in the Central Sudan. It should be noted here that this summary is more like an updating of a previous account on the same issue (Magid, A., 1982: 4-8).

Generally, archaeoethnobotanical research is not yet fully recognized in the field of archaeological studies in the Sudan. This is probably due to the following factors:

1) The scarcity of evidence in the recovered archaeological material for specialized tools and equipment associated with the different activities of exploitation of food-plants, e.g. gathering, cultivation.
2) The poor preservation of the plant remains, e.g. grains, seeds in the archaeological sites; the older the site is, the lesser are the chances of preservation of these remains.

These factors are most likely the reason for the general tendency among archaeologists to underestimate the essence of such type of evidence in the study of the past economic strategies and their development as regards exploitation of food plants. However, despite these problems a number of contributions were made in this field of research through many but uncoordinated efforts. An attempt will be made in this account to bring these contributions together, with the main emphasis on the studies which deal with the Central Sudan in the period between c. 9000 to 1700 B.P.

A.J. Arkell can be considered the founder of the archaeoethnobotanical research in the Sudan generally and in the Central Sudan in particular (Magid, A., 1982: 4). The first botanical evidence from the Central Sudan which consisted of seeds of *Celtis integrifolia*[1] Lam. (Ar. *mahagai*), was recovered by Arkell during his excavations of the site of the Early Khartoum, (Arkell, A.J., 1949: 108-109, Pl.45, fig.no.3). This was one of the first sites to be excavated in this area. He interpreted the presence of these seeds to be an indication that the fruits of *Celtis* sp. were gathered and brought to the settlement as food (ibid). He added that the gathering of wild fruits such as *Celtis* and wild grass-seeds was an activity practised during the Early Khartoum tradition c. 9000- 6000 B.P. probably as a supplementary source of food which was mainly obtained from aquatic resources, e.g. fish, molluscs, and wild game (ibid). However, despite his statement that seeds of wild grasses were gathered as food (ibid) Arkell did not functionally relate the preparation of these and the grinding tools recovered from the same site. As he interpreted the function of these grinders to be used mainly for grinding ochre (Arkell, A.J., 1949: 51, 52-53). New information is now available on gathering practices and the artefacts used in these food-quest activities during the Early Khartoum tradition as shall be discussed in chapters IV, V and VIII. Arkell also listed the presence of seeds of *C. integrifolia* Lam. (together with other relevant faunal evidence) as indicators for a higher rainfall, e.g. at least 500 mm at the time of the occupation of the site of Early Khartoum (ibid: 110) (the present day rainfall is c.160 mm).

During his excavations of the site of the Shaheinab, Arkell also recovered seeds of *C. integrifolia* Lam. (Arkell, A.J., 1953: 103). While the majority of these seeds are desiccated a few specimens exhibit carbonized exteriors indicating their probable exposure to fire. Beside these seeds, one *Acacia* sp., *Ziziphus* sp. and a carbonized shell of *Elaeis guineensis* Jacq. (Ar. *shagarat zeit el-nakheil*) were also found (ibid: 80, 103). Arkell suggested that *E. guineensis* Jacq. was most likely brought from elsewhere because the area of Khartoum was not then a natural habitat for its growth (ibid). On the basis of the recovered plant remains from the site of the Shaheinab, Arkell concluded that cultivation activities were not yet initiated but gathering of food-plants was still practised. Since no evidence of cereals was found, he also stated that grinding equipment were used for purposes other than preparation of food-plants (ibid: 54). More detailed analysis of new data from sites of the Shaheinab tradition and re-evaluation of the previously analysed data from this site were recently made. The results of these provided both new evidence and new interpretations on the economic strategy related to the exploitation of food-plants (this will be discussed in the following chapters). It also seems that Arkell's findings of plant remains were chance discoveries picked by hand while excavating and screening.

Since the publication of the results of Arkell's excavations of the Shaheinab site, in 1953, the archaeoethnobotanical research in the Central Sudan apparently came to a halt

[1] Seeds of *C. integrifolia* Lam. were also found at the site of Gereif, near Khartoum, at the same time when the site of Early Khartoum was excavated (Arkell, A.J., 1949: 110).

for two decades until a report on the excavations of the sites of Shabona, Jebel Tomat and Jebel Moya along the White Nile and the southwestern fringe of the Gezira (cf. fig. no.1) was published by J.D. Clark in 1973. This report included some remarks on the recovery of some plant remains and the possible association of certain artefacts with the exploitation of these food-plants (Clark, J.D., 1973: 55-64). An engine-operated froth flotation machine was used for the first time in the field with the intention of retrieving plant remains from the site of Shabona, which is a site of Early Khartoum tradition (ibid: 61, 62). Some plant remains have been retrieved and the results of their identification show that almost all these remains belong to xerophytic species (ibid: 62) but since they are similar to those present today they were considered to be later intrusives because it was uncertain if they represent the flora during the occupation of the site (ibid). From the Jebel Tomat, which is a post-Shaheinab site, a number of carbonized and desiccated plant remains were recovered from this site and identified as: *Acacia* sp., *Salvadora persica* L. (Ar. *arak*), *Ziziphus* sp. (probably *Z. spina-christi* Lam. (Ar. *nabag*) and *Ficus* sp. (Ar. *gameiz*) (Clark, J.D., and A.B.L. Stemler, 1975: 588). Some of these species have edible fruits, e.g. *Ziziphus spina-christi* Lam. and the others are evidence that the area was about as arid during the period of occupation of the site as it is today (ibid). But the most important botanical material is the finding of large quantities of carbonized grains of *Sorghum bicolor* (L.) Moench (Ar. *aish/durra*). These were recovered from a storage pit but were also found scattered in the settlement debris. This evidence, which is dated to A.D. 245±60 represents the earliest direct evidence of domesticated sorghum from the Central Sudan (ibid: 589). On the basis of these finds, faunal remains and the remains of material culture, it has been suggested that the economy was one of mixed farming in which sorghum was cultivated during the rainy season (ibid: 589-590).

Shinnie has, however, mentioned in an earlier work, in 1967, that the geographer Strabo has reported, after his visit to Egypt in 25-19 B.C., that the Meroites (the population of Meroe in the Central Sudan) used millet[2] and barley, tender twigs, reed roots, etc. as food (Shinnie, P.L., 1967: 18-19). He also reported a rock-engraving of the Meroitic king Sherkarer (A.D. 12-17) at Jebel Qeili (fig. no.1) where heads of sorghum appear in the scene (ibid: 50-51; cf.pl.no.1). The presence of barley as food in the Central Sudan round 25-19 B.C. indicates either that it was cultivated by means of regular irrigation or that it was imported from elsewhere, e.g. Egypt, because it could not have been grown by means of rainfed cultivation in the Central Sudan which at that time received a maximum of 250 mm of summer rain, while this plant species requires at least 300 mm of winter rain. No plant remains were reported from Jebel Moya. However, since the material culture and faunal remains are similar to those from Jebel Tomat, it is

[2] Millet is most likely meant to denote sorghum (Clark, J.D., and A.B.L. Stemler, 1975: 589). The term is still often used in the literature to denote sorghum.

Plate no. 1

Representation of sorghum inflorescences on a Meroitic rock engraving from Jebel Qeili.

(After Shinnie,P.L.,1967: 51)

assumed that the occupants of Jebel Moya had a similar economy, e.g. mixed farming, to that of Jebel Tomat (Clark, J.D., 1973: 59, 60).

The 1970s have, undoubtedly, witnessed the most active operations of archaeological research in the Central Sudan especially in the area of Khartoum (Haaland, R., 1987a: 16-20; Mohammed-Ali, A., 1982: 10-21). Most of these have also included some archaeoethnobotanical studies. Among these are the earliest successful attempts of extracting positive casts from impressions of plants in pottery. Although some attempts were previously made to find evidence of plant impressions, e.g. the attempts made by Arkell (Arkell, A.J., 1949), M. Klichowska was probably the first palaeoethnobotanist to find such an evidence from this area. In a preliminary report, Klichowska stated that a number of plant impressions were found in pottery recovered from the Kadero, which is a site of a typical Shaheinab tradition. Most of these impressions were identified as being of domesticated species such as *Sorghum vulgare* L. (Ar. *aish/tdurra*), three species of millet, e.g. *Pennisetum* sp. (Ar. *dukhun*) and *Eleusine coracana* Gaertn. (Ar. *telebun*), one species of *Setaria* but its domestic status is not certain and finally domesticated barley, *Hordeum* sp. (Ar. *shaeir*). (Klichowska, M., 1978: 42-43). Almost all these identifications have, unfortunately, proved to be incorrect. For instance, the re-examination of what has been identified as domesticated sorghum, *S. vulgare* L. turned out to be an unripe seed of *C. integrifolia* (Haaland, R., 1981: 197). The impression of the grain of domesticated barley *Hordeum* sp. in the Central Sudan, 6000 to 5000 years ago raises a number of questions which all indicate that it is very doubtful that such an identification is correct (Magid, A.: *in press*). Clark stated that if these identifications are confirmed then these grasses must have been cultivated as early as the fourth millennium B.C. (Clark, J.D., 1984b: 33). Thus it is obvious that incorrect identifications of botanical evidence can lead to an inaccurate cultural historical interpretation. Credit should, however, be given to the successful method of making and extracting the positive casts of these impressions. The wrong identification of these was unfortunate but it draws our attention to the following important requirements:

1) The importance of building up a proper reference collection of fresh, carbonized plant materials and their impressions in pottery (for more details see chapter IV).
2) Acquirement of as much knowledge as possible about the area under the archaeo-ethnobotanical investigation. For instance, literature and field knowledge about the past and present vegetation, rainfall, soils, topography, etc. of the area under study and its neighbouring northern and southern areas. This will enable the researcher to have a good control on his data, its identification, interpretation, determination of exotic finds and elimination of the intrusives.

3) Confirmation of the results of identification by cross-checking and exchange of these identifications and other relevant information with other specialists working on similar problems and/or similar geographical areas/zones.

Further examinations of plant remains and their impressions from the same site, Kadero, beside the re-examination of the previous results revealed the presence of desiccated seeds of *C. integrifolia* Lam., *Hyphaena thebacia* (L.) Mart. (Ar. *dom*) as well as impressions of grains of wild sorghum, two species of millets and seeds of other wild grasses, e.g. *Setaria* (Krzyzaniak, L., 1978: 160-166). On the basis of these results it has been suggested that the occupants of the Kadero site practised both plant gathering and cultivation (ibid). As regards the environment, he suggested that a savanna type with patches of forest, shrubs, and grasses prevailed during the occupation of the site (ibid). Krzyzaniak has also taken soil samples to be treated for pollen analytical study, but no pollen grains were obtained both due to unfavourable preservation conditions and the chemical treatment of the soil samples (Krzyzaniak, L., 1979: 69). Haaland reported in 1981 that desiccated seeds of *Celtis integrifolia* Lam. were found at all the excavated sites, all of Shaheinab type, on the east bank of the Nile river north of Khartoum (cf. fig. no.1) except the site of Kadero II (Haaland, R., 1981: 195). She pointed out the use of this plant species as an indicator of the annual rainfall (ibid). Moreover, she presented the recovered evidence of grain impressions of wild sorghum, *S. verticilliflorum* (Steud.) Stapf. on potsherds from the Shaheinab type of sites of Zakiab (5 impressions), Um Direwia (4 impressions) and Kadero I (2 impressions) (ibid). Correlating this evidence with the large number of grinders found at these sites, Haaland hypothesized that the occupants of the site practised the cultivation of food-plants, namely sorghum (Haaland, R., 1981: 197, 213; 1987a: 221). This hypothesis gave further support to Krzyzaniak's interpretation that gathering and cultivation were practised during the occupation of the Kadero (Krzyzaniak, L., 1978: 160-166). The morphological similarities between the cultivated sorghum species and the free breeding population were seen to be related to the methods of sowing and harvesting (Haaland, R., 1981: 211-215; 1987a: 202-204) as I shall discuss this in detail later.

Mohammed-Ali published in 1982 the results of identification of the floral remains he recovered from the site of SRB-1 (cf. fig. no.1), an Early Khartoum type of site, situated on the west bank of the Nile river north of Khartoum. These fossils included *Ziziphus* sp., *Salix subserrata* Willd. (Ar. *safsaf*) and an impression of a leaf-fragment of an unspecified monocotyledonous species (Mohammed-Ali, A., 1982: 82,172). He also stated that although there is no direct evidence of domesticated sorghum from the Central Sudan between the 6th and 5th millennia B.P., yet it appears that cereals such as *Sorghum* sp. and *Pennisetum* sp. were harvested as food (Mohammed-Ali, A., 1984: 69).

The present writer completed in 1982 an M.A. dissertation based on the study of archaeoethnobotanical material dated to c. 8000-4000 B.P. in the area of Khartoum (Magid, A., 1982). A review of previous works in this field of research (ibid: 4-8) was presented with new data from the sites of the Nofalab and Islang on the west bank of the Nile river north of Khartoum (cf. fig. no.1). In addition, plant-impressions found on potsherds from the site of Shaheinab which was re-excavated in 1979 were also included (ibid: 82:92). An abstract of these results is included in the present work (chapter IV). A preliminary report on these was published in 1981 (Magid, A., 1981: 42-45).

Flotation of soil samples from the site of Saggai [an Early Khartoum type of site situated on the east bank north of Khartoum (cf. fig. no.1)] did not yield any results as regards retrieval of plant remains (Costantini, L., 1983: 47). Moreover, the examination of 56 potsherds from the site for plant impressions gave negative results (ibid). Since the size of the processed soil sample(s) is too small and the number of the examined potsherds is very few, it is obvious that the obtained results are not conclusive and that further examination of much larger samples both of soil and potsherds should be done.

As for the hinterlands east and west of the Nile river, there are no published work so far as regards the latter. However, a number of papers were published since the beginning of the 1980s from the hinterlands east of the Nile. Among these are the results of the identified carbonized plant remains from the prehistoric sites of K I and II in Kassala area (cf. fig. no.1). These results show that three different species were identified: *Hordeum* sp., *Ziziphus* sp., and (?)*Leguminosae* (Costantini, L., et al, 1982: 33). One year later, in 1983, more results from the sites of JAG 1 and SEG 9 (cf. fig. no.1) were published but these consisted of plant-impressions (Costantini, L., et al, 1983: 18). The majority of these impressions were identified as domesticated sorghum, *Sorghum bicolor* (L.) Moench. The others belong to *Setaria glauca* Beauv. and less frequently *Paspalum* sp. On the basis of the typology of pottery, these finds were dated to the second millennium B.C. (ibid). What has been identified as *Hordeum* sp. from this area and dated the second millennium B.C. (ibid) raises a number of questions as regards whether this identification is correct (Magid, A.: *in press*) because the area was not a natural habitat of *Hordeum* sp. 3000 B.P. and no evidence of artificial irrigation is found to indicate that it was cultivated during the winter season. Furthermore, since the area of Kassala apparently was part of the most eastern limits of the initial belt of naturally growing sorghum (Harlan, J.R., and A.B.L. Stemler, 1976: 473), the presence of wild sorghum in this area is not unexpected, but its occurrence in a domesticated form as early as around the second millennium is unlikely, as will be explained in the following chapters. Nevertheless, the results of identification of the other plant remains from this area indicate that gathering and perhaps cultivation of wild food-plants were practised.

The present writer published a preliminary report on the initial results of identifying the plant remains retrieved from the cave-site of Shaqadud (Magid, A., 1984: 27-28). Carbonized grains of *Pennisetum* sp. (Brum.) Stapf. & Hubbard, wild *Sorghum* sp., *Panicum turgidum* Forsk. (Ar. *tumam*), *Setaria* sp., *Grewia tenax* (Forsk.) Fiori. (Ar. *guddeim*) and desiccated seeds of *Ziziphus spina-christi* Lam. and many other species are summarized in chapter IV in this thesis. It was not certain, then, whether the *Pennisetum* sp. was domesticated (ibid). Further examinations and consultations have now shown that the retrieved carbonized grains do not retain external morphological features which are well enough preserved to allow identifying the form (wild or domesticated) of these species (see chapters IV and VIII). Thus, the previous implicit assumption that the size and shape of these is similar to the domesticated form (ibid) is too hasty a conclusion. More samples with well preserved exterior morphological features are necessary to clarify the present uncertainty. In this connection, it should be mentioned that it has been suggested that the earliest evidence of domestic millet in the Sahel was recovered from the same cave- site at Shaqadud and dated to about 2500 B.C. (Marks, A.E., et al, 1985: 275). Since it is the earliest evidence of domesticated millet in the entire Sahel at present, it would have been very useful, in fact necessary, to state the diagnostic criteria and to present illustrations that show on which basis the identification was made.

There are many studies on the macroflora and the microflora mainly from archaeological sites in northern part of the Sudan dated to different periods such as that of Wendorf's (Wendorf, F., 1968: 10, 1051), Adams' (Adams, W.Y., 1982: 30), Rowley-Conwy (Rowley-Conwy, P., 1986: 110-111), etc. Since these studies are outside the geographical area being studied[3] here (the Central Sudan), detailed discussion is beyond the scope of this work. In short, this survey on the history and state of archaeoethnobotanical research in the Central Sudan shows that although this field of research is not yet fully established in the archaeological studies it certainly has undergone noticeable development during the last two decades. Furthermore, with the increasing attention to problems of the domestication and exploitation of food-plants in the Central Sudan, there is also a widespread consent among archaeologists and biologists that more attention should be paid to an interdisciplinary approach in archaeological research.

---

[3] Wickens has published some speculative reconstruction of the vegetation history and the past climatic conditions in the Sudan (Wickens, G.E., 1975: 45-65; 1982: 23-50).

CHAPTER III

The Geographical Setting of the Sites

3.1 Introduction.

In this chapter I shall try to bring together the basic information concerning the geographical setting of five archaeological sites; Aneibis, Abu-Darbein in Atbara, Shaqadud, Jebel Tomat and Rabak (cf. fig. no.1). The sites of Aneibis and Rabak are situated,in their respective order, at the most northern and southern boundaries of the area being studied here (cf. fig. no.1). They are separated by some 550 kilometres and this vast area embraces several dissimilar zones. In order to understand the physical setting of this very heterogeneous territory, I have divided it into three areas on the basis of similar geological, geomorphological and vegetational criteria. These are:

1) The Aneibis/Abu-Darbein (Atbara) area in the north.
2) The Shaqadud area in the centre. Mention should be made here of the fact that the sites in the Khartoum area, another area in the centre (cf. fig. no.1), will not be discussed in this thesis. Since the geographical features of this area have been described elsewhere (Arkell, A.J., 1953: 1; El-Mahi, A.T., 1982: 18-24; Haaland, R., 1981: 30-46; Magid, A., 1982: 9-17), it wilnot be included in this description.
3) The Jebel Tomat/Rabak area in the south.

The geographical setting of these areas can generally be outlined as follows:

1) The geological formation of the areas of Aneibis/Abu-Darbein and Jebel Tomat/ Rabak is the Basement Complex. The Shaqadud area along the west fringe of the Butana plains is characterized by its Nubian Sandstone geological formation (cf. fig. no.1).
2) With the exception of the clayey silt plains along the Nile and the Atbara rivers, the types of soils found at the areas of Aneibis/Abu-Darbein and the Shaqadud share some basic similarities in that they are generally sandy. The present dry climate with its high temperatures, erratic precipitation, high evaporation rates and surface winds, plays a major role in producing the incompletely developed soils of these two areas. The types of soils found in the Jebel Tomat/Rabak area are markedly different, being mainly cracking clay assumed to have been deposited down in the past by permanent water (Tothill, J.D., 1948: 133). In contrast to the gently sloping clay type, soil

found in the Jebel Tomat/Rabak area, the areas further north at Shaqadud and Aneibis/Abu-Darbein are characterized by their series of gravel-ridges, more frequent inselbergs and more prominent seasonal water courses of various sizes.

3) Climatologically speaking, the areas presented here are situated within the dry tropics. All experience relatively high temperatures (c. 41°C in summer and c. 15°C in winter) and consequently high evaporation rates. But differences also occur. Among these is the variability of the mean annual rainfall experienced in each of these three areas. The Aneibis/Abu-Darbein area, being at the very southern fringe of what is currently classified as the "Desert zone", receives only 65 mm of annual rainfall. The Shaqadud has at present an annual rainfall of 100 mm. The Jebel Tomat/Rabak area differs considerably from the preceding areas in that it is situated within a more vegetated region "Acacia Thornland alternating with Grassland" where the mean annual rainfall (c. 350 mm) is sufficient and its duration is long enough to bring to maturity many of the grasses, perennials and herbs. The water-retaining, clayey soils of this area are also a positive factor in keeping the water long enough until the plants reach their flowering and reproductive stage. During the last few years, however, these three areas have witnessed a decline in rainfall.

4) The floristic composition is markedly different and changing from one area to another as we move from north to south with vegetation that range from almost non-existent in the Aneibis/Atbara area, through sparse scrubs and annuals in the Shaqadud area and savanna thornland and to grass in the Jebel Tomat/Rabak area. The annual rainfall, topography, soil types and the biotic factors mainly humans and animals undoubtedly affect the prevailing distribution and composition of the plant communities in these areas. The role of each of these factors varies considerably from one area to another. For instance, the effect of the biotic factors is less prominent in the areas of Aneibis/Atbara and Shaqadud if compared with its very marked effects in the White Nile area between Jebel Tomat and Rabak. On the other hand, the physical and natural factors (e.g. soil, topography, rainfall, temperature/evaporation, etc.) have great influence in determining the distribution and the floristic composition of the plant communities both in Aneibis/Abu-Darbein and Shaqadud area. These factors are of secondary importance in the area of Jebel Tomat/Rabak.

5) Mixed farming with emphasis on the cultivation of cereal crops and vegetables is the basis of the economy in both Aneibis/Abu-Darbein and Jebel Tomat/Rabak areas. The mixed farming at Aneibis/Abu-Darbein is mainly dependent on artificial irrigation while that at Jebel Tomat/Rabak combines both artificial and rainfed cultivation. The inhabitants of Shaqadud adopt a type of sedentary pastoralism with infrequent resort to rain cultivation of sorghum.

6) Throughout the recent history of all these areas, the existing environmental and economic adaptations seem to have remained the same (Burckhardt, J.L., 1822: 212-215; Mubarak, M.O., et al, 1982: 143-144). Furthermore, the amount of rainfall was (and to a large extent is) the main factor determining the adaptations of the majority of the inhabitants and the life of the biological resources in these areas.Success or failure of rain is almost the sole determinant of the success or failure of crop yields, the survival and reproduction of livestock and consequently the adaptation of the nhabitants.

I shall, in the following pages, give more detailed information about the geographical setting of each of these areas.

3.2 Description of Aneibis/Abu-Darbein Area.

3.2.1 Geographic location.

3.2.1.1 The site of Aneibis (17° 58' N, 33° 59' E).

The site of Aneibis is located c.22 kilometres north of the city of Atbara (cf. fig. no.1), about 2 kilometres east of the present course of the Nile river. It is 7000 m^2 in extent. It is situated on an old river bank of the Nile, elevated c.2.5 m above the surrounding plains (cf. fig. no.1).

Four different concentrations of cultural material are distinguished and test pits have been dug[1] in each of these. A total of 136 m^2 were excavated. The cultural deposit is generally rich; and one metre deep at the most but no stratigraphy is distinguished in the cultural layer. Among the radiocarbon dates obtained, 7730±110 B.P. is the earliest and 7290±150 B.P. is the youngest date (cf. appendix no.1). According to the chronological chart of the culture-history of the Central Sudan, these dates (as well as the recovered material-culture) place the site of Aneibis within the Early Khartoum tradition.

3.2.1.2 The Site of Abu-Darbein (17° 40' N, 33° 58'E).

The site is located c. 23 kilometres south of the site of Aneibis and c. 2 kilometres southeast of Atbara town (cf. fig. no.1). It is situated on a steeply cut gravel-terrace which is part of the east bank of the Atbara river. This gravel ridge rises c.10 metres above the present flow of the river. Towards the east, the site rises only c. 3.8 m above its plain vicinity (cf. fig. no.6). The area of the site is c. 1200 m^2 of which 50 m^2 have been excavated. The site is badly eroded; yet, at its maximum the cultural debris is 0.6 m thick. The recovered cultural material is homogeneous (from top to bottom of the culture-layer) and no stratigraphy was recorded from the site. The oldest radiocarbon date

---

[1] The Aneibis, Abu-Darbein, Jebel Tomat and Rabak have been excavated by Professor Haaland, Dr. El-Mahi and myself. Analysis of data is not yet completed.

Fig.no 6 Abu Darbein; Contour map and excavated area in season 1985.

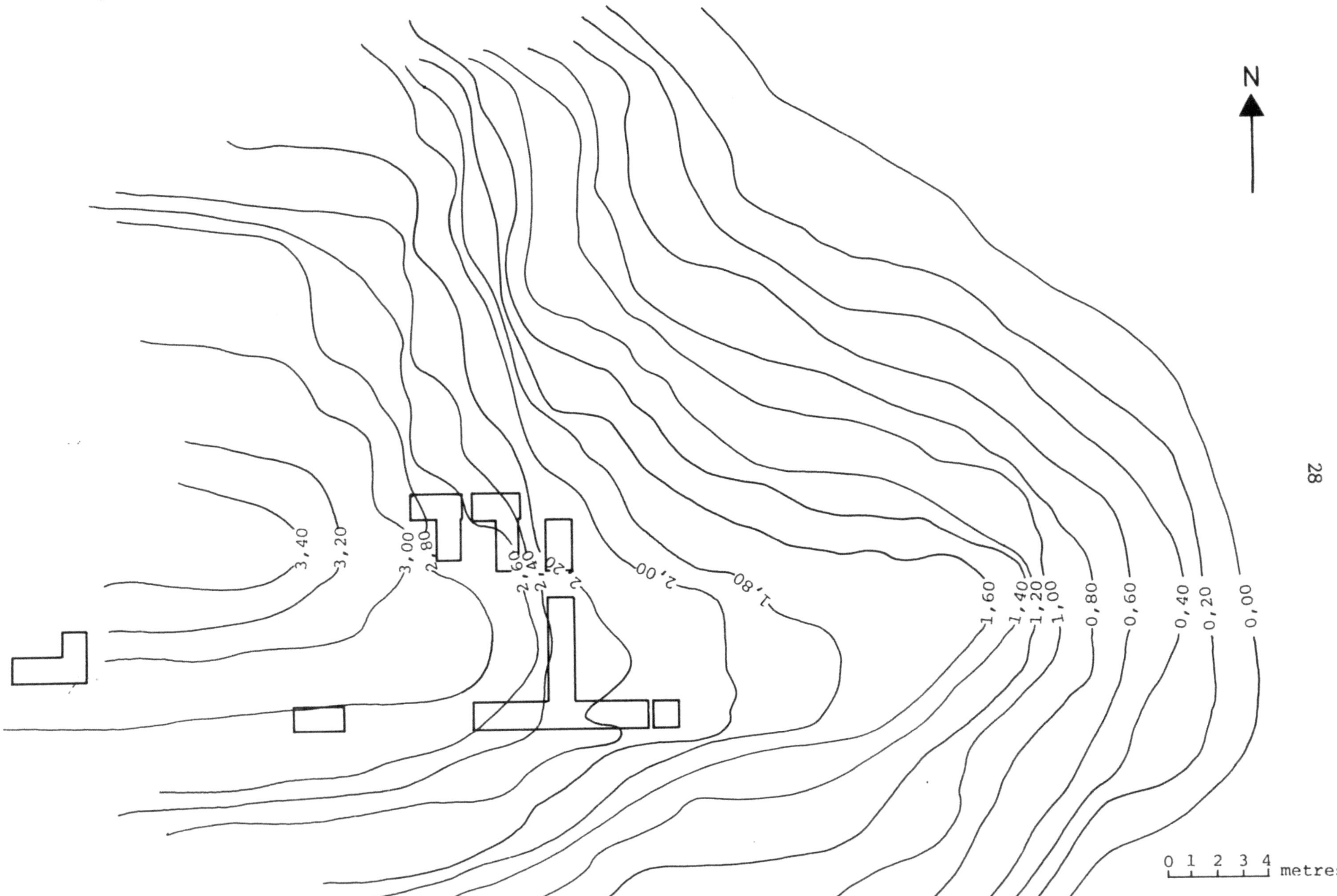

obtained from this site is 8500±100 B.P. and the youngest is 7700±140 B.P.(cf. appendix no.1). On the basis of both the archaeological material (e.g. wavy line pottery and bone harpoons) and these dates, the site of Abu-Darbein is also classified as belonging to the Early Khartoum tradition.

3.2.2 Geology.

The available information on the geology of the area of Aneibis/Abu-Darbein is rather meager. However, it has generally been described as being part of the Basement Complex (Andrew, G., 1948: 112; Whiteman, A.J., 1971: 6-7, 39) which is in this area often covered by superficial deposits.

According to a rather rudimentary classification, the Basement Complex includes all igneous and sedimentary rocks whether they have undergone metamorphosis or not. Their present arrangement has, however, been interpreted as being a result of metamorphism (Andrew, G., 1948: 92) of which granites, schists and gneisses contribute the more important classes of rocks to the Basement Complex. It has been proposed that this Complex is mainly of Pre-Cambrian age, but the probability that it contains younger formations than these has also been considered (Whiteman, A.J., 1971: 11).

3.2.3 Topography.

There is no major published source of information on the landforms of the Aneibis/Abu-Darbein areas, so the following account is based on the observations made by Drs. Haaland and El-Mahi and by the present writer during their field work in the area. In addition, some fragmentary information has also been collected from other sources.

3.2.3.1 The site of Aneibis.

Three overlapping topographical units are distinguishable. These are, a) the gravel-ridge series where the site is located, b) the flood plains of the Nile to the west, and c) the desert plateau stretching east of the site.

The gravel ridge is c.2.5 metres above the old flood-plains. This ridge seems to have been part of an old bank of the Nile. This interpretation is based on the following observations:

1) The test pits which were excavated along the western slopes of the ridge revealed clayey silt (starting from a depth of 20-30 cm and downwards) which is similar to that along the bank of the present water course.
2) Several test pits were also excavated at the centre and on the eastern slopes of the ridge. These revealed pebbles and ash-like gray soils which are different from the clayey silt along the western edge of the ridge. This may be taken as an indicative that

the western edge of the ridge was the border between this area and most likely the high water level of the Nile.

3) A similar gravel ridge stretches for several kilometres north and south of the gravel ridge area of the site. Despite the discontinuity of these ridges, the uniformity of their distribution in a north-south line parallel to the river does support the interpretation that they were an eastern terrace for an old course of the Nile.

West of these units is an area of sandy clay and cracking clayey plains. This area stretches from the area of the sites all the way (c.2 kilometres) to the present bank of the river. The correlation of the present course of the Nile, the gravel ridge series to the east and the fertile soils in between clearly indicate that there was a sizable flood-plain during the occupation of the Aneibis site.

The area east of the Aneibis site is characterized by its loose sandy desert soils (Lebon, J.H.G., 1965: 43-45) which stretches in flat plateau interspersed by infrequent low-lying gravel and sand mounds. A very few beds of seasonal gullies taking their courses towards the Nile in the west traverse this area from the east.

3.2.3.2 <u>The site of Abu-Darbein (Atbara)</u>.

The Atbara river is the most prominent feature. It is the largest main tributary of the Nile north of Khartoum. Thus it is essentially the principal contributor to the Nile, both of water and of sediments. It contributes 12.5 km^2 of annual flow that furnishes 14 million tons of sediments during the flood season, May to September (Williams, M.A.J., and D.A. Adamson, 1982: 119).

The site of Abu-Darbein is situated within an area which has three overlapping topographical units. These are generally comparable with those in the area of the Aneibis site; but differences in details between them do exist and will now be described in the following.

The three overlapping units of Abu-Darbein are: a) the gravel ridge on which the archaeological site is situated, b) a narrow flood-strip and the course of the Atbara to the west, and c) the desert plateau towards the east.

With regard to the first unit, the site of Abu-Darbein is situated on a gravel-ridge which at the present is part of the east bank of the Atbara river (unlike the old bank at Aneibis). The western edge of this gravel-terrace is characterized by its abrupt cut and almost vertical profile (c.10 metres above the present river course). This seems to be due to a heavy erosive action by the river during its period of high water level. The river is characterized by its very strong eroding water-current during the flood period. The course of the river in this area bends slightly eastwards so that the effect of erosion by the strong current is greater on the east bank. This creates a very narrow flood-strip (denoted as the second unit) along the east bank, while a wider flood-plain with gently sloping deposits is developed on the west bank.

The desert plateau (the third unit) exhibits a surface coated with mixtures of sandy and calcareous soils spotted with pebbles, conglomerates and strips rich in ferrous gravels. Traces of dry water courses and beds of gullies of different sizes are also present in this area.

3.2.3.3 Comparison between the topographic features of Aneibis and Abu-Darbein.

There are some basic similarities between the topographical features of the areas of Aneibis and Abu-Darbein, but there are also some major differences. The similarities between the areas may be summarized as follows:

1) The two localities have three similar overlapping topographical units; the gravel ridges (where the sites are situated), the flood-plains of the Atbara and the Nile rivers and the desert plateau.
2) The types of soil in the two areas have been derived mainly from weathering, transportational and depositional processes. Some have classified these as "semi-desert" type of soils (El-Tayeb, G., 1972: fig.3; Lebon, J.H.G., 1965: 45). Others have classified them as of a desert type (Barbour, K.M., 1964: 53; Greene, H., 1948: 145-147; Whiteman, A.J., 1971: 130). If we consider the nature of the arid origin of these types of soil co-related with other physical and biological features of the area, I support the latter classification (cf. figs. no.7 and 8).
3) The upper layers of the immature[2] poor soil-deposits are to a large extent natural products of the prevailing dry climate, but this is not necessarily true of the older type of soil that are underlying these. The abrading action of these generally sandy type of soils, the contrasting temperatures (e.g. between day and night, summer and winter) are some of the factors which do not favour preservation of plant remains.

The differences between the two areas are as follows:

1) The series of the gravel-ridges at Aneibis is limited to a narrow strip running north-south and is less than a kilometre wide at its broadest. The Abu-Darbein gravel-ridge is much wider as well as discontinuous.
2) A distinction can easily be made between the three overlapping units in the Aneibis area whereas they are rather obscure in the Abu-Darbein area; for instance the gravel-ridge and the area east of it (the desert plateau).
3) The two kilometres wide present flood-plain in the Aneibis area is at least six times broader than the narrow strip at Abu-Darbein. Accordingly, one may argue that during the occupation of the Aneibis site, the flood-plain was considerably wider;

---

[2] Immature soils are described as being the type of soils which are recently deposited and lack what is defined as soil horizon(s) (layers) and soil profile (Greene, H., 1948: 145).

Fig. no. 7   Types of soil in the Sudan

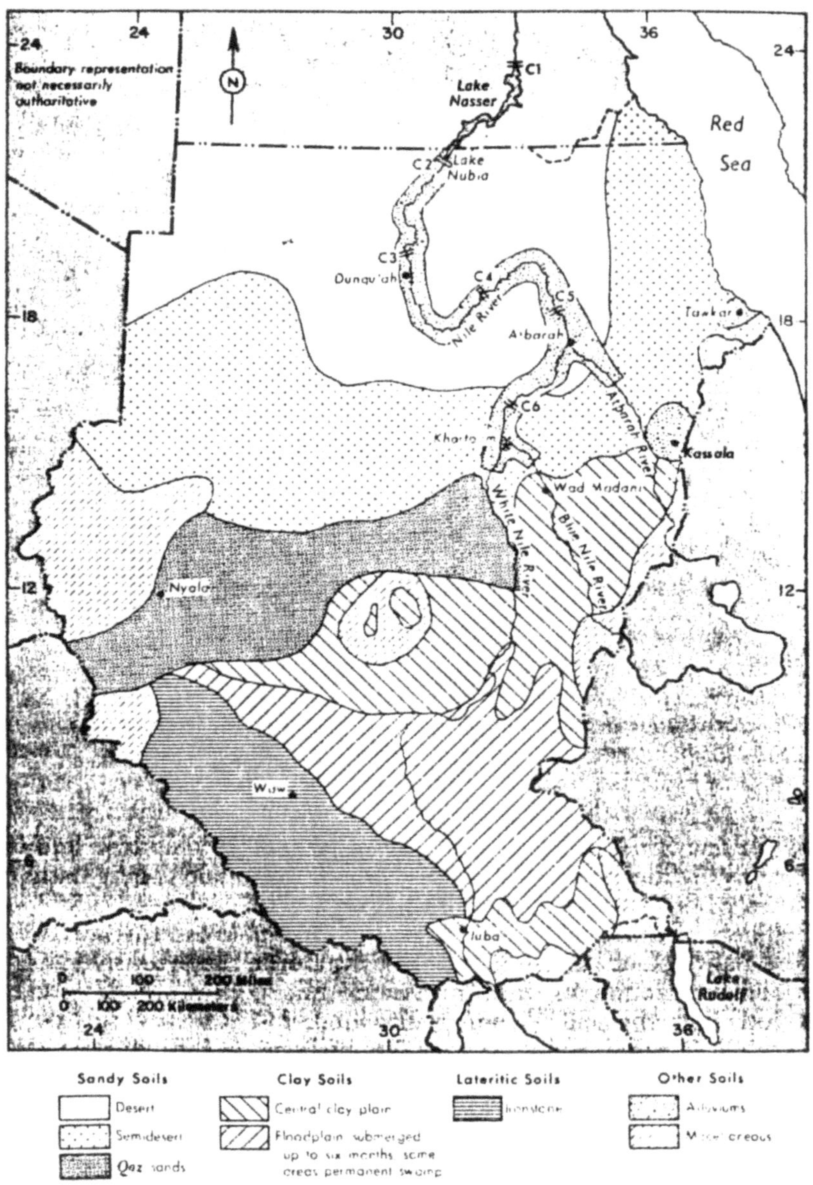

After Whitaker, D.P., 1982:150

Fig.no 8 The Central Sudan:
Past and Present vegetation.

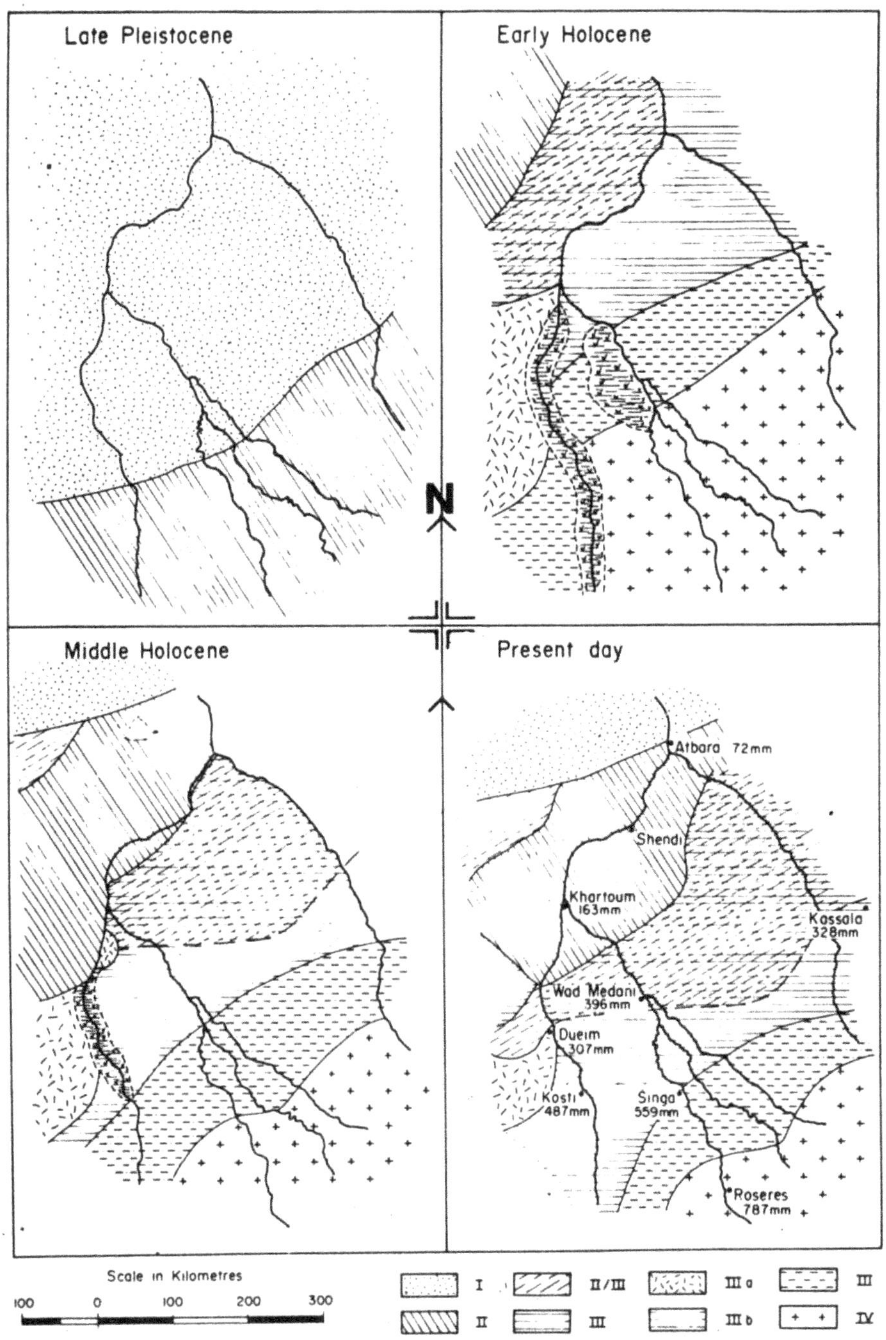

Legend:

I. Desert
II. *Acacia tortilis* – *Maerua crassifolia* semi-desert scrub.
II/III. Semi-desert grassland.
III. *Acacia mellifera* thorn savanna.
IIIa. *Acacia senegal* thorn savanna.
IIIb. *Acacia mellifera* – *Commiphora* thorn savanna.
IIIc. *Acacia seyal* thorn savanna.
IV. *Anogeissus leiocarpus* – *Combretum hartmannianum* deciduous savanna woodland. Swamp symbols denote swamp vegetation along White Nile and 'torch' vegetation west of Blue Nile.

After Wickens,G.E.,1982; 40-41.

while there was hardly any flood plains during the occupation of the Abu-Darbein site. This is further supported by the information obtained when comparing the types of soils from the excavated pits at the Aneibis site with the present types of soil adjacent to the Nile river.

3.2.4 The present climate.

The Aneibis/Abu-Darbein area is situated within the dry, tropical, continental climatic zone which is markedly arid (Lebon, J.H.G., 1965: 5; Whiteman, A.J., 1971: 1-5). The dominant feature of the climate is the north-south migration of the intertropical zone.

There are two main seasons in this area; determined primarily by the movement of the sun and secondarily by the prevailing seasonal winds: These seasons are winter and summer; and they are more or less equal in length. Winter is generally cloudless, warm and dry, and lasts from November to March. Northerly and, infrequently, northeasterly winds and high pressure dominate during this season. Mean temperatures drop to their lowest point (14.4°C and 15.7°C) during January and February.

As the intertropical convergence zone moves north towards the end of March, and the cool winter winds cease to blow, the effect of the high pressure is considerably weakened; and solar radiation increases both in intensity and duration (Barbour, K.M., 1964: 42-44). These features mark the beginning of the summer season. This season is fairly hot and the highest mean temperature reaches 42.4°C to 42.9°C during May and June. The highest evaporation rates concur with the maximum temperatures, recording 23.3 mm and 23.7 mm respectively.

Southeasterly and southwesterly winds dominate throughout the period from July to September. Passing over the Indian ocean, these winds bring summer rains to the Sudan. Almost the whole load of moisture of these winds is discharged further south. Since this area is situated at the northern fringe of the monsoonal regime, it hardly receives any rain; the mean annual rainfall is barely c. 60 mm (cf. histogram no.1; Williams, M.A.J., et al, 1982: 130). The rainfall occurs irregularly both in space and time. We should, however, bear in mind that the mean annual precipitation which is calculated over a fairly long period will undoubtedly mask many rainless years, as has been the case in this area between 1981 to 1987.

Histogram no.2 clearly shows that the precipitation rate, low humidity compared with the high temperatures and the great potential evaporation all account for the prevailing aridity in the area.

Histogram no 1

RAINFALL TRENDS IN CENTRAL SUDAN :
From three stations between 1941- 84

ATBARA

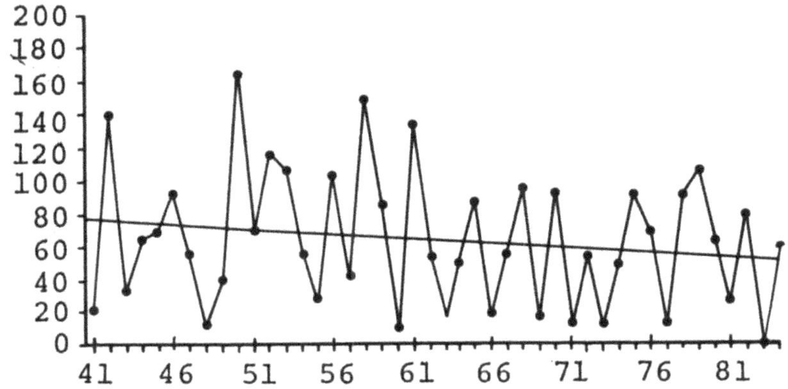

Mean 63,6mm.

KHARTOUM

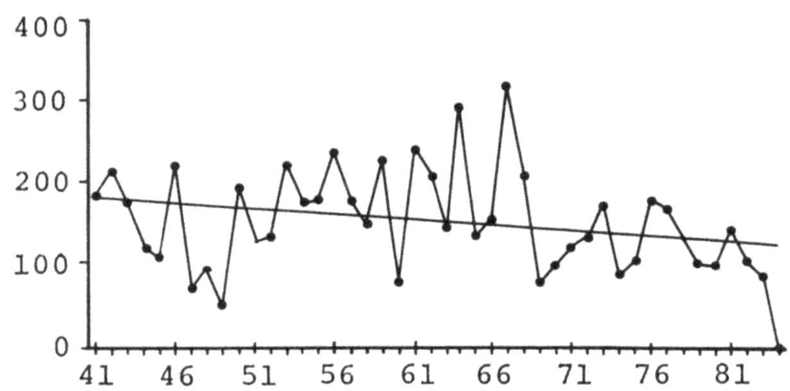

Mean 150,5mm.

KOSTI

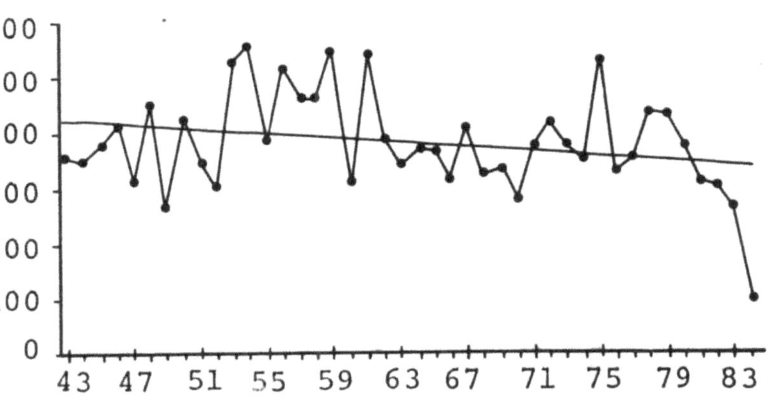

Mean 383,8mm.

Histogram no 2

ATBARA

AIR TEMPERATURE (the last 30 yrs.: 1941–84)

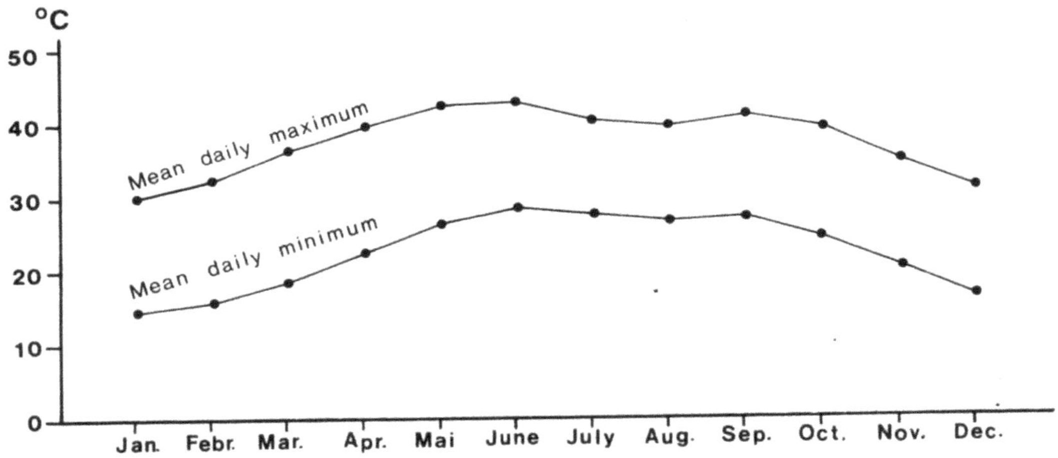

EVAPORATION (the last 30 yrs.: 1941–84)

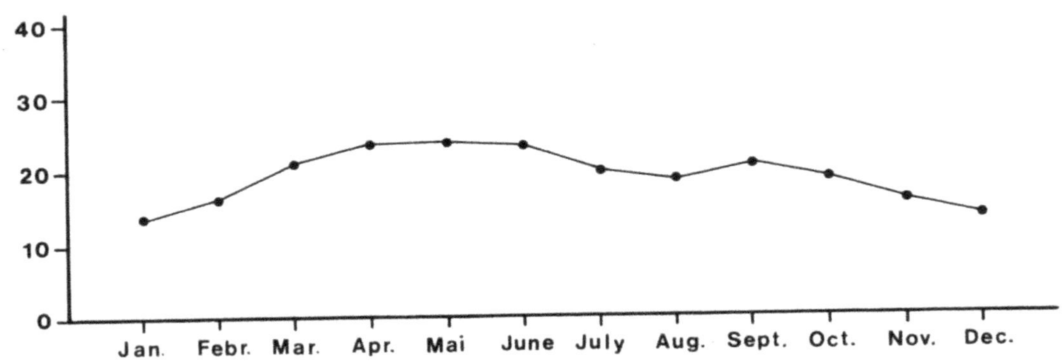

RAINFALL (the last 30 yrs: 1941–84)

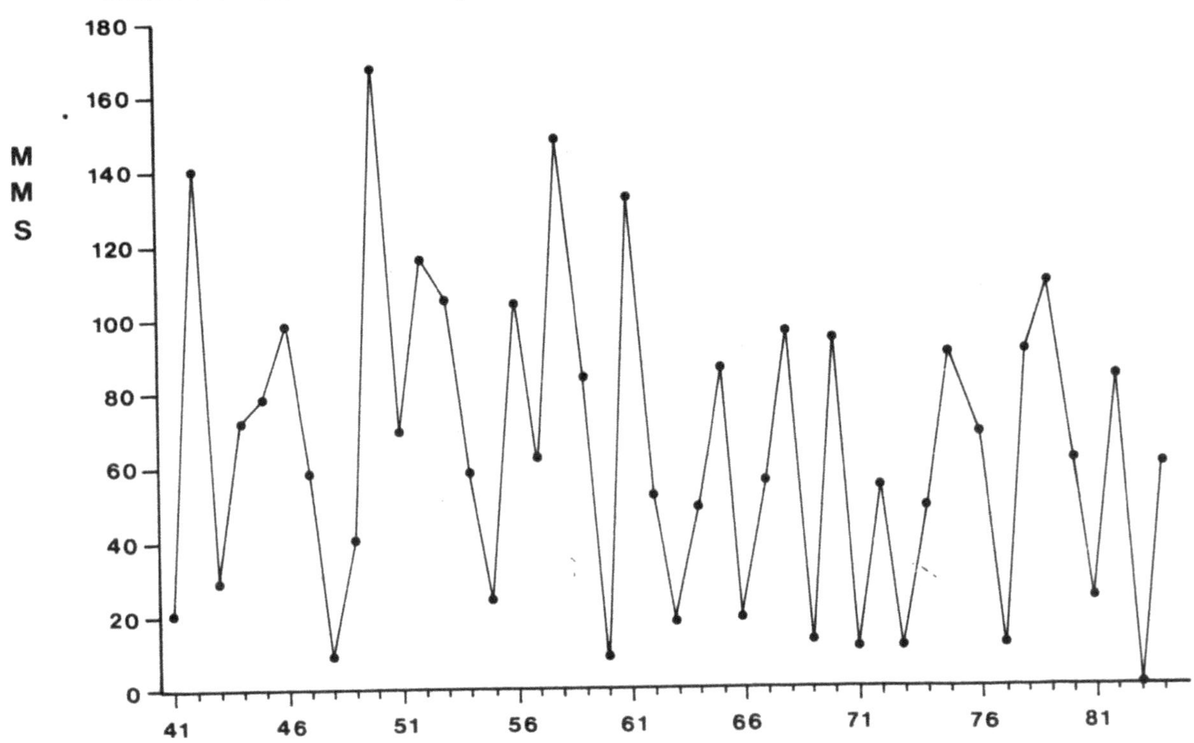

3.2.5 The present flora and fauna.

3.2.5.1 Flora.

3.2.5.1.1 Floristic composition.

The chief floristic character of this area is the deeply rooted *Acacia* species. Along the alluvial plains of the Nile and the Atbara rivers there grows a fair variety of trees and scrubs of *Acacia* spp. Among these, the most common are *Acacia arabica* Willd. (Ar. *sunt*) and *Acacia albida* Del. (Ar. *haraz*). The former is a very good source of fire wood and its leaves are browsed by camels. The fleshy-orange coloured flat pods of the latter are most favoured by goats and sheep and its leaves by camels (cf. table no.1). Sufficiently watered, low-lying plains form the optimum habitat for the growth of *Capparis decidua* Pax. (Ar. *tundub*) and *Ziziphus spina-christi* (L.) Willd. (Ar. *sidir, nabag*). Both species have edible fruits. Both these species and the above mentioned *Acacia* sp. have dispersed distribution patterns. A fair number of *Phoenix dactylifera* Linn. (Ar. *nakhal*) and a few *Hyphaena thebacia* Mart. (Ar. *dom*) also grow here. Both species are heavily exploited (cf. table no.1). Many herbs and grasses grow along the alluvial plains. Among these are *Desmostachya cynosuroides* (L.) Stapf. (Ar. *halfa*), and several species of *Aristida* and *Panicum turgidum* Forsk. (Ar. *tumam*); the latter grows sparsely in the drier strips. Species of wild *Sorghum* Stapf. (Ar. *adar*), *Echinochloa colona* Link. (Ar. *difra*) and *Cenchrus biflorus* Roxb. (Ar. *haskanit*) are among the many grasses growing in this area. All these (except *Cenchrus* sp. with its unpleasant, spiny fruits) provide very good fodder for livestock (cf. table no.1).

As we proceed eastwards away from the alluvial plains of the Nile and Atbara rivers, there is a quite observable decrease, if not an abrupt decline, in the vegetation cover. The soil consists - as mentioned previously - of either loose sand plateau or gravel ridges and compact pebble mounds. A few seasonal gullies and water courses traverse the area. It is along these that *Acacia tortilis* (Forsk.) Christensen (Ar. *samr*) predominates. Besides these, there are a few bushes of *Acacia mellifera* Benth. (Ar. *kitr*). Both are palatable for camels and goats, but the latter is an inferior browse feed. The otherwise dormant seeds of the unpalatables, *Cassia senna* L. and *Cassia italica* (Mill.) Lam. (Ar. *senna*), germinate in sandy depressions and the fans of *wadi*s after the infrequent showers moisten the soil.

Most of the above mentioned plant species have special attraction for human beings, animals or both (cf. table no.1).

3.2.5.1.2 Discussion.

I have presented in the foregoing only a descriptive account on the present vegetation of the Aneibis/Abu-Darbein area. Now, I shall discuss critically some of the main classifications proposed for the vegetation of this area.

Table no. 1: An example for the plant-species exploited and their uses in the Aneibis/Abu-Darbein area.

Species name		Use of plants								
							Construction & other uses			
Latin	Arabic	Sheep	Goats	Cattle	Donkeys	Camels	Houses	Fences	Fire-wood	Other uses
Acacia arabica	sunt	0	0-2	0	0	1-3	0	0-1	5	0
Acacia albida	haraz	1	5	0	1	1-4	0	1	1-3	0
Acacia tortilis	samr	0-1	2-3	1	1	5	0	2-3	5	3 animals shelter
Acacia mellifera	kitr	0	0-1	0	0	2-3	0	5	0-1	0
Capparis decidua	tundub	0	0	0	0	0	0	0-1	0	0-1 edible fruits
Z. spina-christi	nabag	1	3-4	0	0	5	0	3-4	0-1	2-3 edible fruits
Phoenix dactylifera	nakhal	1-2	3-4	0	2-3	0-1	5	0-1	1-4	5 edible fruits
Hyphaena thebacia	dom	0-1	1-3	0	0	3-4	4	0-1	1-2	1-2 edible fruits
Panicum turgidum	tumam	5	5	5	5	5	0	0	0	0
Sorghum sp.	adar	5	5	5	5	5	0	0	0	0
Cenchrus biflorus	haskanit	0-1	2-3	0	0	2-3	0	0	0	0
Echinochloa colona	difra	5	5	5	5	5	0	0	0	0

Legend:
0 = not grazed and/or used
1 = little grazed and/or used
2 = grazed and/or used
3 = fairly grazed and/or used
4 = well grazed and/or used
5 = intensively grazed and/or used.

According to Andrews' system of classification of the vegetation in the Sudan, the Aneibis/Abu-Darbein area is situated within the desert vegetation zone (Andrews, F.W., 1948: 34). It should be noted here that the records show a relatively low rainfall and dry phase during the period between 1939 - 1948/49 (Ibrahim, F.N., 1984: 123; histogram no.1). There are other classifications that have been proposed but are not quite in conformity with Andrews'. Lebon, for instance, suggested that there are five principal vegetation zones in the Sudan (Lebon, J.H.G., 1965: 20). He assigned the vegetation of the Aneibis/Abu-Darbein area to zone no. II, the "Semi-Desert Acacia Scrubs and Short Grassland" lying approximately between latitudes 14° - 17° N (ibid: 20-25). Although the period between 1950-1965 was markedly wet (Ibrahim, F.N., 1984: 123; histogram no.1) and accordingly the vegetation cover was denser (Lebon, J.H.G., 1965: 20-25) than the previous dry phase (1939-1949), yet Lebon's classification seems to contradict itself when applied to the area being studied. That is to say, differences in latitudinal location are essential in determining the rainfall and consequently the general vegetation cover (Wickens, G.E., 1982: fig. no.3.3, 34). But according to Lebon's system (ibid: 20), this area is geographically located within the desert zone, no. I; north of latitude 17° N; at the same time he referred to its vegetation as being of zone no. II type which he geographically confined to the area between latitudes 14° and 17° N.

In his hypothetical reconstruction of the late Quarternary vegetation regions in the Central Sudan, Wickens has included only Atbara (Abu-Darbein) area in his second category of vegetation: "Semi-desert Scrub". To be more specific, Atbara was included within the semi-desert vegetation but was located right on the border-line between this vegetation region and the desert vegetation (Wickens, G.E., 1982: 32-40; cf. fig.no.8). Although the Aneibis is only 23 kilometres to the north, yet the implication of his classification is that this area is part of the desert vegetation zone (ibid). On the basis of the nearness of the Aneibis and Abu-Darbein from each other, their basic similarities in e.g. the types of soil, topography, rainfall, etc. and from the point of view of a person in the field; these two areas should rather be related to one zone and not two different ones.

It is obvious from what is mentioned above that such definitions of desert and semi-desert zones or regions were made but with emphasis on different criteria. Another secondary reason may be related to the recently accelerated drought and the alternating dry and wet phases in the past, for instance (as implicitly cited above) different classifications of the vegetation in this area were proposed by different authorities at different points in time only during the last forty years. In an attempt to clarify these issues, the present writer (having recently been doing surveys in both zones) has tried to recombine these criteria and will put forward a more comprehensive definition bearing in mind the following:

1) Admittedly, the transitional territory from one zone to another, i.e. from desert to semi-desert is gradual and there is no clear or abrupt line to divide them. It should, however, be stressed that conformity of the prevailing vegetation (in its broadest sense) with the mean annual rainfall (calculated over c.50 years) approximate the optimum criteria for the classification of the vegetation zones.
2) There are also three other factors which should be considered. These are:
   a) the variability of the physical and chemical composition of the different types of soil.
   b) the nature of the dominant physical feature(s) of the region i.e. mountains, plains, drainage pattern, etc. These two (a) and (b) may impose local differences in the vegetation composition even within the same area.
   c) the border zone between for instance the desert vegetation and the semi-desert vegetation may fluctuate during different periods of time due to a few years with an exceptionally high rate of rainfall when the desert zone retreats and the semi-desert zone expands or vice-versa in case of having some years with low or no rainfall.

Accordingly, the present writer proposes that both the Aneibis and Abu-Darbein (Atbara) areas are situated within the desert region where the mean annual rainfall is 70 mm or less (histogram no.1). Since this area is situated within the desert region the vegetation is virtually non-existent except for localities with ample of water for instance along the Nile and Atbara rivers and along the seasonal water courses where emphemeral herbs and grasses grow after infrequent rain showers. This classification is further confirmed by a very recent study on the vegetation of the Sudan (Fadlalla, B., 1987).

3.2.5.2 Fauna.

Since Atbara is one of the big towns in the Sudan, its inhabitants, including those of Abu-Darbein rarely bother to keep animals; the people from the neighbouring villages, however, normally come with animals to this town either as their means of transport or to sell them. If we consider the whole area, the majority of the inhabitants are farmers; a few reside in the vicinity of Atbara, and the remainder are natives of Aneibis and its neighbouring villages. The most important domestic animals kept are goats (*Cappra aegagrus* f. hircus), secondly sheep (*Ovis ammon* f. aries), thirdly cattle (*Bos primi-genius* f. taurus), fourthly donkeys (*Equus africanus* f. asinus) and least important are camels (*Camelus* sp.). Chicken, pigeons and a few ducks were kept but they were not economically significant until recently when the increasing market demand for them has changed the attitude of the people towards a more commercial investment in these species. Many of the inhabitants keep watch dogs (*Canis lupus* f. familiaris) specially in the town of Atbara.

Both goats and sheep are of typical desert species. The sheep with their roman nose, hornless head and generally brown or infrequently spotted (black and white or brown and white) hairy coat. The goats are characterized by their long legs, long black hair (very few with brown hair) and medium length horns (Lebon, J.H.G., 1965: 108-110). The one-humped camel of the desert type is the most suitable animal for such dry habitat.

The goats are kept mainly for milk; sheep are raised basically for their meat and to a lesser degree for their milk. Cattle are the plough-animals; but their milk, and, rarely their meat are also exploited. Donkeys are the principle means of transportation for the villages.

There are also different species of birds such as weavers, kites, tits, egrets, etc. and quite a number of fish species are also available in both the Nile and the Atbara rivers. The most favoured species are the Nile perch and the catfish.

The female Anophalis and house-flies are two serious threats of the inhabitants' health particularly during summer. Small flying insects (local Arabic name: *nyimitti*) disturb both human beings and animals. They also destroy the seasonal vegetables grown along the cultivable plains especially during winter.

3.3 Description of the Shaqadud Area.

3.3.1 Geographical location: The cave-site of the Shaqadud (16° 15' N, 33° 45' E).

The complex of sites at Shaqadud is situated c.140 kilometres northeast of Khartoum. It is c. 50 kilometres east of the Nile at Wad Ben-Naga (Marks, A.E., et al, 1980: 30-35; 1982: 47-50) and about 12 kilometres east of the archaeological monuments of el-Naqa (cf. fig. no.1). This complex of sites consists of four different sites (cf. fig. no.9) but the main emphasis of the present work is on the cave[3] site (S1A). The area of the site (S1A) is c.36 m^2 but due to the very low roof of the cave, only c.9 m^2 were excavated. The cultural layer is c.3.50 m deep. Four radiocarbon dates were obtained from the site (appendix no.1); the oldest of these is 4123±86 B.P. and the youngest is 3615±88 B.P. (Marks, A.E., et al, 1984: 32). Since no other site of this period has been excavated in the Central Sudan the material culture e.g. pottery is new to the archaeology of this area (Marks, A.E., et al, 1985: 273).

[3] The complex of sites of Shaqadud were excavated in the seasons of 1981 and 1982 by a joint team of Sudanese (from U. of K.) and Americans (from S.M.U.). I participated in the last season-excavations and I am given permission to take samples from 0.25x0.25x3.50 m for analysis of macrofossil plant remains.

ig. no. 9 The complex-sites of Shaqadud:contour map.

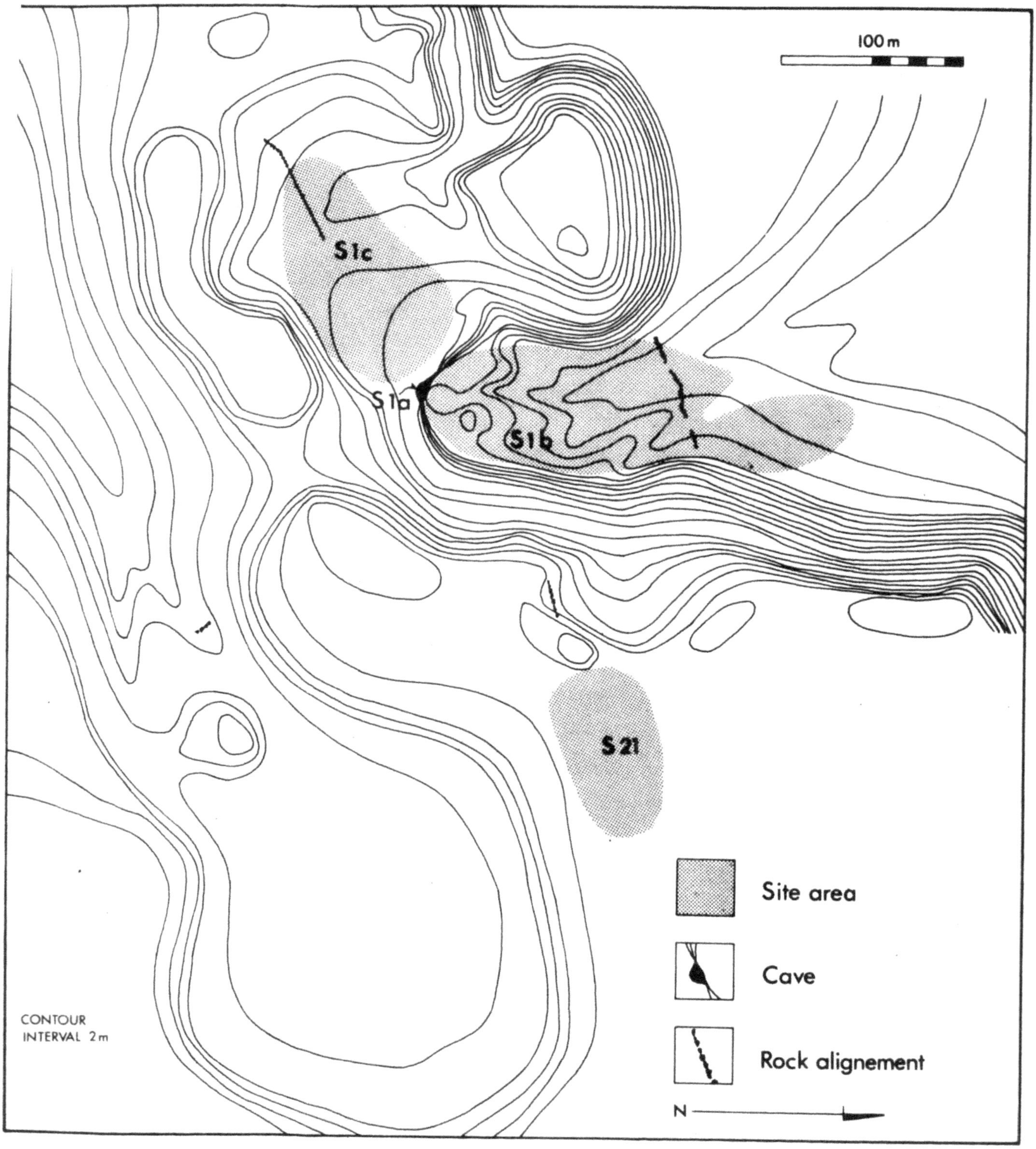

3.3.2 Geology.

The Nubian Sandstone is the main geological formation of the Shaqadud area (Al-Horr, A., 1983: pers.comm.). This formation is usually bedded with horizontal conglomerates, sandstones and sandy mudstones (Whiteman, A.J., 1971: 54). The beds of the formation in this area are horizontal or slightly dipping and are characterized by their discontinuity and variable thickness (Al-Horr, A., 1983: pers. comm.). The thickness of the exposures in this formation ranges between 30 and 35 kilometres. Their dominant colour is chocolate brown, but dark-brown to blackish-brown ferruginous crusts cap them in some localities (ibid). As described by Whiteman,

> 'often the Nubian Sandstone Formation is iron-rich and in a number of localities beds of iron ore occur. These are well developed in the Shendi district and were worked in Meroitic and ? later times. They are now of little economic value' (Whiteman, A.J., 1971: 62).

The Nubian Sandstone Formation in the Shaqadud area (and in the Central Sudan at large) is believed to be of a Mesozoic age (ibid: 52).

The water-table in this area is estimated to lie between 90 and 120 metres (300-400 feet) below the surface (ibid: 198); but a detailed record for the Nubian Sandstone Formation of the Sudan as such, and for this area in particular has not yet been made.

3.3.3 Topography.

Unfortunately, the modern landforms of the Shaqadud area are nowhere adequately studied, yet, the present writer has tried to record some of their main features. Generally speaking, the Shaqadud area can be delineated as being part of the western fringe of the Butana plains. These plains form the most easterly part of the Sahel zone; the Nile and the Atbara rivers delimit the east-west barriers of the plains but the southern boundary is not yet well defined. The area generally comprises flat plains consisting of sandsheets with some patches of sandy-silt along the beds and banks of the seasonal streams (Ar. *wadi*), seasonal watercourses (Ar. *khor*) and gullies (Ar. *eiddyiat*). It seems that deflation, among other geomorphic processes, has resulted in the formation of these plains, sandsheets and silt deposits.

The soils of this area are not completely developed due to many factors but mostly to climatic effects e.g. high temperature, erratic precipitation, high evaporation rates, strong surface winds and sandstorms. Lebon included these soils within his category of immature soils (Lebon, J.H.G., 1965: 45).

The drainage system of this area can be related to the second category of Lebon's classification, i.e. the central zone of seasonal watercourses which only flow during the short rainy season (ibid: 13) between July and September and remain dormant most of the year. According to local informants some of the streams in this area e.g. Wadi Al-Awateib for instance used to flow as far as the Nile (50 kilometres to the west) until one and a half decades ago, and hold water all year round. These streams, then, made an invaluable contribution to securing the inhabitants' food-supply for the reason that they were flowing through fertile silt areas which were used for rain-cultivation of sorghum.

The Shaqadud area is dissimilar to the area of Aneibis/Abu-Darbein in that it is situated in the hinterland away from rivers. The only permanent source of water are the dug-out wells of which some are believed to date back to the late Meroitic period. The seasonal water courses are, however, more frequent in the Shaqadud area. This may partly be attributed to the mountainous terrain when compared with the relatively flat landscape at the Aneibis/Abu-Darbein area. Furthermore, the soil of the Shaqadud area is characterized by its coarse loose sand, except the fine silts along and within the beds of the seasonal water courses.

3.3.4 Climate.

The Shaqadud area is situated within the zone of the dry tropics (Obeid, M., and Mahmoud, A., 1969: 140; Whiteman, A.J., 1971: 1-5). The prevailing climate is generally arid.

Similar to Aneibis/Abu-Darbein area, the Shaqadud experiences yearly two main seasons: the cool dry winter and the hot rainy summer. The lowest mean temperatures are 14.4°C and 16°C respectively during January and February. The highest mean temperatures are 41.1°C and 41.2°C in May and June. The summer rains are erratic; the precipitation is variable both temporally and spatially. The area has been placed within the 100-200 mm rain isohyets (cf. fig. no.10); this has, to a large extent, been taken as a differential criterion between this area and the desert region directly north of it (Lebon, J.H.G., 1965: 8). However, for the last two decades it has experienced an accelerating decline in the rate of rainfall, it is most likely that the present annual rainfall is c.100 mm. August is the wettest month of the rainy period; the relative humidity during this month is 30-50% while it is only 18-38% during the hottest months in summer (May-June).

The potential evaporation is much greater than the calculated precipitation rate throughout almost the entire year; this, together with other physical and biotic factors, accounts for the prevailing predominantly hot and arid climate in the area.

3.3.5 The present flora and fauna.

3.3.5.1 Flora.

3.3.5.1.1 Floristic composition.

The chief floristic component is *Acacia* sp. Other scrub species are also sparsely represented. The vegetation cover is generally dispersed and areas entirely bare

Fig.no.10 Average annual rainfall in the Sudan (1930 - 1960)

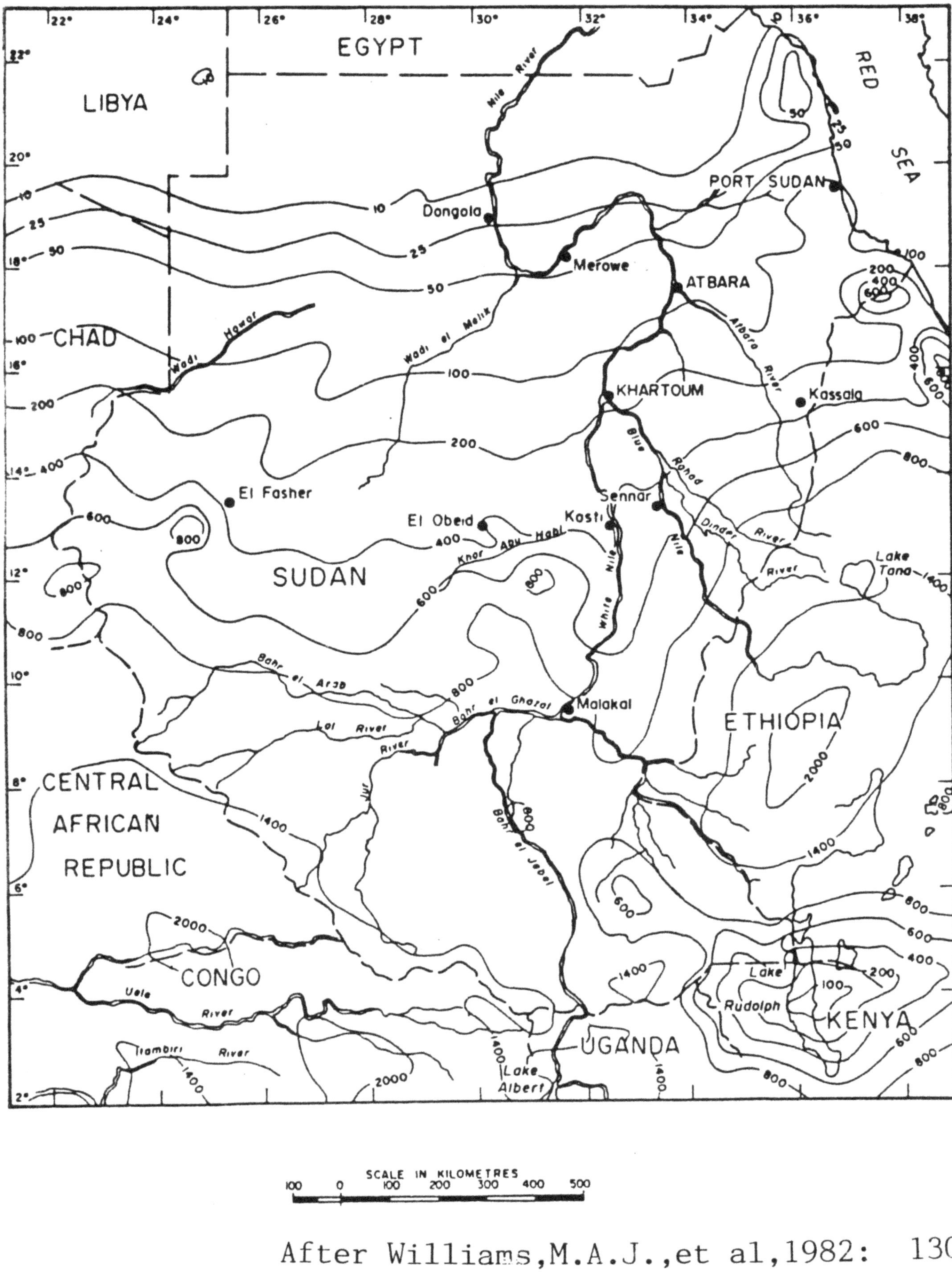

After Williams,M.A.J.,et al,1982: 130

of vegetation may occur. Other areas which are comparable to Shaqadud in terms of annual rainfall, soils were described as being of:

> 'a varying mixture of grasses and herbs, either without any woody vegetation at all, or more usually with variable scatter of scrub bushes .... interspersed with bare areas ....' (Lebon, J.H.G., 1965: 20).

Acacia raddiana Savi. (Ar.*seyal*) and *Acacia tortilis* (Forsk.) Christensen (Ar.*samr*) are the two dominant species among the scrubs growing in the area at present. Both of them were heavily browsed, and intensively exploited for fire-wood, charcoal and hut construction (cf. table no.2). Consequently, these species have been drastically reduced both in number of population and in the size of each individual. In addition, there are *Acacia flava* (Forsk.) Schweinf. (Ar.*sallam*), *Acacia mellifera* Benth. (Ar.*kitr*). The former is browsed by animals and the latter is exploited by the local people (cf. table no.2). The following species *Ziziphus spina-christi* (L.) Willd. (Ar.*sidir/nabag*); *Maerua crassifolia* Forsk. (Ar.*sarih*); *Balanites aegyptiaca* Del. (Ar.*heglig*); *Combretum aculeatum* Vent. (Ar.*shaheit*) and *Grewia tenax* (Forsk.) Fiori. (Ar.*guddeim*) are used for a variety of purposes by humans and/or animals as illustrated in table no.2. The dry light wood of the unpalatable *Calotropic procera* Ait. (Ar.*ushar*) is used for making enclosures for newly born goats and sheep.

The distribution of these species is mainly concentrated along the seasonal streams and gullies and other low depressions, provided that the soils of the latter contain some elements of clay and silt.

Herbs and annuals grow in this area during the short rainy season. These include several species of the gen. *Sorghum* Stapf. (Ar.*adar*); *Echinochloa colona* Link. (Ar. *difra*) and *Panicum turgidum* Forsk. (Ar.*tumam*). Being highly favoured by all animals, these species are intensively grazed (table no.2); *Cenchrus biflorus* Roxb. (Ar.*haskanit*) is a palatable forage grass but has an unpleasant spiny fruit. *Citrulus colocynthis* L. (Ar. *handal*) also grows during the short rainy season but it is an unpalatable species. Edible annuals that grow in the this area are *Hibiscus esculentus* L. (Ar. *bamia*) and *Corchorus terristeris* L. (Ar. *melukhyia khalaweyia*).The local people use both as food-plants; thick stews are cooked from the fruits of the former and the leaves of the latter. *Citrulus vulgaris* Schrah. (Ar.*battikh*) which grows during the short rainy season, bears, at maturity, edible fleshy and juicy fruits that provides the local people with food and a sweet alternative for water. *Cymbopogon proximus* Stapf. (Ar. *mahareib*) which grows in sandy depressions and protected sandy depressions on the tops of the mountains, is well known as a medicinal plant for stomach-aches and it is also commonly used to flavour the tea.

Table no. 2: An example for the plant-species exploited in the Shaqadud area.

Species name		Uses of plants								
							Construction & other uses			
Latin	Arabic	Sheep	Goats	Cattle	Donkeys	Camels	Houses	Fences	Fire-wood	Other uses
Acacia raddiana	seyal	0-1	1-2	0	0	2-4	5	0-1	5	2 storing of fodder
Acacia tortilis	samr	0-1	2-3	0-1	0-1	5	5	0-1	5	4 storing of fodder
Maerua crassifolia	sarih	0-1	2-3	0-1	0-1	5	5	0-1	5	0
Z. spina-christi	nabag/ sidir	0-1	3-4	0	0	4	0-1	3-4	0-1	1-2 edible fruits
Balanites aegyptiaca	heglig	0	0	0	0	3	3	0	1-2	0
Grewia tenax	guddeim	0	0-1	0	0	0-1	0	0	0-1	1-2 edible fruits
Sorghum sp.	adar	5	5	5	5	0	0	0	0	0
Echinochloa colona	difra	5	5	5	5	0	0	0	0	0
Panicum turgidum	tumam	5	5	5	5	0	0	0	0	0-1 seed processed and eaten
Hibiscus esculentus	bamia	0	0	0	0	0	0	0	0	2-3 fruits cooked and eaten
Corchorus terristeris	meluk-hyia	0-1	2-3	0-1	0-1	2-3	0	0	0	2-3 leaves cooked and eaten
Citrulus vulgaris	battikh	1-2	2-3	0-1	0-1	2-3	0	0	0	5 edible fruits

Legend:
0 = not grazed and/or used
1 = little grazed and/or used
2 = grazed and/or used
3 = fairly grazed and/or used
4 = well grazed and/or used
5 = intensively grazed and/or used

Fairly frequent showers can bring out the blossoms of these herbs and annuals within few weeks. If the rainy season is adequate, the scenery of the Shaqadud area turns (by late August and the beginning of September) into a good sward of short grasses and green foliage covering the thorny scrub bushes and trees. This green scenery does not last for long, however, once the dry season begins, plant species start to yellow and eventually, as the season advances, most of them die.

3.3.5.1.2 Discussion.

As for the area of Aneibis/Abu-Darbein, different classifications have also been proposed for the area of Shaqadud. Andrews classified this area as an Acacia Desert Scrub vegetational zone (Andrews, F.W., 1948: 35). Lebon classified the same area as being part of "zone no. II; the Semi-Desert Acacia Scrub and Short Grassland" lying approximately between latitudes 14° and 17° N (Lebon, J.H.G., 1965: 20). Among the recent classifications there is one proposed by Wickens, in which he related the vegetation of the area to the "Semi-Desert and Grassland" type (Wickens, G.E., 1975: 52; 1982: 35). The most recent ecological review on the vegetation of the Sudan is also in favour of Wickens' classification of the vegetation of this area in particular and in the Sudan at large (Fadlalla, B., 1987: pers.comm.). Furthermore, Wickens' classification (ibid) has the following advantages,

a) Keeps an up to date general vegetation register.
b) Proposes a combination of Andrews' (1948) and Lebon's (1965) definitions for the vegetation cover while taking into consideration the difference in time between them.
c) It serves - perhaps - as a temporal datum for reconstructing the past vegetational history; studying the present vegetation cover and provide the basis for predicting the near future vegetation if the conditions of the recent drought and desertification remain unattested.

3.3.5.2 Fauna.

The inhabitants of the Shaqadud area are sedentarized pastoralists[4] who have retained the option of nomadism as means of adaptation to the prevailing environment. Their livestock include sheep (*Ovis ammon* f. aries) which is the predominant animal; a lesser number of goats (*Cappra aegagrus* f. hircus), a few camels (*Camelus* sp.) and a few cattle (*Bos primigenius* f. taurus). Donkeys (*Equus africanus* f. asinus) are the most common means of transport. Similar to those of Aneibis/Atbara area, the sheep and goats here are typical desert species, and so are the camels whose distribution covers all the semi-desert and desert zones north of latitude 13° N (Lebon, J.H.G., 1965: 110).

---

[4] The fast-maturing *Sorghum vulgare* (Pers.) var. *caudatum* Stapf. is the favoured rainfed-cultivated crop. The cultivated plots are confined to the low-lying plains with their fine sand-silty soils which are situated along the beds of the water courses and headwaters and fans of seasonal streams.

The wild game only consists of some few rabbits (*Lepus* spp.) and a very few herds of gazelle (*Gazella dorcas*) which visit the area during the short rainy season. It has been more than twenty years ago since jackals (*Canis mesomelas*), antelopes and other big wild game were (frequently) seen in this area.

The bird-species present in this area are similar to those in the Aneibis/Atbara area, but the former have a migratory cycle while those of the latter are present the year round. This migratory cycle comprises movement towards the west (river Nile) during the dry season where greens, seeds and water are available and return movement during the rainy season.

3.4 Description of the Jebel Tomat/Rabak Area.

3.4.1 Geographical location.

3.4.1.1 The site of Jebel Tomat (13° 36' N, 32° 44' E).

The site of Jebel Tomat (also J. et-Toweimat) is situated some 50 kilometres north of Rabak town and c.10 kilometres east of the White Nile river at Esh-Shawal village (cf. fig. no.11). The site is situated between two inselbergs[5] which seemed to have been part of an old bank of the White Nile (cf. fig. no.11). The cultural deposit is rich; in some places it is 1.2[6] m deep, e.g. storage pits (Clark, J.D., and A.B.L. Stemler, 1975: 589) but in average it is 0.5-0.6 m deep in the 8 m^2 we excavated. Unfortunately, due to road-building activities and continuous blasting of the two inselbergs during our excavations, it was not possible to measure the extent[7] of the site or its elevation above the surrounding plains. However, the material culture is in part pre-Meroitic, e.g. pottery, and in other is contemporary with the late Meroitic period. Close similarities are observed between the material culture from the later period of this site and that of Jebel Moya .We obtained two radiocarbon dates[8], the oldest being 4140±90 B.P. and the youngest 1770±80 B.P.

3.4.1.2 The site of Rabak (13° 09'N, 32° 46'E).

The site of Rabak is situated on an old bank of the White Nile (cf. fig. no.1) which is now 3 kilometres east of the present course of the river; similar in that sense to the site of Aneibis some 550 kilometres north (cf. fig. no.1). The site is situated on a ridge which stretches east-west about c.3.5 m higher than the surrounding flood plains

---

[5] The Ar. name Tomat is derived from these two inselbergs and it means "twins".

[6] J.D. Clark has also conducted some excavations in the early 1970s and stated that 80-120 cm is an average depth of the cultural layer (Clark, J.D., and A.B.L. Stemler, 1975: 588).

[7] The size of the site was estimated to be over 12 acres (ibid).

[8] J.D. Clark has also published a series of radiocarbon dates from this site (Clark, J.D., and A.B.L. Stemler, 1975: 589) and presented in appendix no.1 in this work.

Fig. no. 11 Past and present extent of the course of the White Nile

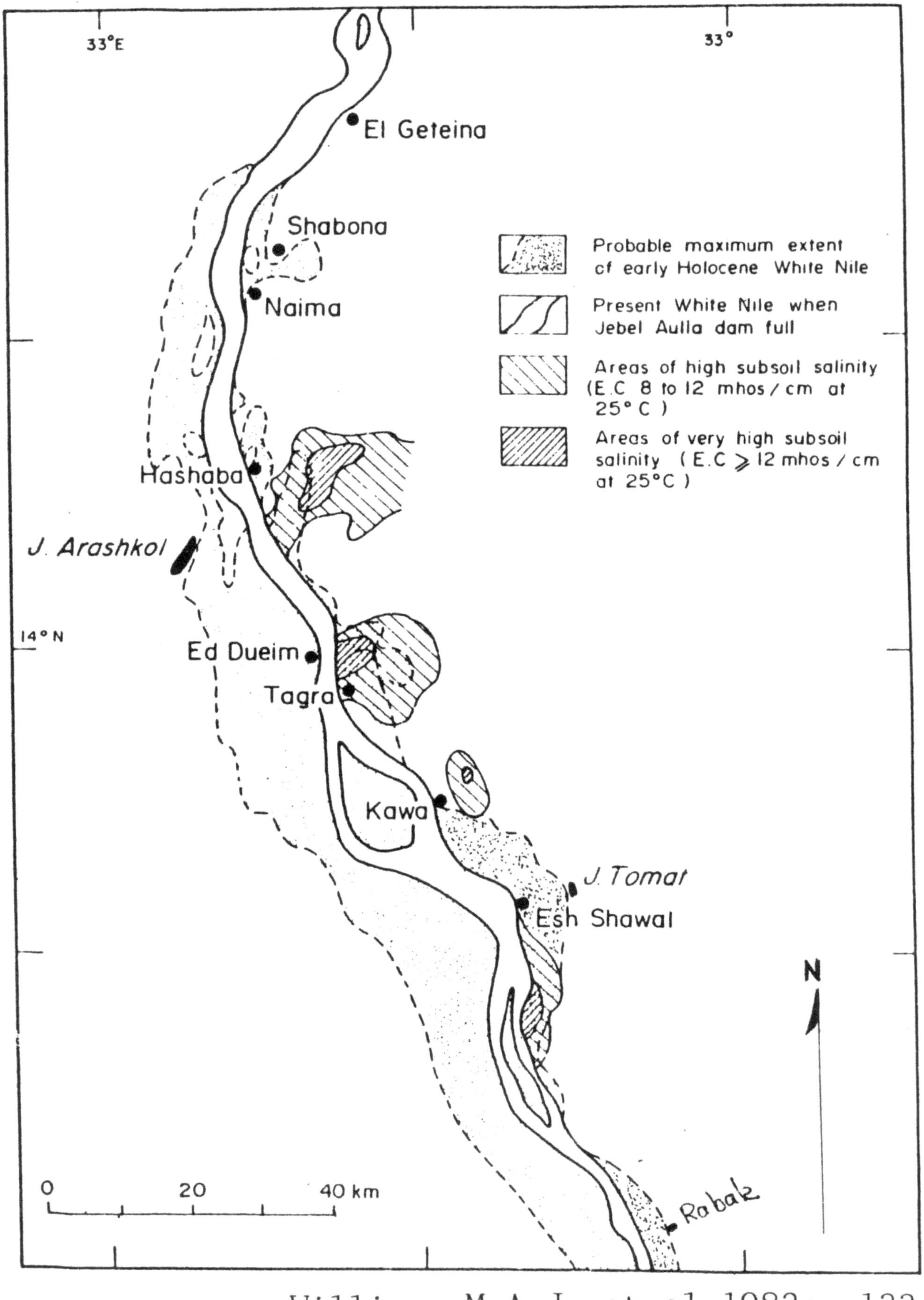

Williams,M.A.J.,et al,1982: 133

(cf. fig. no.12). The extent of the site is 200x80 m (16000 m^2) and the excavated area is 18 m^2 (cf. fig. no.12). The cultural debris generally varies in depth between 0.5 to 0.8 m and one trench has a deposit of 1.5[9] m (interpreted as kitchen midden) but none of these deposits showed any stratification. Pottery recovered from the bottom levels of this site is almost identical to the Khartoum Neolithic tradition and that from the upper levels is related to the Jebel Moya tradition . The diagnostic tool, the gouge, of the Khartoum Neolithic is, however, absent at this site and a very few standardized lithic tools (scrapers, lunates, groovers) were found (El-Mahi, A.T., and R. Haaland, 1984: 30). The results of radiocarbon dating of the site are 4490±100 B.P., 6050±100 B.P. and 6020±130 B.P. (cf. appendix no.1).

3.4.2 Geology.

The area of Jebel Tomat and Rabak (located 50 km to the south) is characterized by the same geological formation; the Basement Complex (cf. fig.13; Vail, J.R., 1982: 52; Whiteman, A.J., 1971: 4-7). Unfortunately, there is not, at present, any comprehensive geological study that provides detail information e.g. on the subdivisions, groups, etc. of this complex in the Sudan (Vail, J.R., 1982: 54). It is the Gezira area (between the White and Blue Niles) which has been studied in more detail. The area studied here has been classified as part of the gneisses and migmatites group (ibid). The twin inselbergs of Tomat are some of the few isolated outcrops of the Basement Complex in this area (cf. Whiteman, A.J., 1971: 116). Similar to Jebel Moya and Saqadi, the Tomat Jebels are metasediments of soda granite (ibid: 30). Information on the Basement Complex has also been presented, in the foregoing (3.2.2) when describing the area of Aneibis/Abu-Darbein.

3.4.3 Topography.

The history of the present landforms of the Jebel Tomat/Rabak area is intimately connected with the history of the White Nile (Adamson, D.A., et al, 1982: 165), which is the main physical feature delimiting the area to the west. Unlike the Atbara river, further north (which contributes 14 million tons to the main Nile), the sediments deposited by the White Nile into the main Nile, is small; some 2 million tons during its annual flow (Williams, M.A.J., et al, 1982: 119).

This area, transected from west to east, is characterized by its gently sloping clayey plains spotted by few isolated inselbergs. Although the gradient is only a few centimetres per kilometre, a distinction has conventionally been made between the "White Nile clays" to the west and the "higher-level clays" towards the east (cf. fig. no.14). This relative distinction is sometimes expressed as "low-land" for the former and the "upland" for the

---

9 This trench 107x/100y was exceptionally deep because it has a down-cutting from a depth of c.80 cm.

Fig.no.12

Rabak site:contour map and the excavated area.

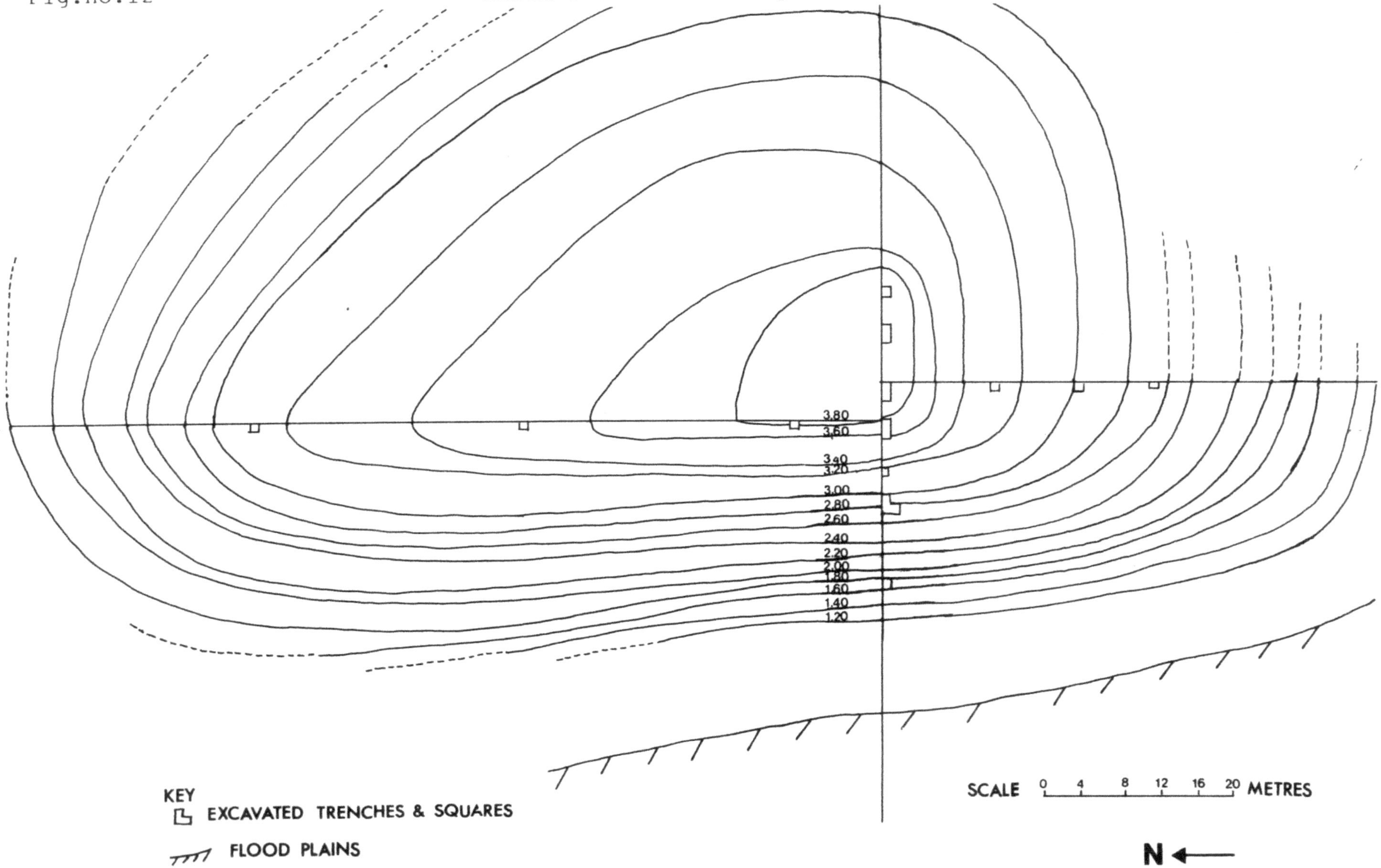

Fig. no 13 Geology and types of soil in the area of J.Tomat / Rabak and further north and east

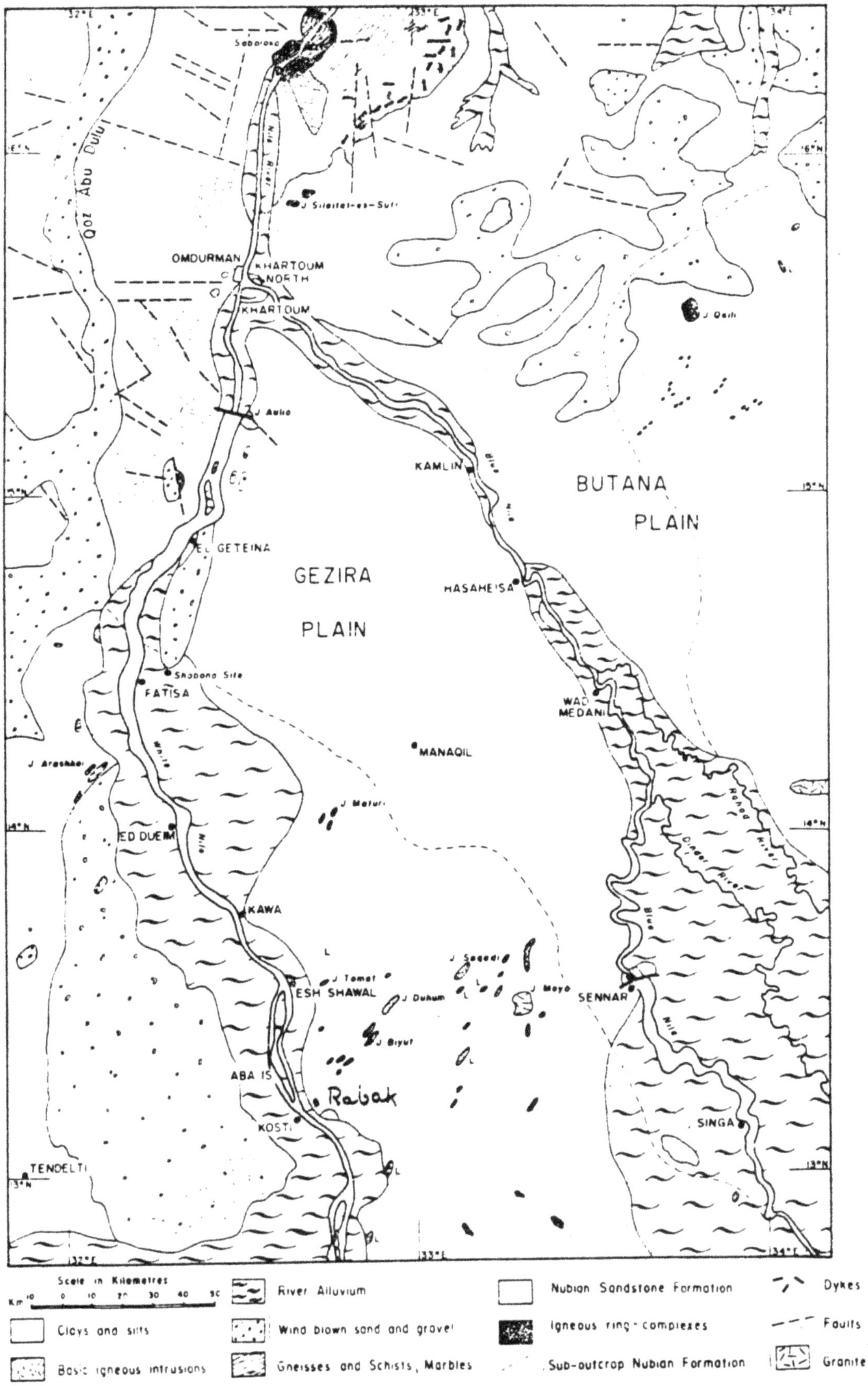

After Vail, J.R., 1952: 52

Fig. no.14 Upland and lowland clays and drainage in the White Nile area

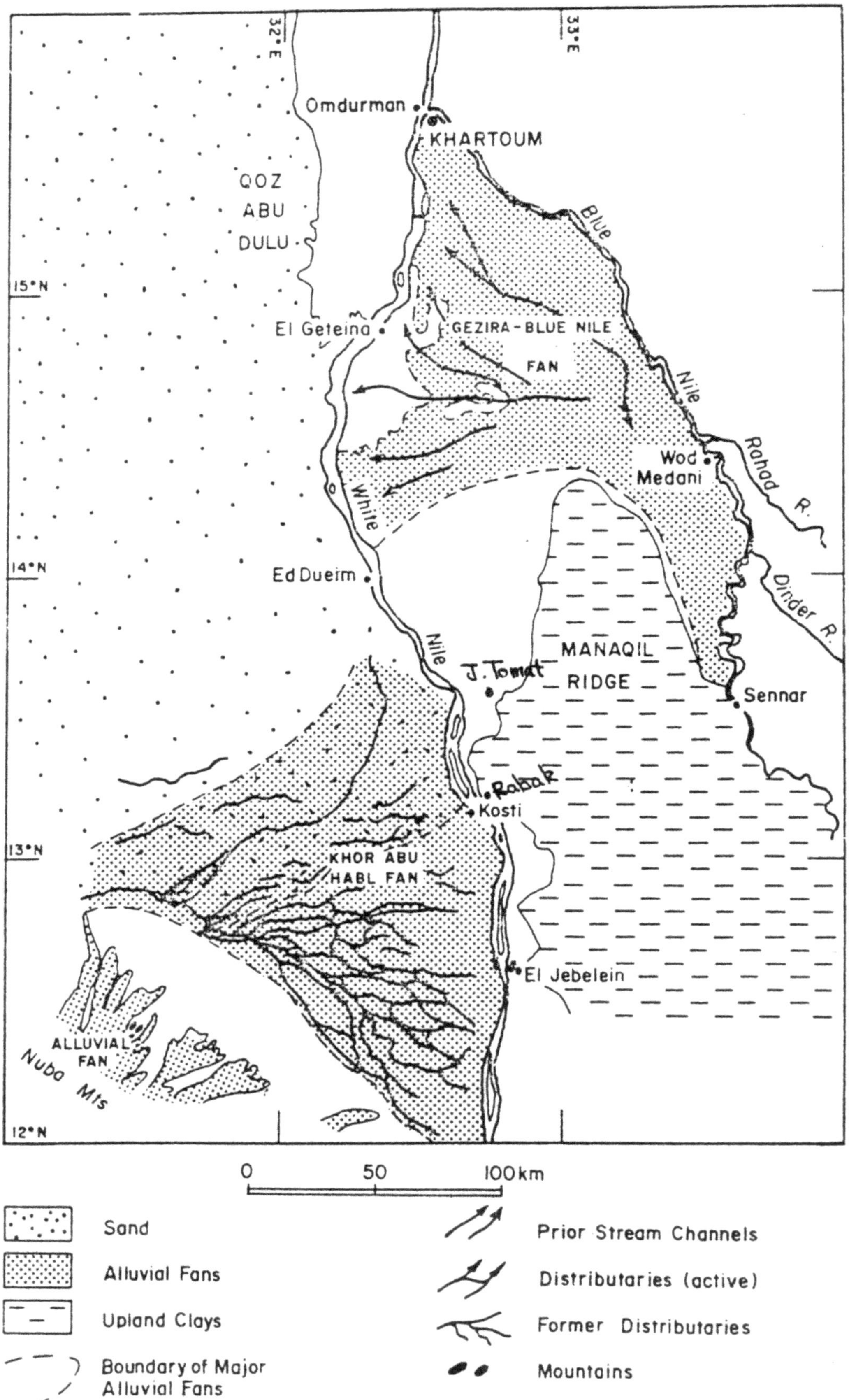

After Williams,M.A.J.,et al,1982: 112

latter (Mubarak, M.O., et al, 1982: 146; Williams, M.A.J., et al, 1982: 124). Both were deposited by former flood waters of the White Nile. It is believed, however, that the "higher-level" or "upland clays" are older and that some of their material was laid down by former channels of the Blue Nile (ibid). The White Nile "low-land" clays are deeper, darker and heavier textured than what is found in the high-level "upland" area (ibid: 124). The distribution of both types of clay is, however, uneven varying between very thin limited strips at Rabak in the south, to relatively wider and thicker segments in the area of Jebel Tomat in the north (cf. fig. no.14). The boundary between these (White Nile and higher-level clays) is set as the mark for the maximum extent of the White Nile flooding at the beginning of the Holocene some 11000 years ago (ibid: 121-127).

The types of soil found in the area are generally of a cracking clay type which were deposited under permanent water (Andrews, G., 1948: 123). This type of cracking clay has been correlated with the previously discussed "White Nile and higher-level clays", and it has been postulated that during the early Holocene a palaeochannel of the White Nile extended as far eastwards as the Jebel Tomat area in the north and the site of Rabak in the south (cf. fig. no.11). Accordingly, the latter site (Rabak) was probably under water during that period.

Salinity and alkalinity are reported to be of markedly different proportions in different parts of this area (cf. fig. no.15). But the greater part of the area is characterized by its dark alkaline cracking clays (cf. fig. no.15) which were laid under permanent water (Mubarak, M.O., et al., 1982: 151). The red and brown coloured soils are associated with the Jebels e.g., Jebel Tomat. The sandy loams and clays and recently deposited alluvial sands (Ar. *gerif*) are observed along the White Nile Valley (ibid).

Seasonal watercourses, streams and gullies are very few in this area (cf. fig. no.16) due to its gentle gradient, but some shallow linear depressions (with or without distinct boundaries are in evidence especially in the vicinity of Jebel Tomat (cf. fig. no.16). These, together with the swampy depressions, e.g. in the area of Rabak, are seasonally filled with water during the wet season either by the trapping of rain-waters or by flooding from the White Nile during the period of the high water level.

3.4.4 Climate.

This area lies within the same tropical continental climate (Ireland, A.W., 1948: 67; Lebon, J.H.G., 1965: 5; Whiteman, A.J., 1971: 1-5) described in connection with the areas of Aneibis/Abu-Darbein and the Shaqadud. It has, like these areas, two main seasons; a hot rainy summer and a cool dry winter. There, are however, differences mainly as regards the annual rainfall. This area generally receives a mean annual rainfall of c. 350 mm (cf. fig. no.10 and histogram no.1) which is almost six times as much when compared with that of the area of Aneibis/Abu-Darbein and three to four times the

Fig.no.15 Salinity and alkalinity of soil in the area of J.Tomat/Rabak

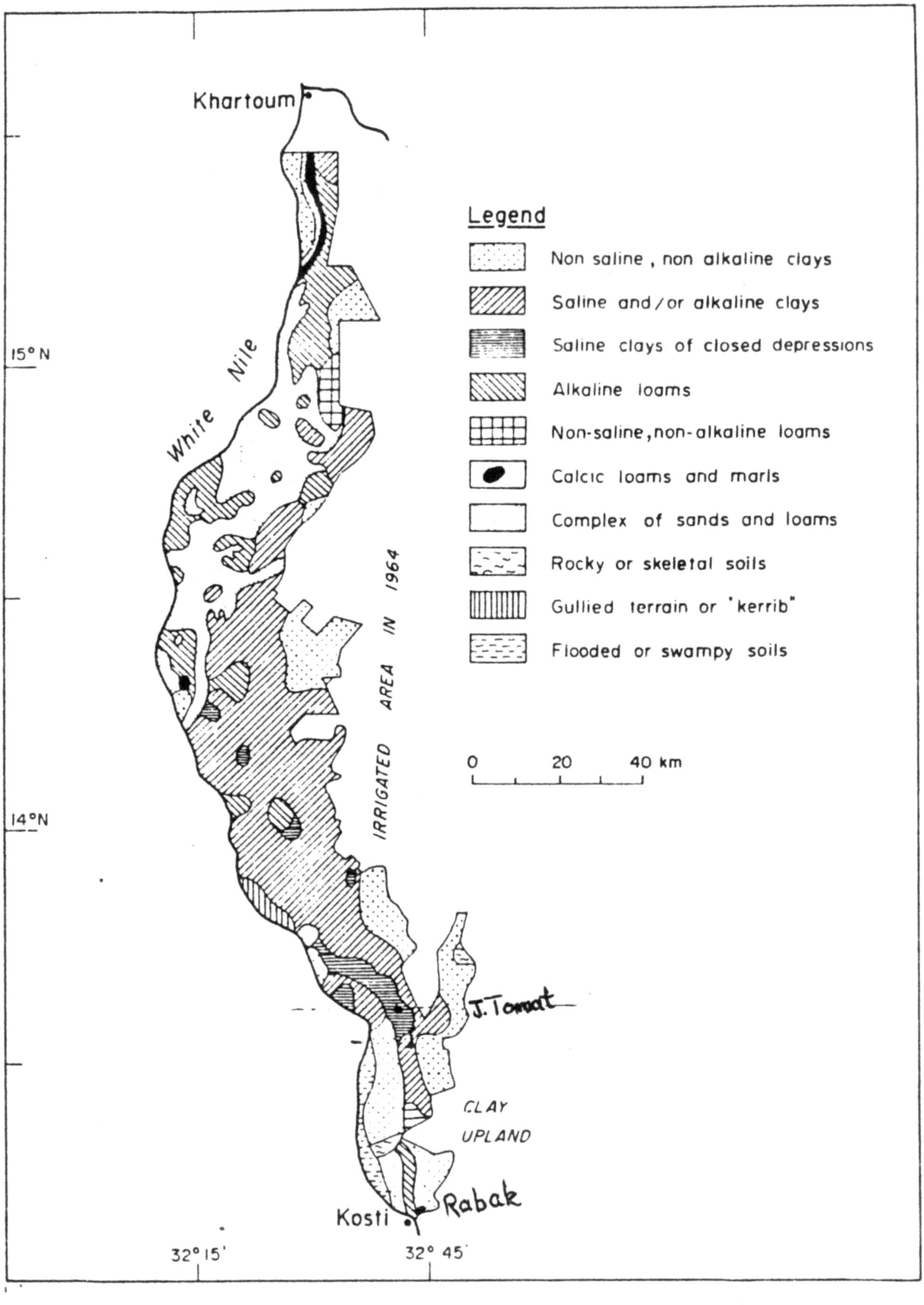

After Williams, M.A.J., et al, 1982: 131

Fig. no. 16 Depressions seasonally filled with rain and/or flood water

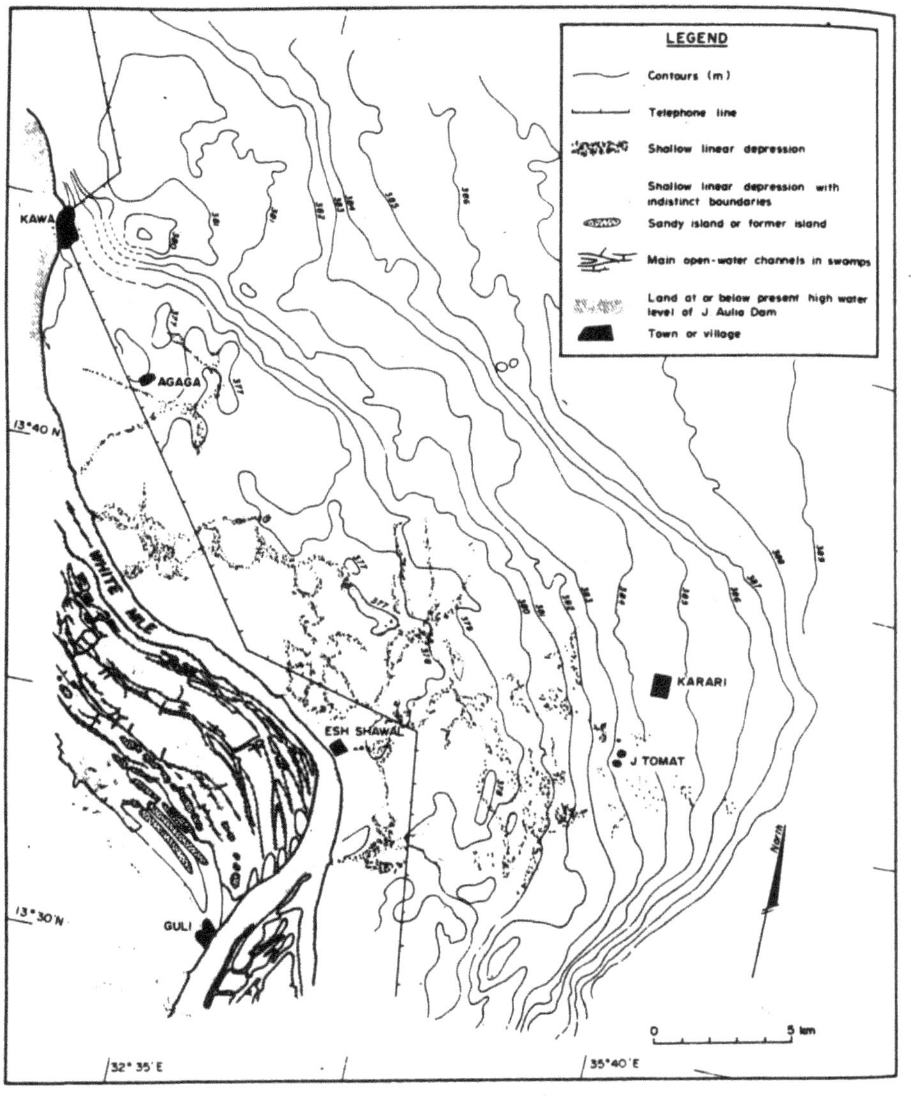

After Adamson,D.A.,et al,1982: 194

amount of the annual rain in the area of the Shaqadud. However, the distribution of this mean annual rainfall, 350 mm, is uneven within this area. A recent study showed that the southern part of the area (Rabak/Kosti) receives greater amount of rainfall than its northern part (e.g. Jebel Tomat); the Rabak/Kosti part receives 406 mm of rain (Williams, M.A.J., et al, 1982: 137) but,

> 'the isohyets swing away north-east from Kosti, so the localities in the White Nile Valley receive less rain.....' (ibid).

August is relatively the wettest in the area of Jebel Tomat/Rabak; similar in that to the areas further north (Aneibis/Abu-Darbein and the Shaqadud). Furthermore, summer and winter temperatures in this area resemble the summer and winter temperatures in the areas of Aneibis/Abu-Darbein and the Shaqadud. The evaporation rate is almost 20 mm throughout the year (Mubarak, M.O., et al, 1982: fig.8:1,144).

3.4.5 The present flora and fauna.

The natural habitat of the vegetation in the vicinity of Rabak is almost completely destroyed as a result of the rapid urbanization of the area. Thus the following account largely concerns itself with the flora and fauna of the central and northern parts of the area.

3.4.5.1 Flora.

Several accounts and maps on the composition of the principle vegetation have been made of this area. Regardless of the slight differences of terminologies each of these used, they agree in general that the vegetation of this area is Acacia scrub and short grass of a low rainfall woodland savanna type (Andrews, F.W., 1948: 37; Barbour, K.M., 1964: 65, 67; Lebon, J.H.G., 1965: 20-21). Recently, more elaborate botanical surveys and research were conducted in the area (Mubarak, M.O., et al, 1982: 143-162). Consequently, a clearer picture based on more comprehensive study has been obtained. "Acacia Thornland Alternating with Grassland" is the current characterization of the vegetation (cf. fig.no.8). The area is part of a broad region that embraces this type of vegetation (ibid: 151).

The distribution of the floristic composition of the area correlated with its annual rainfall, temperatures, evaporation rate and types of soil, has been described as follows,

> 'the vegetation .... can be subdivided into riverain associations; and Acacia Thornland on dark clays alternating with grass areas' (ibid).

The riverain associations include *Acacia nilotica* (L.) Willd. ex Dc. (Ar. *garad/sunt*), *Acacia seyal* Del. (Ar. *talih*), *Ziziphus spina-christi* (L.) Willd (Ar. *nabag/sidir*) and

Balanites aegyptiaca (L.) Del. (Ar.*heglig*). These species are found along the flood plain of the White Nile and in its vicinity (ibid), but they are variable in density and distribution. The presence of *Capparis decidua* (Forsk.) Edgew. (Ar. *tundub/humbog*) and *Acacia nubica* Benth. (Ar. *la'ot*) are relatively dense towards the northern part of the area. *Phoenix dactylifera* L. (Ar.*nakhal*) is observed near Esh-Shawal. *Mimosa pigra* L. (Ar. *habit al-ain/habas*) and *Salix subserrata* Willd. (Ar. *safsaf*) are water demanding species. They grow along the bank of the White Nile close to the water and are in evidence all the way from Kosti to Esh-Shawal. Furthermore,

> 'along the lagoons and banks of the White Nile *Eichlornia crassipes*, *Cyperus rotundus*, *Cyperus* spp., *Eichinochloa* spp., and *Typha domingensis* are common. On the higher ground of the banks, *Chrozophora plicata*, *Leonotis africana*, *Tribulus terrestris*, *Heliotropium* spp., *Trianthema petandra* and *Aerva javanica* occur. On watercourses which drain the higher ground and where sandy deposits prevail, tussocks of *Panicum turgidum* are common' (ibid: 152).

The local people grow a variety of vegetables and fruit-trees in orchards along the east bank of the White Nile.

With regard to the Acacia Thornland on the "higher-level clays", the list of vegetation includes *Acacia mellifera* (Vahl.) Benth. (Ar. *kitr*) confined to depressions, *A. nubica* Benth., and *Capparis decidua* (Forsk.) Edgew.; these are the predominant species. A few dispersed species of *Acacia seyal* Del. (Ar. *talih*), *Cadaba rotundifolia* Forsk. (Ar. *kurmut*) and *Balanites aegyptiaca* (L.) Del. are also reported from this area (ibid). Unlike the *Balanites* sp. which are hard to cut and also fire resistant, the stands of *Acacia seyal* are subject to intensive wood-cutting for charcoal and fire-wood. Table no.3 is set only as an example to show how the inhabitants exploit these plant species. In areas where grasses alternate the thorny *Acacia* species, there is an abundance of grass species which flourish during the rainy season. These include several species of the genus *Sorghum* (Ar. *adar*); *Setaria verticillata* (L.) Beauv. (Ar. *luseig*); *Cymbopogon nervatus* (Hochst.) Chiov. (Ar. *hamareib*); *Cenchrus ciliaris* L. (Ar.*haskanit*); *Sehima ischaemoides* Forsk. (Ar. *garéz*) and *Hyparrhenia rufa* (Nees.) Stapf. (Ar. *umm silela*).

Sorghum spp. are the grasses most favoured by sheep, goats and cattle. *Setaria* sp. and *Cenchrus* sp. are less grazed due to the stickiness of the former and the spiny fruits of the latter. The most common species among the herbs are *Cassia* spp. (Ar. *senna/senna-makka*) and *Solanum dubium* L. (Ar. *gubbein*).

Many trees, herbs and grass species grow between the two inselbergs (Ar. *jebel*) of Tomat and in their vicinity (cf. table no.4). It was rather a surprise to find *Combretum*

Table no. 3: An example for the plant-species exploited and their uses in the Jebel Tomat /Rabak area.

Species name		Uses of plants								
							Construction & other uses			
Latin	Arabic	Sheep	Goats	Cattle	Donkeys	Camels	Houses	Fences	Fire-wood	Other Uses
Acacia nilotica	garad/ sunt	0	0-1	0	0	2-3	0	0	5	1-2 seeds to cure and flue
Acacia seyal	talih	0	0-1	0	0	2-3	0	0	5	0
Acacia nubica	la'ot	0	0	0	0	0-1	0	1-2	2-3	0
Z. spina-christi	nabag/ sidir	0	0-1	0	0	1-2	0	2-3	0-1	2-3 edible fruits
Balanites aegyptiaca	heglig	0	0	0	0	1-2	0	0	0-1	0
Capparis decidua	tundub	0	0	0	0	0	0	0-1	0	0-1 edible fruits
Phoenix dactylifera	nakhal	0	0	0	0	0	5	0	2-3 leaves	5 edible fruits
Solanum dubium	gubbein	0	0	0	0	0	0	0	0	2-3 fruits are milk ferminter
Cymbopogon nervatus	hamareib	0	0	0	0	0	0	0	0	3-4 stalks medicine & tea flavour
Panicum turgidum	tumam	5	5	5	5	5	0	0	0	0
Cenchrus ciliaris	haskanit	0	1-2	0	0	0-1	0	0	0	0
Sorghum sp.	adar	5	5	5	5	5	0	0	0	0

Legend:
0 = not grazed and/or used
1 = little grazed and/or used
2 = grazed and/or used
3 = fairly grazed and/or used
4 = well grazed and/or used
5 = intensively grazed and/or used

Table no. 4: A list of plant species at Jebel Tomat.

Family	Gen./species	Eastern Jebel	Western Jebel	Between the two jebels
Capparidaceae	*Cadaba rotundifolia*	6	7***	
	Cleome viscosa	6	6	
	Maerua sp.		1	
Amaranthaceae	*Aerva javanica*		1	
Combretaceae	*Combretum aculeatum*	1	1	
	Combretum hartmannianum		1	
Tiliaceae	*Grewia tenax*	5	7**, 6	
Leg. Mimosoideae	*Acacia mellifera*	5	7*	6
	Acacia nubica	4	1	6
Leg. Papilionoideae	*Tephrosia* sp.	1	3	
Simaroubaceae	*Balanites aegyptiaca*			6
Asclepiadaceae	*Calotropis procera*	2^{L}		2, 6
Solanaceae	*Lycium persicum*	4	4	
	Solanum dubium	1		
Pedaliaceae	*Rogeria adenophylla*	6		
Gramineae	*Beckeropsis uniseta*	4^{H}	4	

Legend:

1 = rare
2 = local
3 = occasional
4 = frequent
5 = common
6 = widespread
7 = dominant
7*, 7**, 7*** = first, second and third order of dominance
L = on lower grounds
H = at higher levels

(After Mubarak,M.O.,et al,1982:156 with modification)

hartmannianum Schweinf. (Ar. *habil/áweer*) growing at the area of the site of Jebel Tomat (in particular) and in this area in general (cf. table no.4). Its presence here is, however, interpreted as being "a northern outlier from its main area of area of occurrence, suggesting possible isolation arising from climatic isolation" (ibid: 157).

3.4.5.2 Fauna.

The following account is meant to describe the natural fauna of the whole area without reference to domestic fauna of Rabak or that of Jebel Tomat for two reasons: First; Rabak is a town whose heterogeneous inhabitants adapt to town life. Thus, the few selected animals found in the town - mainly to transport people and goods - are not representative of the fauna of the area.
Second; the purpose of this account (as we also intended with the previous ones) is to concentrate on ethnoarchaeological issues that we hope will assist us in our subsequent interpretation of the archaeological data. Accordingly, it is more beneficial to stress all the traditional practices including those related to animal husbandry, exploitation of wild game and aquatic resources. These activities are better examined in rural than urban settings.

Like most of the fauna in other rural regions of the Sudan, this part of the country is basically an area of domestic animals. The livestock include the so-called Nubian goats (*Cappra aegagrus* f. hircus) (John, E.R., 1948: 643). But their physical description (ibid) is very much similar to that of the desert goats (Lebon, J.H.G., 1965: 108-110) found in the areas further north (Aneibis/Abu-Darbein and Shaqadud). It, however, seems that the sheep (*Ovis ammon* f. aries) in this area are somehow different from the desert-sheep present in the areas further north. The former is classified as being of Northern Riverain type (John, E.R., 1948: 641). It is believed to be most probably an intermediate breed between the "Desert" and the "Nilotic" sheep because it has a medium size, medium length of legs, fairly long ears and relatively short coat (ibid). Like the areas further north camels (*Camelus* sp.) and donkeys (*Equus africanus* f. asinus) are also present in this area and until recently they were the prevalent means of transport. Considerable numbers of cattle (*Bos primigenius* f. taurus) are also raised in this area.

Except for donkeys, all the above mentioned domestic animals - but rarely camels - are raised for their milk and its by-products, meat and hide. Their presence in the area, in order of dominance, is that of sheep/goats, cattle, donkeys and camels.

Wild game has not been reported for the last two decades in the history of this area due to the expansion of mechanized farming and urbanization projects. But similar to the Aneibis/Abu-Darbein area, the aquatic resources, mainly various fish species, are both plentiful and can be exploited with traditional fishing methods.

Since the area provides a habitat with ample of water and plenty of food-plants throughout the year, birds of different species are attracted there either as transients or as permanent residents. Shooting with slings and/or trapping with traditional tools is performed mainly by children.

CHAPTER IV

Plant Remains: Nature of Evidence, Methods and Techniques of Its Recovery

4.1 Background.

Plant remains may be preserved in archaeological sites for thousands of years by such diverse conditions as extreme dryness, extreme cold or water-logging (Stemler, A.B.L., and R.H. Falk, 1981: 191). They can also be preserved due to the very nature of the remains, e.g. carbonized and/or desiccated material with hard or resistive exterior wall-structures and impressions in pottery (Krzywinski, K., 1988: pers.comm.). They are retrieved in a variety of sizes, shapes and different states.

On the basis of the size and shape of these remains, they can be classified into two main categories. However, this categorization should be understood merely as a practical procedure to enable a consistent presentation of the data. One of these categories is microfossil[1] plant remains, e.g. pollen grains, which cannot be seen, after their retrieval, without using microscopes. The traditional study of fossil-pollen cast light mainly on vegetation history, palaeoclimatic conditions, etc. Currently, pollen analysis is also applied in the study of past economic adaptations related to the exploitation of plants and the study of cultural landscape (ibid; Magid,A. and K. Krzywinski,1988). The other category comprises macrofossil[2] plant remains. These are remains which can be seen by the naked eye, e.g. cobs, kernels, seeds, nuts, and leaves, but the study of their external morphology often requires microscopic magnification. Although some macrofossils can sometimes serve as climatic indicators, they are basically useful in the study of former economic activities such as the gathering and the cultivation of wild and domesticated food-plants. They would also be useful in obtaining information on the utilization of plants for purposes other than dietary ones such as medicine, ritual, and household utensils. The centre of attention in the following archaeoethnobotanical account will be on the latter category: the macrofossil plant remains, with the main emphasis on seeds/grains

[1] See definition of the term "fossil" presented under heading 4.2.1.

[2] Other categories of fossils, e.g. diatoms are beyond the scope of the present study.

and their impressions in pottery (potsherds) as indicators of past economic adaptations in the Central Sudan.

4.2 Macrofossil Plant Remains.

4.2.1 Nature of evidence.

Although there is no satisfactory definition for the word "fossil" it is certain that it is derived from the Latin verb *fodre* which means 'to dig' (Arnold, C.D., 1947: 14). Thus it originally referred to anything removed from the earth, but its modern usage is restricted to organic remains removed from earth and defines fossils as 'the remains or traces of organisms that lived prior to historic times' (MacFall, R.P., and J.C. Wollin, 1983: 11) and also any remains of beings that lived in the past, as such, be it the historic or prehistoric times (Krzywinski, K., 1988: pers.comm.). Accordingly, macro-fossil plant remains may be defined as remains which are dug out of earth (Bohrer, V.L., and K.R. Adams, 1977: 5) regardless of the state in which they are found, e.g. whether it is petrified or not. Modern usage also extends the term fossil to inorganic matters, e.g. fossil landscape and fossil lakes. Most of the macrofossil plant remains which I found in the archaeological sites in the Central Sudan are carbonized and/or desiccated (mummified), yet they retain some or all of their basic external morphological features such as size, shape, and surface structure.

One of the reasons for carbonization of plant remains is thought to be spontaneous action by a slow natural process which takes place at normal temperatures (Renfrew, J.M., 1973: 9) but no experiments have been conducted - so far - to confirm this suggestion (Krzywinski, K., 1988: pers.comm.). Biffen believed that carbonization is a result of the action of anaerobic bacteria and that such an action takes place fairly rapidly as the bacteria obtain their oxygen from the cellulose in cell walls and thus reduce them to charcoal (Renfrew,J.M.,1973: 9). But Renfrew noted that Biffen's statement does not explain the often encountered occurrences of both charred and uncharred grains in the same context, pit, pot, etc. (ibid). The relatively widely accepted interpretation is that which attributes carbonization of plant remains to their having been exposed to intense heat just beyond the roasting point (Bender, B., 1975: 50; Dimbleby, G.W., 1967: 101; Renfrew, J.M., 1973: 9-10; Stemler, A.B.L., and R.H. Falk, 1981: 192). Exposure to intense heat could happen if the remains were near to a hearth or a house fire in the area where they were stored (ibid). Direct contact with heat and/or fire beyond the roasting point for a longer time would reduce them to ash. This is particularly true with cereal grains which have too little cellulose and too much decaying starch that can easily be burnt to ash (Renfrew, J.M., 1973: 10). Recently, Krzywinski and Soltvedt suggested that 'carbonization occurs best under relatively low temperatures and in the absence of or with low input of oxygen' (Krzywinski, K., and E.C. Soltvedt,1988: 4).

Desiccated (mummified) remains are those which are not carbonized and to a large extent look like their extant counterparts except, however, as regards their colour which may be darker or brown (Stemler, A.B.L., and R.H. Falk, 1981: 191). One of the most likely factors for the desiccation of plant remains is chemical weathering of minerals such as silica, especially during the rainy season. When these minerals are dissolved, they wash into the plant remains buried in the earth and eventually consolidate on their exterior (Krzywinski, K., 1988: pers.comm.). Repeated weathering of minerals, their dissolution, precipitation and consolidation on the exterior of plant remains created silicified exterior structures on them and hence protect them from destruction and/or decay (ibid). This process could as well be an explanation for the desiccation of plant remains I recovered from sites in the Central Sudan where chemical weathering is very active. It should be stressed here that while a natural process of preservation by desiccation is slowly taking place, there are also destructive processes and factors which can be effective in a fairly short time after the plant remains are discarded. Among these the main ones are destruction by microorganisms and/or decay of the soft parts of the plants . This implicitly means the surviving desiccated remains ought to have had hard exterior surfaces. These surfaces resisted the destructive and decaying factors long enough for the slow process of their desiccation[3] to have furnished them with an extra surface-resistance which would probably be ever-lasting. Although details of desiccated plant remains (such as small hairs and spicules on the kernels, husks and internodes) can be well preserved, the proteins in the endosperm are completely disorganized and decomposed and the embryo is incapable of germination (Renfrew, J.M., 1973: 7, 18, 19; Stemler, A.B.L., and R.H. Falk, 1981: 191). Thus desiccated seeds and seeds of berries can retain all the details of the pericarp (fruit/seed wall) but will be either hollow or only have brown powder inside, e.g. *Celtis integrifolia* Lam. indicating the former contents (Renfrew, J.M., 1973: 7).

4.2.2 Preservation.

Mention has been made, in the foregoing, of the fact that the best conditions of preservation of macrofossil plant remains are in general extreme dryness, extreme cold and water-logging. Furthermore, the nature of the specimen itself, be it carbonized and/or desiccated or an impression in pottery, helps preserving it regardless of the environmental conditions in which it exists.

With regard to the conditions of preservation of plant remains in archaeological sites in the Central Sudan, the only optimum preservation is that of impressions of plants in pottery. The preservation of carbonized plant remains in such an arid (between latitudes 16°-18°N) and semi-arid (between latitudes c. 13°-16°N) area with its variable

[3] It has also been suggested that extremely arid conditions would lead to desiccation of plant remains (A-Bari, E., 1986: pers.comm.).

environmental and microenvironmental conditions, depends on many factors. Among these the most important are the type of soil[4] (e.g. alkaline or acidic; sandy, clayey, loam, etc.), the depth at which the plant remains are buried, moisture content and temperatures in the soil, contrast between seasons (and perhaps contrast between day and night), and the presence or absence of microfauna, insects, rodents, etc. In general, the conditions of preservation are not particularly favourable for the survival of plant remains in archaeological sites, especially carbonized remains, e.g. cereal grains which were buried for thousands of years. The only exceptions to this are remains with hard woody structure e.g. seeds of *Ziziphus* sp., *Celtis* sp., etc. and remains which are retrieved from naturally protected sites such as the cave-site of Shaqadud (cf. fig. no.1) from which relatively well preserved carbonized remains were recovered (cf. table no.8). Since ants, rodents, etc. are very active in the Central Sudan, there are possibilities that these plant remains are not contemporary with the cultural debris. These possibilities and others are dealt with in the presentation of the plant remains retrieved from the excavated sites in the area.

4.3 Methods and Techniques.

Apart from the chance discoveries of sizable seeds which were carefully picked (and stored after being properly labelled) while excavating and/or during sieving, different methods and techniques were used to collect the material studied here.

4.3.1 Sampling in the field.

It is often very difficult to detect by the naked eye alone plant remains in archaeological deposits even if froth flotation shows they are abundantly found in those deposits (Jarman, H.N., et al, 1972: 39). Furthermore, indicative archaeological features do not always successfully guide one to where plant remains were deposited in the site. H.N. Jarman and others referred to Helbaek's experience that 'dark patches, ashy layers and other promising or doubtful occurrences yielded negative as often as positive result' (ibid). Therefore, in order to ensure fairly representative sampling of the full range of the excavated area on the archaeological site and at the same time to eliminate errors that can happen due to localized concentration of seeds, samples of constant size were systematically taken from the trenches and squares in the different sites we excavated in the Central Sudan. A regular procedure was followed, while excavating, by dividing the trenches into four equal units and alternatively taking soil samples from each of these. In

---

[4] For more information on soils, see chapter III.

this manner, four soil samples were regularly taken from each arbitrary level[5]. The site of each sample was 50x50x2.5 cm of soil, except on one site (namely the Shaqadud site) where the excavated trenches were 0.5 m^2 and hence the size of each soil sample taken from this site was 25x25x2.5 cm. Besides these, samples of soil of different sizes and/or weights were always taken wherever ashy patches and concentrations of charcoal or grinders were spotted during the process of excavation (Magid, A., 1982: 87). Thus, all cultural deposits, features and stratigraphic units were sampled.

As regards the transportation of the samples, it is certainly practical, efficient, easy and cheap to process the collected soil samples in the field during the excavation period rather than to transport a heavy load (sometimes a few tons) of soil and process it somewhere else. But in the case of the Central Sudan the situation is quite often the contrary. The main problem there is the provision of water at the sites regardless of their being near or at a distance from the sources of water, i.e. rivers, water-traps, etc. Due to this basic problem, there was practically no alternative other than to transport the soil samples from the site(s) to where water could be obtained without difficulty. Although transportation of the samples is often impractical, difficult and expensive; it is not without some advantages. One of these is that any unforeseen risks of damage of retrieved plant remains are avoided. Another advantage is that, since the transported soil samples were sometimes stored for a couple of weeks before they are processed by flotation, the period of storage often contributed to reducing and/or eliminating the exposure of the embedded carbonized plant remains to rapidly fluctuating temperatures (specially in an arid and semi-arid region such as the Central Sudan). I observed that when the soil samples are processed by froth flotation directly after their excavation and then packed and left for some time in normal air temperature during the excavations, such rapidly fluctuating temperatures sometimes led to damaging the fragile carbonized plant remains retrieved.

4.3.2 Retrieval of macrofossil plant remains by froth flotation.

. Flotation is basically a liquid, e.g. water-separation technique used to collect light organic materials (Bohrer, V.L., and K.R. Adams, 1977: 13). Different improvements and modifications have been added and apparatuses invented, all of which have developed this basic simple technique. Among these is the flotation cell-apparatus, which I used to retrieve macrofossil plant remains from the soil samples collected. The apparatus consisted of a chamber, granulometry sieves, a valve, water pipe, and regulated compressed air pump (cf. fig. no.17). Liquids and chemicals used are water, terpinol, paraffin and soap powder. In this method of froth flotation some catalysts (additions or

[5] Arbitrary levels of 10 cm were taken in the excavation of all the sites except the Shaqadud cave-site where it was 5 cm.

Fig.no. 17 Different parts of the flotation apparatus

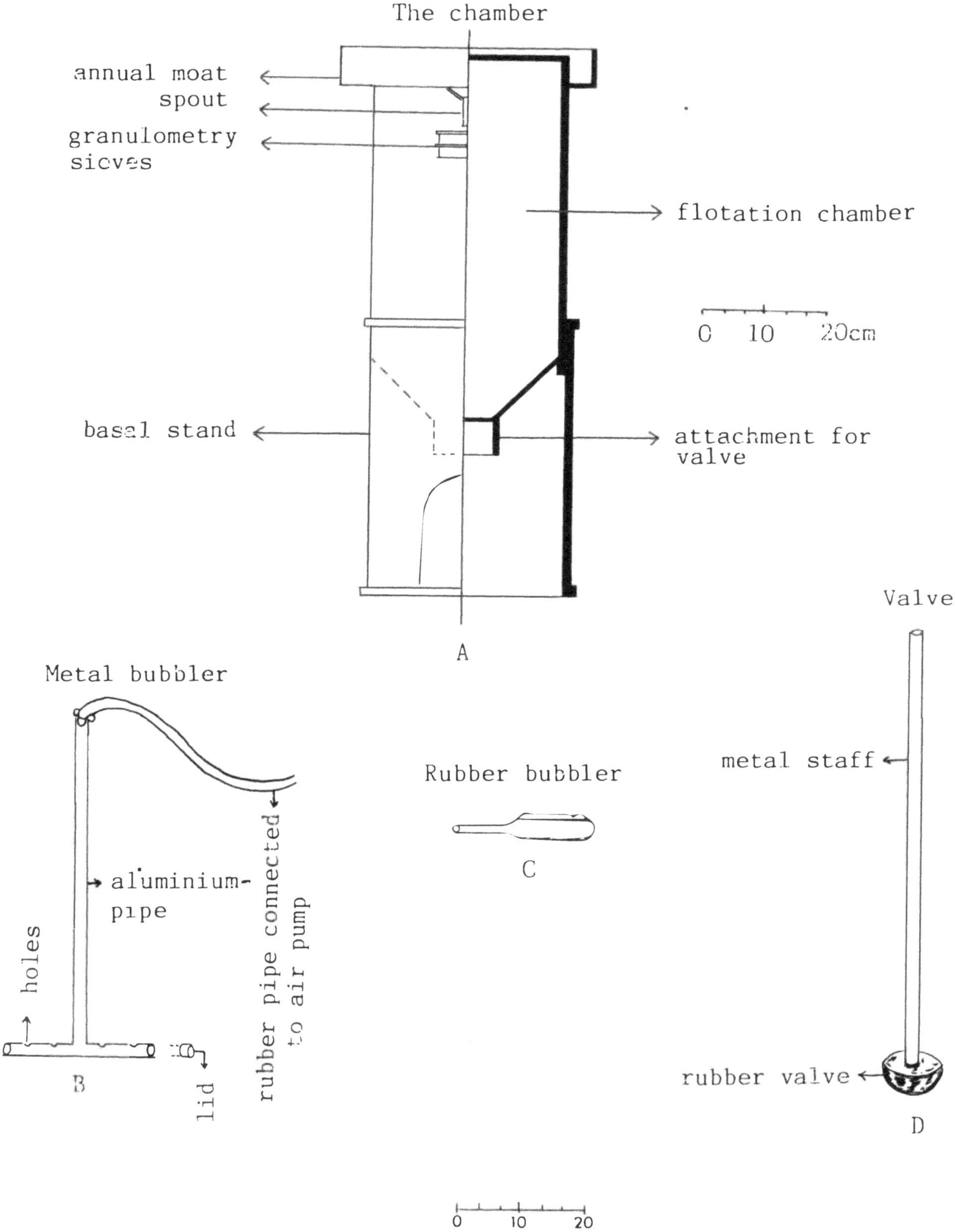

collectors) are added to help suspend the light, mixed solids in water by coating their surfaces. Thus, they become more water repellent and air avid than, e.g. the hydrophobic and aerophillic particles they are mixed with (Bender, B., 1975: 50; Jarman, H.N., 1972: 40; Renfrew, J.M., 1973: 15).

> 'The suspension is searched by rising air bubbles and the formation of short-lived froth at the surface is facilitated by making additions of a frothing agent .... which lowers the air/water surface tension so that bubbles may approach one another without coalescing. When suitably coated particles [seeds] meet an air bubble they attach to it and are lifted into the surface froth for removal' (Jarman, H.N, et al, 1972: 40).

Thus, with a continuously supplied liquid-stream in the chamber, the light[6] plant remains are the first to be raised to the surface and drained through the spout in the sieves, then heavier remains follow, then still heavier ones and so on. Terpinol and (mostly) soap-powder were used as frothing agents. The paraffin oil being efficient and also available and cheap, served the purpose of coating and hence collecting most of the seeds which otherwise could have sunk to the bottom of the flotation cell, or floating (Jarman, H.N., et al, 1972: 40-41) without being trapped and driven to the set of granulometry sieves (Magid, A., 1982: 87-88). Regulated compressed air was pumped into the contents of the cell in order to disturb the soil samples and to generate air bubbles which trapped the carbonized plant remains (and other light remains) and lift them to the surface.

Removal of large stones and objects by means of coarse sieving before the flotation of the soil samples is found to be most essential to avoid risks of destroying the internal components of the apparatus (Jarman, H.N., et al, 1972: 44) and also the fragile carbonized and desiccated plant remains. The soil samples should also be gently passed through large mesh sieve e.g. 2.5 and 1.00 cm mesh, after removing large objects as another measure to preserve the retrieved plant materials (ibid). It is also important to start the air pump before starting to pour a steady and gentle stream of soil sample into the cell. In this manner, creation of clay lumps is avoided and a thorough processing of the sample is completed. Some of the seeds are frequently trapped and lifted to the surface even before the soil sample reaches the bottom of the cell.

The retrieved plant materials were gently washed[7] with water (to avoid damage), then covered with damp tissue-papers, labelled and left to dry out slowly away from any source of direct heat (because rapid drying often results in damage to the material). Then

---

[6] This explicable in terms of the theory of dynamics diameter (Krzywinski, K., 1988: pers.comm.).

[7] Container with fine mesh sieve (0.5 mm) to let the water drip through, can also be used for drying out the remains if the number of the available granulometry sieves is limited.

they were carefully removed from the sieves and wrapped with soft tissue-papers or cotton and stored in labelled appropriately sized boxes[8].

The type of froth flotation-apparatus used in this work (cf. fig.no.17) has some major disadvantages. One of these is the rubber bubbler unit (cf. fig. no.17-C) which failed to provide an even distribution of the air pumped into the chamber. The soil samples could not be thoroughly disturbed to effect good separation of plant remains without the continuous manual effort of mixing the samples in water and hence longer time was wasted moving the bubbler in the chamber by means of a wooden rod attached to it. Since better units than this (see Jarman, H.N., et al, 1972: 43) were not accessible, some modifications were made to the available one by replacing the rubber piece by two hafted, T-shaped, aluminum pipes with several holes and removable lids at the ends (cf. fig. no.17-B). Better disturbance of soil and hence better separation of plant materials was achieved. This hafted aluminum bubbler could also be cleaned by removing the lids whenever needed. Other defects of the apparatus were both the valve (cf. fig. no.17-D) and the chamber attachment used to discharge deposits after retrieving the plant remains. The valve was vertically mounted in a discharge attachment at the bottom of the flotation chamber (cf. fig. no. 17-A,D); hence, it did not provide control when releasing the mud deposits. Therefore, all the water in the cell was wasted. As I previously mentioned, the provision of water for flotation purposes is frequently a major problem in many parts of the Sudan. For this reason any type of regulated valve, such as the controllable butterfly valve, is most needed in the Sudan because it will save time, normally, wasted to refill the chamber. More importantly, since a controllable valve allows a given volume of soil to be processed with the loss of only half of its volume of water (Jarman, H.N., et al, 1972: 44), water can be saved and even reused in situations where supplies are inadequate. A third disadvantage of the apparatus I used was that the flotation chamber was made of hardened plastic, that proved to be breakable under the field conditions in the Sudan, e.g. transportation and exposure to high air temperatures for a long time while in use. These factors resulted in serious damage and cracks that reduced its efficiency in holding water. Unfortunately, there was not any available alternative or effective means of repair.

Despite of all these major disadvantages and other minor ones, this froth flotation apparatus provided faster and more efficient large-scale processing of soil samples and retrieval of plant remains with minimum handling of the samples than would have otherwise been possible to achieve by other methods[9].

4.3.3 The method of positive casts of plant impressions in pottery.

---

[8] Soft brush and tweezers can be used when removing and/or identifying the specimens.

[9] Exceptions are concentrations of, e.g., seeds found as cobs or inside pots, pits, etc.

Pottery, mainly the hand-made, often contain evidence of impressions of small plant materials such as seeds/grains, husks, leaves, twigs, etc. Although these plant materials might have been originally in the clay or accidentally incorporated while shaping the pots/vessels, it is more common for them to have been added intentionally as a tempering medium and sometimes as decoration[10]. The dry plant materials absorb a certain amount of water from their moist clay surroundings, and the passage of water from the clay deposits a layer of fine clay particles around them. This fine cast often reproduces minute morphological details of the grain surface (Helbaek, H., 1955b: 653). During the firing of the pottery, these plant materials are been burnt out leaving depressions which retain more or less the same external morphological features as the original plant remains (Renfrew, J.M., 1973: 8).

On the basis of these facts, I examined the impressions of plant materials in pottery[11] (mainly potsherds) recovered from sites in the Central Sudan with the main emphasis on impressions of seeds and spikelets of grains. The procedure of examination proceeded as follows:

1) Preliminary gross-sorting of the potsherds by visual inspection and hand-held magnifying glass (x10) and selection of all those which bore any cavities or depressions which were characterized by regular shapes (e.g. rounded, elongated, oval, etc.) or showed obvious impressions of, e.g., fragments of the leaves of grasses. About 1200 kilograms (over 105600) potsherds[12] of different sizes were examined in the current work. These were recovered from the sites of Zakiab, Um Direwia, Kadero, Jebel Tomat, Rabak, Abu-Darbein, Aneibis and Shaqadud.
2) Reflected light stereo-microscopic examination (max. x40) was the next step which was intended to cross-check those potsherds which were visually inspected and thought to bear impressions of plants. Since the chemicals used in making the positive casts of the these impressions are not always easy to get, this step of microscopic examination led to dropping out a few of the sherds selected during the finest gross inspection and hence avoided needless waste of chemicals.
3) All the potsherds which were still thought to bear plant impressions after the microscopic examination, were subjected to a final treatment in which positive casts were taken from the original impressions (Magid, A., 1982: 87). This treatment consisted of the following procedure:

---

[10] Grains and sometimes fruiting cobs may be used for the decoration of pottery vessels. For instance, whole maize cobs were used for decorating large ancient Peruvian pottery from Mochia (presently Chile & Peru) dated to A.D. 900-1000 (Schery, R.W., 1954: 368), but evidence of this nature was not encountered in the type of pottery being studied here.

[11] The collection of pottery included that recovered from sites in the excavation of which I participated and pottery which was recovered previously but which I was given permission to examine.

[12] Previously examined collections included 150 kilograms from the sites of Kadero, Um Direwia and Zakiab (Haaland, R.: pers.comm.) and from the Nofalab, Islang and the Shaheinab sites about 200 kilograms of potsherds were examined by the present writer in 1982.

a) Careful cleaning of the potsherds with distilled water and diluted acetic acid (Stemler, A.B.L., and R.H. Falk, 1981: 195), then rinsed in ultrasonic-bath (with distilled water) for 10 minutes and leaving them to dry. This step is essential if the impressions are filled with dirt and/or dust. Otherwise distilled water or even blowing air in the impressions will be sufficient.
b) Each impression was then smeared with a mixture[13] of vaseline (9 grams), paraffin (1 gram) and toluen (100 ml) in order to facilitate the extraction of the positive casts without damaging them (Magid, A., 1982: 89).
c) Methyl polysiloksan (R.T.V.M. 400) with its elastic filling compounds was used as casting material. A hardener[14] (T.37 T42) which is a mixture of organic components, tin and silica ester (Krzywinski, K.: pers.comm.), iron oxide[15] (FeO) and/or India ink was quickly but thoroughly mixed with the casting material and poured into the cavity exposed on the surface of the potsherds and then set up to dry[16] in normal room temperature.
d) The positive casts were then carefully extracted from the potsherds using two pairs of forceps.
e) The cast specimens can be directly examined with a light microscope. But if they are to be examined by a scanning electron microscope (abbreviated SEM), they must be prepared in sputter coater, e.g., mounted (with silver paint) on aluminum stubs and coated with gold palladium, etc. (cf. appendix no.2) before their examination.

Some problems may be encountered when employing the method of positive casts. The basic ones are:

1) If the potsherds with plant impressions are not well fired, it will be very damaging to the cavities to attempt removing the dirt and dust often deposited in these by using, e.g. diluted acetic acid; for the potsherds tend to disintegrate if treated by such acids (Stemler, A.B.L., and R.H. Falk, 1981: 193). Sonication is found to be the most appropriate alternative (ibid); but it has to be carefully, patiently and properly done.
2) If the casting components are not mixed and freed from air bubbles, these air bubbles will destroy the casts and hence produce distorted and/or incomplete positive casts of the impressions. An ordinary vacuum pump is found most useful in reducing and

---

13 This mixture gives far better results than those obtained by using i.e. paraffin, alone (Krzywinski, K.: pers.comm.)

14 Iron oxide and India ink were added to provide a dark-grey, light absorbent surface to facilitate both identification and photography. However India ink is more efficient as regards the thinning of the casting material and hence ensures its thorough penetration into the cavity, no matter how small it may be.

15 There are two types of hardeners, fast, which help consolidating the casting material in less than an hour and which requires that the mixture be prepared and poured quickly; and a slow hardener, which takes 3 to 4 hours to consolidate the casting material. Both, however, if excessively used or left for longer periods of drying would involve risks when extracting broken casts.

16 See footnote number 15.

eliminating this disadvantage. In this connection, it is also quite possible, if fast and thorough pouring is not observed, to produce incomplete positive casts. This often happens in the case of deep impressions (Stemler, A.B.L., and R.H. Falk, 1981: 193) and or of such as have a narrow opening.

These problems are to a large extent minor and avoidable. In addition, they are related to the technical application rather than the final results and objectives of the method of positive casts.

The value of this method can be summarized in the following points:

1) Once the pottery is burnt, the cavities (impressions) of the plant materials would be, like the pottery, permanently preserved. Accordingly as many positive casts (replicas) as desired can be made of each individual impression without destroying it or distorting and/or reducing the quality of the casts.
2) Since the method succeeds in duplicating casts from the original impressions of the plants it provides an accurate and reliable representation of the former plant morphology (ibid). That is to say, less distortion is expected during the formation of the plant impression than that expected during the process of carbonization. In this connection, unlike the doubtful survival of carbonized/desiccated, the survival of evidence of plant impressions is almost certain.
3) There is a direct association between the plant impression and the material culture as represented by the sherds. Thus the dates of these impressions and their positive casts are secured, if the dates of the types of pottery and the context in which they are found are established (Renfrew, J.M., 1973: 16).

The information obtained from a single impression will therefore often be more reliable and less biased than a big sample of carbonized seeds. It can also be referred to if need arises to make an assessment on the carbonized/desiccated plant remains. But, if several samples of carbonized/desiccated specimens were recovered from different contemporary levels of a site, they would certainly give a better picture than a few of grain impressions (ibid).

4.4 The Identification Procedure.

The identification procedure was based on comparisons of external morphological features (e.g. surface pattern, size, and shape) of carbonized and desiccated (mummified) seeds/grains and their impressions in pottery with a reference collection of extant plants. Manuals for identification of seeds, e.g., Martin, A.C., and W.D. Barkley, 1973, and comparison with existing prehistoric and historic evidence of plant impressions and carbonized/desiccated specimen (plus their magnified pictures) were also used. In addition, I continuously depended on consultation with the plant taxonomist Ekhlas A.

Bari (University of Khartoum), Mahammoud Ahmed (the Savanna Research Project, Sudan) and Knut Krzywinski (University of sergen).

4.4.1 Comparison of the archaeological specimens with a reference collection of extant plants.

4.4.1.1 The reference collection.

Since the theme of the present work is the exploitation of food-plants mainly *Sorghum* sp. and *Pennisetum* sp. in the Central Sudan, and since most of the archaeological specimens seemed to belong to the grass family *Gramineae*, I conducted an experimental study on seeds/grains of several grass species. Special emphasis was laid on both wild and domesticated species of *Sorghum* and *Pennisetum*. sp. This experimental study involved carbonizing of extant seeds/grains and making impressions of these in pottery. The basic objective of these experiments was to provide some of the basic requirements for the process of identifying the archaeological specimens. These are:

1) To provide a reference collection that consists of carbonized specimens and positive casts of impressions in pottery in addition to the traditional reference collection of fresh extant plants.
2) To examine the change in the external morphology of the carbonized specimens and their impressions in pottery and to measure statistically the percentage of change in their size as compared with the size of the same specimens prior to their treatment. The results obtained are thought to be essential for understanding 'the changes in shape and size proportions caused by heat in order to visualize the original appearance of the deformed and often mutilated carbonized grains [or their impressions in pottery], to be able to identify' (Renfrew, J.M., 1973: 11).

4.4.1.1.1 Experiments and results of carbonization.

Thirty to fifty seeds/grains each of *Sorghum vulgare* L., *Sorghum arundinaceum* (Desv.) Stapf., *Pennisetum americanum* (L.) K.Schum., *Setaria verticillata* (L.) Beauv., *Echinochloa pyramidis* (L.) Hitchc., *Triticum aestivum* L. and *Hordeum* sp. L. were randomly selected from different positions in their inflorescences i.e. from the apex, middle and distal parts and were measured (tables from 1 to 7 in appendix no.3). The seeds/grains of each species were then wrapped in aluminum foil and gently[17] and evenly heated in a sand bath to between at least 100 and 110°C[18] until they were properly carbonized. The time needed for carbonization depended mainly on the size of the seeds/grains. The specimens were then left in the sand-bath (after turning the heating off) until the temperature dropped to normal room temperature, then the specimens were

[17] Direct exposure to high temperatures leads to explosion and/or greater distortion of the seeds/grains.

[18] Although it required a long time, (e.g. 3 to 4 hours), 100-110°C was found to be the optimum range of controllable heating in sand-bath during the carbonization process of, e.g., grains of grasses.

carefully removed and left in their aluminum foil for some time before they were finally taken out. The exposure of the specimens to normal room temperature directly after carbonization often leads to their explosion and damage. Further, if the specimens were exposed to heating beyond their charring point, they might also be destroyed, even turning into ash. The carbonized specimens were measured, and the results compared with those before their carbonization as shown on fig. no.18 and tables from 1 to 7 in appendix no.3. However, it can be seen in all of the specimens that some distortion[19] and changes have occurred in the following pattern (cf. fig. no.18 and tables from 1 to 7 in appendix no.3):

1) The length decreases after carbonization. The extent (i.e. in percentage) of such decrease varies from one species to another.
2) The breadth increased in general, however, the extent of this increase varying from one species to another. The only exception to this general tendency is the decrease in the breadth of *Sorghum arundinaceum* (cf. table no.3 in appendix no.3). The most likely cause of such a decrease is that the grains are tightly enclosed in their lemma(s). A similar explanation was given for the decrease in breadth after carbonization of hulled barley (Renfrew, J.M., 1973: 13).
3) The thickness of the grains generally increases, but the extent of this varies from one species to another. Results similar to this have been obtained after carbonizing grains of *Triticum aestivum*, *Hordeum vulgare*, *Avena sativa* and *Secale cereale* (ibid: pp.10-13).

Other general observations on the results of these experiments include the following:

1) Grains with soft structure, e.g. wild species and/or primitive ones, shrink more than those with hard structures, e.g. domesticated ones.
2) Grains with greater breadth than their length (e.g. in *Sorghum vulgare* L. (cf. table no.1 in appendix no.3) have a tendency to reverse the general pattern of shrinkage and expansion in the foregoing. That is to say, the breadth shrinks and the length would either expand or remains without change after carbonization.
3) The length and breadth may not change, but a slight increase in thickness may occur after the carbonization of some of the grains. It may also happen that the length remains constant after carbonization while both breadth and thickness puff. It may also happen that while the length decreases and the thickness increases after carbonization, the breadth remains the same or is only very slightly changed (0.1± mm).

[19] In cases of obvious distortion of the grains, it is recommended to avoid placing too much emphasis on their measurements while identifying the archaeological specimens (Renfrew, J.M., 1973: 11).

Fig.no18 PLOTS SHOWING CHANGES IN DIMENSIONS (OF GRAINS OF SEVEN SPECIES) DUE TO CARBONIZATION

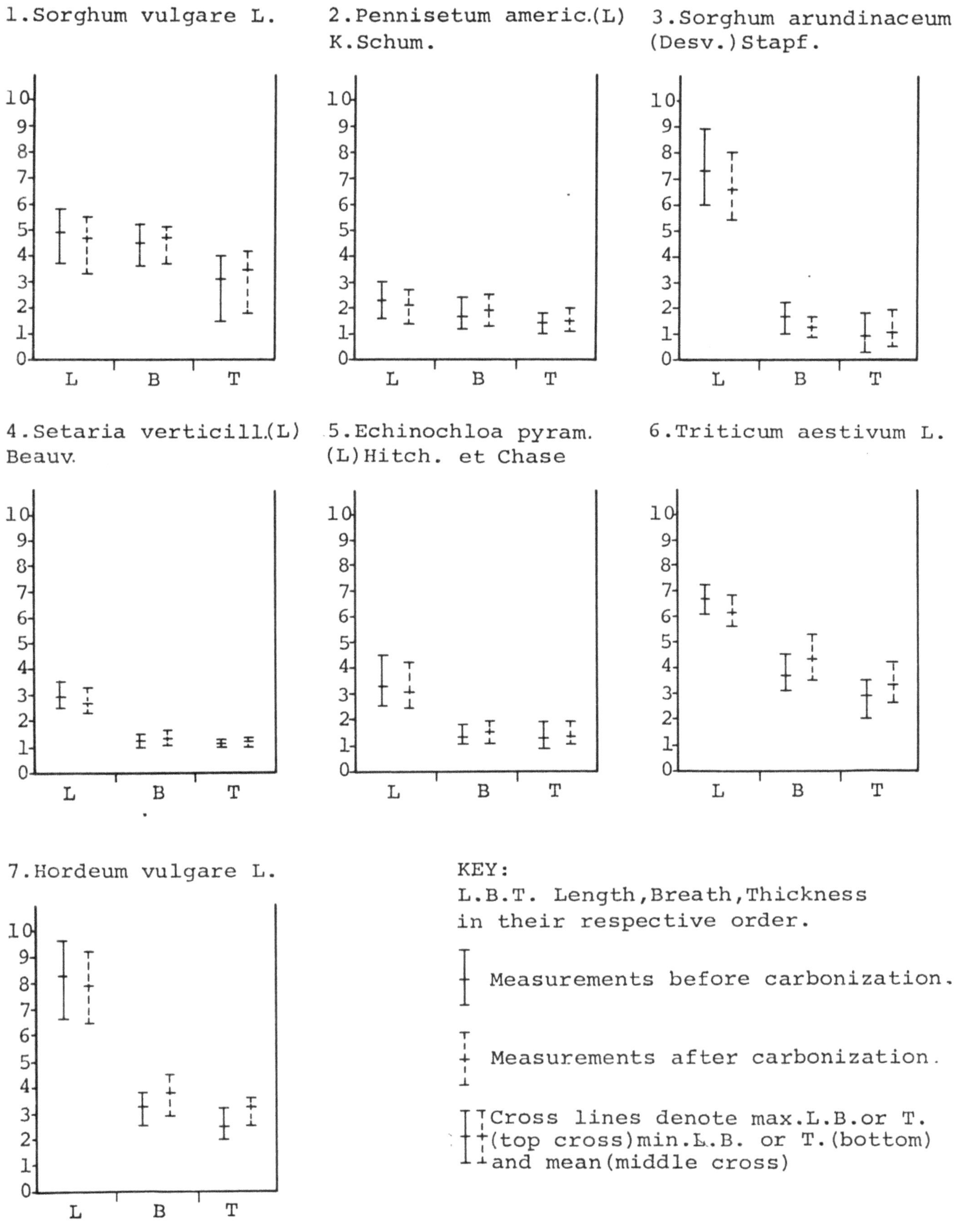

4) The differences in the sizes of the individual grains before and after carbonization are attributed to a number of reasons. These are the position of the grain within the inflorescence, e.g. those picked from the apex are generally bigger than with those picked from the distal end of the inflorescence. Furthermore, the size of grains of different inflorescences, e.g. of sorghum differ from one another due to differences in the amount of water received and where each of the inflorescences was growing in the field.

These individual observations and the general results obtained from the experiments were found to be of great help when identifying the archaeological specimens. They also stimulate further such experiments in future to include a wider list of different species/genera, immature seeds/grains, kernels (stones) and seeds both of ripe and unripe fruits, etc.

4.4.1.1.2 Experiments on and results from impressions in pottery.

Experiments on making impressions of fresh seeds/grains in pottery used seven different grass species including some species in both wild and domesticated forms. These species are *Sorghum vulgare* L., *Pennisetum americanum* (L.) K.Schum., *Sorghum arundinaceum*(Desv.) Stapf., *Setaria verticillata* (L.) Beauv., *Echinochloa pyramidis* (L.) Hitchc., *Pennisetum pedicellatum* Trin., *Sorghum verticilliflorum* (Steud.) Stapf. and *Sorghum aethiopicum* (Hack.) Rupr. A sample of 60 grains (omitting *P. pedicellatum* and *S. aethiopicum*[20]) was randomly selected from each of these species, labelled and measured (cf. tables from 8 to 15 in appendix no.3). A mixture of Nile silt, coarse sand and plant materials[21] (mainly of grass origin), donkey and cattle droppings was treated with water in the same manner as the clay of the traditional hand-made pottery is mixed and treated[22]. Then flat rectangular blocks were made from the clay. The following step was the insertion of the measured grains of each species in each of the clay blocks according to their labels which were now duplicated on the clay. Each population[23] of the grains was divided into two equal groups (predominantly 30 grains in each). One of these groups was further divided into two equal groups (each group 15 grains). The idea behind such divisions and subdivisions was to have a population of grains, inserted with their apex ends in the wet clay, equal to that inserted with their

---

20 It was not possible to provide for these experiments more than 15 grains of *P. pedicellatum* and *S. aethiopicum*.

21 The coarse sand and plant materials were used as tempering material.

22 Detailed information on the production of hand-made pottery (e.g. in Darfur) in the West Sudan was published in the work of Haaland, R. 1978: 47-61).

23 Generally speaking, the more the number of specimens involved in the experiments the better; however, the minimum number of any population in experiments should not be less than 30 specimens (A-Bari, E.: pers.comm.). This minimum number was doubled in these experiments because there was no other way of making positive casts of the ventral and dorsal surface of the grain impression than using a double number of specimens to fulfill the minimum population required in experiments.

dorsal/ventral surface embedded. The specimens were left until completely dry, then placed in an electric oven and kept there at 700°C[24] until the clay was finally baked into pottery and nothing of the grains was left except their impressions. It was only in the case of *Sorghum vulgare* that the grains were left to soak[25] in the clay and were then measured (cf. table no.8 in appendix no.3). After firing the pottery, positive casts of the impressions were made in the same manner as described in the foregoing discussion. These impressions were measured and the results were compared with those of the original seeds/grains (cf. tables from 8 to 15 in appendix no.3). The following general observations were recorded from these comparisons (cf. fig.no.19 and tables from 8 to 15 in appendix no.3):

1) The lengths, breadths and thicknesses of the impressions of all the different species in the pottery (as seen from their positive casts) generally decreased. The extent (percentage of this decrease varied from one species to another; but, generally speaking, the smaller the grain the less the extent of shrinkage. Thus, the size of impressions in pottery, showed different proportions as compared with the size of the original grains. These results confirm J.M. Renfrew's statement that impressions 'should not be compared directly with those obtained for carbonized grains, or grains preserved by any other means' (Renfrew, J.M., 1973: 16).
2) The lengths, breadths and thicknesses of the grains (e.g. *S. vulgare* L.) soaked in wet clay increased noticeably (cf. table no.8 in appendix no.3) as compared both with the original size of the grains and with their impressions in the fired clay. This increase is mainly related to the absorption by the grains of moisture from the wet surroundings (clay).
3) A certain amount of shrinkage was recorded during both the drying and firing of the clay. The extent of shrinkage increased with the increase in the amount of clay and organic temper (e.g. plant materials) and it decreased with the decrease of these in pottery and/or with the increase of the amount of inorganic temper, e.g. mica, coarse sand. For instance, the pottery I made in my experiments predominantly consisted of fine-grained clays (Nile silt) and organic temper. It shrank about 8% of its original size after drying and firing. But it has also been reported that if the components of pottery are dominated by coarse clay and inorganic temper, it will shrink only 5% of its original size (Renfrew, J.M:, 1973: 16). There is no doubt that such variable ranges of shrinkage in pottery would proportionally affect the size of the grains and hence their impressions in pottery.

---

24 700°C was found to be the approximate degree of firing the prehistoric pottery recovered from sites in the Central Sudan (Haaland, R., 1981: appendix no.3).

25 Due to the limited number of grains of all the other species, I did not risk soaking them because I had to experiment with soaking the grains in water before the insertion in clay and also with soaking them by insertion in wet clay.

Fig.no19 PLOTS SHOWING CHANGES IN DIMENSIONS OF GRAIN-IMPRESSIONS OF EIGHT SPECIES AS SEEN FROM THEIR POSITIVE CASTS.

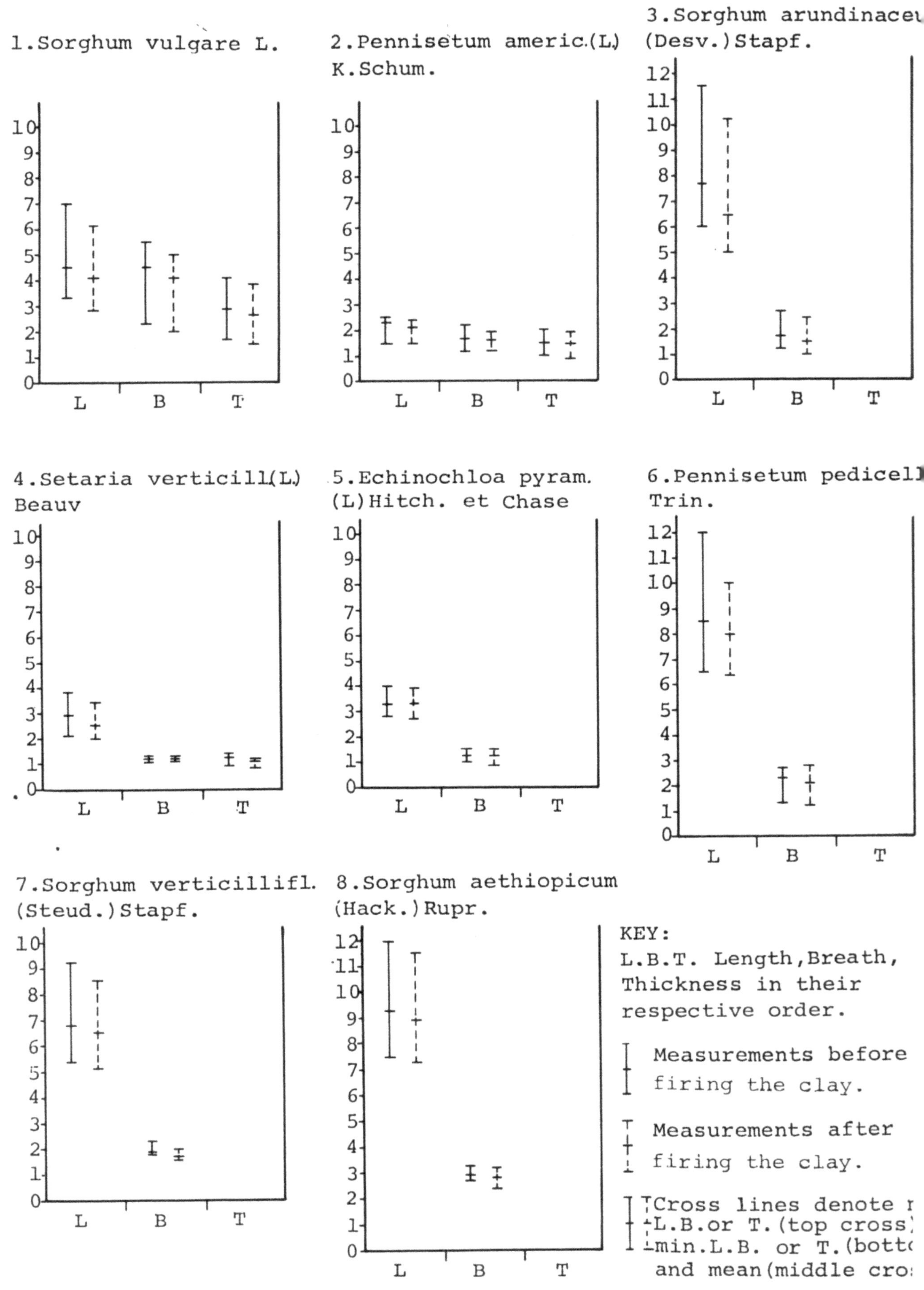

4.4.1.1.3 Results of identifying the archaeological specimens[26].

The conventional light microscope and the scanning electron microscope (SEM) were found satisfactorily efficient in reflecting the minute external morphological features of both carbonized specimens and positive casts of their impressions in pottery. The SEM has, however, a more marked effect on the identification procedure due to its enormous depth of field (power of magnification), resolution and the ease with which the specimens can be examined and at the same time photographed. Even stereomicroscopic photographs can be made by SEM if desired. Thus, in practise the SEM serves as a means of obtaining and/or storing accurate information which would otherwise necessitate both a number of apparatuses and a series of time-consuming procedures (Magid, A., 1982: 90).

The results of the identification of the retrieved specimens[27] (both carbonized/desiccated and impressions in pottery) attested the presence of a number of genus/species as shown on tables nos.5 to 15, and appendix no.4. But before I proceed in the interpretation of these results[28], it is essential to evaluate the methods of sampling, and to examine the reliability of the retrieved evidence, the preservation conditions and the context in which they were found. Any interpretation of botanical remains from archaeological sites (specially in arid and semi-arid regions) which does not include an assessment of these issues and their probable limitations would certainly be incomplete if not biased and misleading.

As regards the retrieval of carbonized/desiccated evidence, the method of froth flotation was found to be the most efficient of the available methods in recovering some representative indications of the species still preserved in archaeological contexts (Jarman, H.N., et al, 1972: 48). However, this method has its limitations (cf.4.3.2). In addition, there are also possibilities that some of the, e.g. seeds/grains may explode and that others may not float or be trapped in the granulometry sieves (ibid; Renfrew, J.M., 1973: 16-17).

· In connection with the temporal relation between the retrieved plant remains and the context (site) from which they were recovered, I found out that not all the retrieved specimens are contemporary with their archaeological contexts. With reference to the period (c. 9000-1700 B.P.) and the sites being studied here, most of the plant remains recovered were actually not contemporary with the contexts they were found in. This is attested in the predominant presence of annuals, particularly among the plant remains

---

26 Illustrations of specimens similar to most of the ones mentioned in this text are presented in the works of A.J.Arkell (1949,1953),D.Clark and A.B.L. Stemler (1975),R.Haaland (1987a) and A.Magid (1982).

27 I was informed that no soil samples were taken for retrieval of macrofossil plant remains from the sites of Kadero, Um Direwia and Zakiab (Haaland, R.: pers.comm.).

28 This study also includes the results of identification of other studies on macrofossils (both carbonized/desiccated and their impressions) from other sites in the Central Sudan.

Table no. 5. Carbonized and desiccated plant remains: Abu-Darbein site.

	Radiocarbon dates B.P.	7700±140			7860±190	8500±100	
	Depth in cm	0-10	10-20	20-30	30-40	40-50	50-60
	Level no.	1	2	3	4	5	6
Seeds - Grains	*Panicum* sp.	c	c				
	Setaria sp.		c		d	d	
	Crotularia sp.		c				
	Capparis sp.	c+d	c+d				
	Ziziphus spina-christi		d	d	d		
	Celtis integrifolia		d	c+d	c+d	d	
	Unidentified	c	c	c+d		d	
Others	Leaves	d(A)	d(A]				
	Twigs						
	Unidentified	c	d				

Legend: c: carbonized; d: desiccated; (A): *Acacia* sp.

Table no. 6. Plant impressions in pottery:
Abu-Darbein site.

	Radiocarbon dates B.P.	7700±140	7860±190
	Depth in cm	0-10	30-40
	Level no.	1	4
Seeds - Grains	Ziziphus spina-christi		+
	Unidentified	+	
Others	Leaves		
	Twigs	+	
	Unidentified	+	

Table no. 7. Plant impressions in pottery: Shaqadud cave site.

Radiocarbon dates B.P.*		3615±88				4123±86			4059±65
Depth in cm BCD		65-70	90-95	125-130	150-155	155-160	205-210	215-220	220-225
Level no.		14	19	26	31	32	42	44	45
Seeds - Grains	*Setaria*		+						
	Pennisetum sp.								+
	bracts ?						+		
	Cucurbitaceae	+							
	Unidentified				+				
Others	Leaves								
	Twigs			+		+	+		
	Unidentified						+	+	

Legend: BCD: abbreviations denoting Below Cave Datum.

* The precise levels from which the material for radiocarbon dates was obtained, are shown in appendix no. 1.

Table no. 8. Carbonized and desiccated plant remains: Shaqadud cave site.

	Radiocarbon dates B.P.	3615±88																					4123±86						
	Depth in cm	75- 80	80- 85	85- 90	90- 95	95-100	100-105	105-110	110-115	115-120	120-125	125-130	130-135	135-140	140-145	145-150	150-155	155-160	160-165	165-170	170-175	175-180	180-185	185-190	190-195	195-200	200-205	205-210	210-215
	Level no.	16	17	18	19	20	21	22	23	24	25	26	27	28	29	30	31	32	33	34	35	36	37	38	39	40	41	42	43
Seeds - Grains	*Setaria* sp.																							d					
	Wild *Sorghum* sp.																												
	Panicum turgidum												c	c		c	c	c			c								
	Pennisetum sp.																	c	c	c	c	c	c	c	c	c	c	c	c
	Sida alba																									c	c	c	
	Cratalaria sp.																						c	c					
	Solanum dubium										c	c							c										
	Grewia tenax				c	c		c+d	c	c	c+d	c			c+d	c	c				c	c+d	c	c	c			c	c
	Ziziphus spina-christi										d				c+d								c+d				d		c+d
	Cruciferae																		c										
	Chenopodiaceae																									c			
	*Leguminaceae**																		c							?c			
	Unidentified				c						c			c				c				c			c				

* Desiccated leaves of *Acacia* sp. were also recovered from levels 38 and 46.

Table no. 8 (cont.):

Radiocarbon dates B.P.																								4059±65				4046±101	
Depth in cm	215–220	220–225	225–230	230–235	235–240	240–245	245–250	250–255	255–260	260–265	265–270	270–275	275–280	280–285	285–290	290–295	295–300	300–305	305–310	310–315	315–320	320–325	325–330	330–335	335–340	340–345	345–350	350–355	355–360
Level no.	44	45	46	47	48	49	50	51	52	53	54	55	56	57	58	59	60	61	62	63	64	65	66	67	68	69	70	71	72
Seeds - Grains																													
Setaria sp.																													
Wild *Sorgum* sp.				c																									
Panicum turgidum																													
Pennisetum sp.	c	c	c	c	c	c																							
Sida alba																													
Crotalaria sp.													c																
Solanum dubium																													
Grewia tenax	c	c	c	c	c	c	c		c				c	c															
Ziziphus spina-christi	d	c							c+d								d			c	c	c	c						
Cruciferae																													
Chenopodiaceae				c													c												
Leguminaceae[*]																													
Unidentified	c																c			c									

Legend: c = carbonized; d = desiccated

[*] Desiccated leaves of *Acacia* sp. were also recovered from levels 38 and 46.

Table no. 9. Plant-impressions in pottery Um Direwia site.

	Radiocarbon dates B.P.*	4950 ± 80 D. 10cms.		5600 ±110 D. 15cms.	6010 ± 90 D. 45cms.	
	Square no.**	100x/100y	102x/100y	104x/ 84y	122x/100y	122x/102y
Seeds - Grains	*Setaria* sp.			+		
	Wild *Sorghum* sp.				+	
	Staked spikelet of ? Wild *Sorghum*	+				
	Lower glume of ? Wild *Sorghum*	+				
	Unidentified		+			+

Legend: D: stands for Depth.

* The R/C date 5600±110 was obtained from square 104x/100y and 6010±90 ws obtained from square 104x/168y (appendix no.1).

** Although the postsherds with plant impressions were not laballed by depth and/or level, it is certain that they are basically associated with the Shaheinab period (c. 6000-5000 B.P.)

Table no. 10. Plant-impressions in pottery:
Kadero site.

Radiocarbon dates B.P.		*
Square no.**		C-23(3)
Seeds - Grains	Wild *Sorghum* sp.	+

* Neither dates nor depth nor level [of C-23(3)] are provided. Thus reference is made to R/C dates obtained from other squares as shown in appendix no.1

** See footnote no.** on table no.9.

Table no. 11. Plant-impressions in pottery: Zakiab site.

Radiocarbon dates B.P.*		5350±90 D. 15 cm		5660±80 D. 35 cm	
Square no.**		100x/100y	102x/100y	102x/102y	102x/104y
Seeds - Grains	*Setaria* sp.		+(?)		
	Wild *Sorghum* sp.	+		+	
	bracts		+		
	Unidentified		+	+	+

Legend: D: stands for Depth.

* The R/C date 5350±90 B.P. was obtained from square 102x/98y.
** See footnote no.** on table 9 and appendix no.1.

Table no. 12. Carbonized and desiccated plant remains: Nofalab site.

Radiocarbon dates B.P.		5290±100				5520±130			
Depth in cm		0 - 10	10 - 20	20 - 30	30 - 40	40 - 50	50 - 60	60 - 70	70 - 80
Level no.		1	2	3	4	5	6	7	8
Seeds - Grains	*Panicum turgidum*	c	c						
	Crotalaria sp.	c	c						
	Capparis sp.	c	c						
	Boerhavia sp.	c	c		c				
	Chenopodaceae								
	Molluginaceae	c							
	Celtis integrifolia		d*	d	d	c+d	d	c+d	d
	Unidentified	c							
Others	Leaves**	d(A)	d(A)						
	Twigs								
	Unidentified			c					

After Magid, A., 1982: 92 with modified presentation.

* The remains consist of a few seed fragments.

** All the leaves are of *Acacia* sp.

Table no. 13. Plant-impressions in pottery:
Shaheinab site.

	Radiocarbon dates B.P.*	5550±90 D.10cms.	5240±80 D. 5cms.	5370±80 D.47cms.
	Square no.	G61	G87	
Seeds - Grains	*Sorghum verticilliflorum*	+		
	Celtis integrifolia		+	
	--			

After Magid, A., 1982: p.90
with modified presentation.

Legend: D: stands for Depth.

* R/C dates 5240±80 and 5370±80 are obtained (in 1979) from square P.87 (cf. appendix no.1).

Table no. 14. Plant-impressions in pottery: Jebel Tomat site.

		1	2	3	4	5	6
Radiocarbon dates B.P.		1770 + 80		4140 + 90			
Depth in cm		0 - 10	10 - 20	20 - 30	30 - 40	40 - 50	50 - 60
Level no.		1	2	3	4	5	6
Seeds - Grains	*Setaria* sp.						+
	Panicum sp.						+
	Sesamum sp.						+
	Wild *Sorghum*				+	+	
	Unidentified	+					+
Others	Twigs			+			
	Unidentified		+				

Table no. 15. Plant-impressions in pottery:
Rabak site.

Radiocarbon dates B.P.			4490±100	5860± 80		6050±100				6020±150
Depth in cm		0 - 10	10 - 20	20 - 30	40 - 50	50 - 60	70 - 80	110 -120	120 - 130	140 - 150
Level no.		1	2	3	5	6	8	12	13	15
Seeds - Grains	Wild *Sorghum* sp.	+				+	+			
	Cucurbitaceae sp.				+				+	
	Unidentified					+				
Others	Twigs		+							
	Unidentified		+							

recovered from the upper levels of the cultural debris (cf. table[29] no.12). It is most likely that these annuals were deposited in a later period. This interpretation is based on the following:

1) The shape, surface pattern, e.g. seed ornamentation, and size of the seeds retrieved are indeed well preserved. This is actually one of the indications that the seeds may have been recently buried; otherwise they would have had - to some extent - an abraded surface or distorted, crumpled shape or ill-proportioned size if they have not been completely decomposed (Magid, A., 1982: 92). This interpretation is further supported by the absence of similar species within the bottom levels (cf. table no.12) of almost all the excavated sites (except for the well protected cave-site of Shaqadud). This indicates that the plant materials at the bottom levels decomposed due to the unfavourable preservation conditions (ibid). These include abrasion by eroded Nubian sandstone pebbles, and coarse sand[30] and the conflicting high temperatures and rainfall during the summer season provided the time that has lapsed (e.g. thousands of years). In addition, despite of any cautious measures taken during the excavation the large amounts of quartz pebbles and small and large stone slabs remained mechanical factors that caused fragmentation of the few surviving *in situ* evidence of plant materials.
2) Similar species are growing at present during the short rainy season in the semi-arid habitat regions including the Central Sudan. Furthermore, there is evidence of leaves of *Acacia* sp. (cf. table no.12) deposited in the same levels as these annuals. This acacia is similar to *Acacia* sp. growing today in the Central Sudan.
3) In those sites where evidence of seeds of, e.g., *Celtis integrifolia* Lam. was recorded, there were no remains of this species within the upper levels as in the bottom ones of the same context (tables no.5 and 12).

Therefore, it is most likely that the presence of these "intrusive" species in the archaeological sites in the Central Sudan is to be attributed to the development of seed deposits by means of physical agents, i.e. wind and sandstorms, which are very active under the prevailing arid and semi-arid conditions in the area being studied here. Seed deposits can also be developed (as mentioned previously) by the action of, e.g., ants Thus, it is very important to consider all these factors with regard to the question of the contemporaneity of these plant materials with the cultural remains in archaeological sites. This is also the case for the development of deposits of carbonized seeds in such deforested arid areas in a period of only 25 years (A-Bari, E.: pers.comm.).

---

29 Detailed information on macrofossil plant remains found within the upper levels was given from the site of the Nofalab only. Finds from other sites were too meagre to demonstrate on tables.

30 In some of the sites i.e. Khartoum and Atbara area, the soil components consist of high calcareous elements.

In conclusion, it is most likely that the well preserved carbonized seeds and leaves which were found within the upper levels of the cultural debris were deposited some time after the Meroitic burials for which dates were suggested to be round the fourth century A.D. (Al-Hassan, A.A.: pers.comm.). This interpretation is supported by the similarities of the soil components (e.g. coarse sand, quartz pebbles, sandstone slabs) of the upper levels in which neolithic debris or finds from the late Meroitic period, e.g. beads, or both are in evidence. Therefore all the superficial finds have been excluded from the final interpretation of the past economic activities related to the exploitation of food-plants.

As regards the impressions of seeds/grains in pottery, the method of positive casts proved indispensable (cf. 4.4.3) for obtaining evidence which is both well preserved and contemporary with the material culture recovered. This method is also found useful in confirming or falsifying the value of evidence of carbonized/desiccated plant remains from the same archaeological context. However, this method demands strenuous, time-consuming effort examining several tens or hundreds of thousands of potsherds while yielding only a few potsherds that bear impression and sometimes none at all as in the case of the sites of Islang island and Aneibis (cf. fig. no.1). One of the reasons for the scarcity of evidence of impressions may be the type of temper used in the pottery, which, in the case of the pottery excavated from these sites, was mainly mica and coarse sand with very little or no organic temper (plant materials) being used. Another reason may be the difficulty of retaining the external morphological features of plant materials as impressions in basically coarse gritted pottery with mica, fine angular quartz grains, and sand as the main components of the temper. It also happened that no plant impressions could be found either on the exterior or interior surfaces of the pottery while evidence could be found embedded within the walls as once was evident in a section of an accidentally and freshly broken potsherd from the site of Jebel Tomat[31]. Thus, it could be possible to find plant impressions in this way but only at the cost of breaking the pots and potsherds to examine their sections.

Quantitative analysis, relative frequencies and other statistical analyses are useless and often misleading in studies of macrofossil plant remains; however, the total number of the seeds recovered or their impressions in pottery are sometimes reported. The present study excludes any statistical analysis of the data recovered for the following reasons:

1) 'Although there is, in principle, no difference between identifying, on one hand, pollen grains and on the other hand, seeds ...., etc, there is yet one important practical distinction: The fossils dealt with in pollen analysis are all comparable; consequently,

[31] The potsherd is labelled JT74-TT2-A2 and was found at a depth of 30-40 cm. An impression of a grain of wild sorghum was found in the section when the sherd broke.

> their relative frequency can be expressed as percentages of a total. This is not possible with macrofossils, since they are not comparable: a sensible sum total can not be made of leaf impressions, seeds, twigs ....' (Fægri, K., and J. Iversen, 1975: 123).

That is to say, as in the case of the remains of material culture recovered macrofossil plant remains are heterogeneous, e.g. twigs, seeds, leaves, etc. and hence are statistically comparable. Furthermore, it is often the case that the number of each of these categories will be too low for any numerical or quantitative treatment.

2) The conditions of preservation may not favour the survival of one type of species (e.g. seeds/grains of grasses); while remains of another species might be well preserved (e.g. seeds of *Celtis* sp. or *Ziziphus* sp.) due to differences, for instance, in their exodermic structure. This is particularly true in the case of the sites from semi-arid Central Sudan.

3) The specimens found on a site do not necessarily represent their actual consumption as foods. For instance, grain-foods such as sorghum would be consumed, and nothing or very little might be left behind, while large numbers of seeds of fruit-foods such as *Celtis integrifolia* Lam. might often be left behind in the same context.

The identification and interpretation of the macrofossil plant remains was completed in the light of these considerations (cf.4.4.1.1.2). The results are as follows:

4.4.1.1.3.1 The sites in the Atbara area.

No evidence was recovered of any macrofossil carbonized and/or desiccated plant remains or their impressions in pottery from the site of Aneibis (cf. fig. no.1). This does not mean there is a complete absence of such evidence; more extended study is required before drawing this conclusion. However, it should not be unexpected to find little and/or fragmentary evidence due to the preservation conditions previously discussed and the type of temper used for making pottery (cf.4.4.1.1.2).

Large numbers of desiccated seeds of *Celtis integrifolia*, L. (col. Ar. *mahagaya*) were recovered from the sites of Abu-Darbein (cf. table no.5 and pl.no.2)[32] and Ed-Damer[33]. A few seeds of *Ziziphus spina-christi* (L.) Willd. (col. Ar. *nabag*) and desiccated seeds of *Setaria* sp. (col. Ar. *danab al-kalib*) were also found at the site of Abu-Darbein (table no.5). It seems quite certain that these plant remains are contemporary with the period of occupation of the sites (cf. appendix no.1). This conclusion is based on the facts that seeds of *C. integrifolia* Lam. and *Z. spina-christi* (L.)

32 Although the tables include all the macrofossil carbonized and desiccated plant remains retrieved, mention shall be made only of those specimens interpreted as being contemporary with the period under consideration in this work.

33 The study of macrofossil plant remains from the Ed-Damer site is not yet finished. Mention is made in this text only of those finds picked out during the excavations.

Willd. are found in sites which are contemporary with these in the area of Khartoum (Arkell, A.J., 1949: 108-110, Pl.45, fig.3; Magid, A., 1982: 95-96). Further-more, these specimens were found *in situ* in the same context as cultural remains associated with the period being studied here, c. 9000-6000 years B.P. Most of them were found within the lower levels of the cultural layer, and no evidence of them was found in the upper levels except in disturbed contexts. Since the seeds of *Setaria* sp. retrieved (cf.pl.no.3) were the only grass-seeds found in the bottom levels with those of, e.g., *Celtis* (cf. table no.5), it is most likely that they are contemporary with the period being studied in this work (c. 9000-6000 years B.P.). In addition, the possibility that they were transported by ants is excluded due to the absence of similar species at present in the area (A-Bari, E., 1986: pers.comm.). If so, their recovery from this site marks the earliest evidence of food grains found so far in the sites of the Early Khartoum type in the Central Sudan. This interpretation needs, however, to be confirmed by further excavations.

As regards evidence of plant impressions in pottery, one impression of a seed[34] of *Ziziphus spina-christi* (L.) Willd. (table no.6) was found in a potsherd from the site of Abu-Darbein. The significance of this evidence lies in its recovery from pottery of an Early Khartoum type and hence in confirming that most[35] of the seeds of *Z. spina-christi* (L.) Willd. recovered are contemporary with the Early Khartoum cultural debris. This is an example that demonstrates the practical value of impressions of plants for cross-checking, the reliability of the carbonized and/or desiccated plant remains recovered, particularly if their presence in an archaeological context is doubted[36] and/or unexpected.

The implications of the recovered evidence can be summarized as follows:

1) Gathering of immediately eaten food-plants and perhaps the beginning of exploitation of food-plants, e.g., of grains that required processing, was practised in the area of Atbara during the Early Khartoum cultural tradition.
2) The rainfall was 6 to 7 times greater (at least 500 mm) during the Early Khartoum occupation than at present in the Atbara area (c. 70 mm mean rainfall). This interpretation is mainly based on the presence of *C. integrifolia* Lam. which grows in habitats with a minimum of 500 mm of annual rainfall (ibid). This plant species grows at present about 600 km south of this area. Evidence of the same kind for a similarly moist climate during the same period was reported from the area of Khartoum (Arkell, A.J., 1949: 108-110; Mohammed-Ali, A., 1982: 60).

---

[34] It is an impression of one half of a split seed.

[35] All the seeds recovered from the lower undisturbed levels are contemporary with the cultural layer.

[36] The reliability of the specimen could be doubted due to its unexpected presence in a particular site at a specific period or due to possibilities of its being present in any context at any time. Thus, in both situations the contemporaneity of the specimen with the context in which it was found would be doubted.

Plate no.2 Desiccated seed of <u>Celtis</u> integrifolia.

Plate no.3 Desiccated seed of <u>Setaria</u> sp.

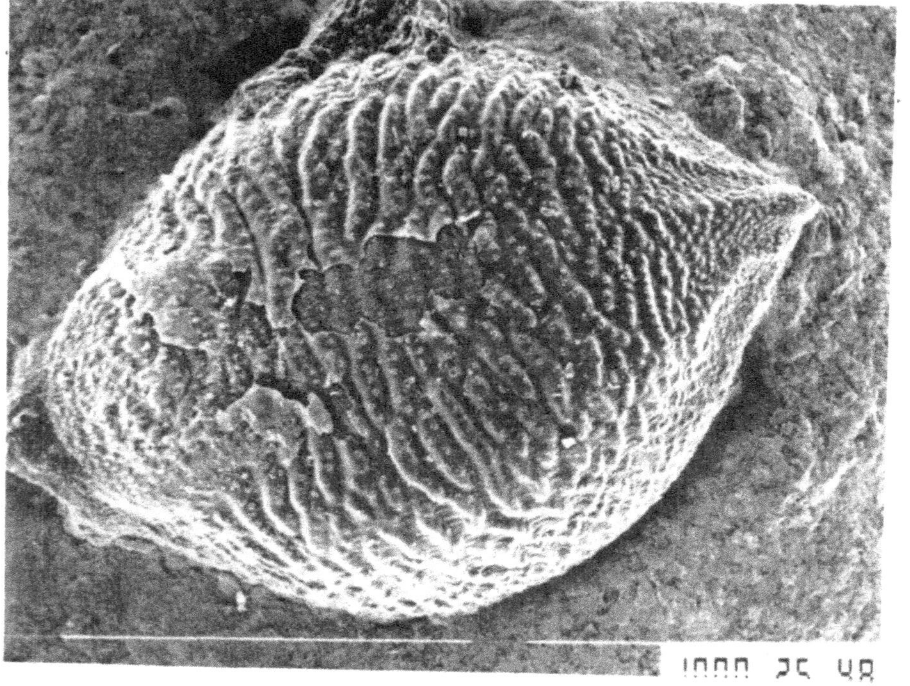

4.4.1.1.3.2 The area of West Butana.
The Shaqadud cave-site.

The Shaqadud cave-site (cf. fig. no.1) represents one of the ideal sites for good preservation conditions for macrofossil plant remains and their survival in such arid and semi-arid regions as the Central Sudan. Being sheltered and protected by a cave, the carbonized and desiccated plant remains (mainly seeds) which were recovered from this site are both *in situ* and remarkably well preserved considering the time that has passed (appendix no.1) since they were deposited there. Large seeds with hard thick exodermic walls and small ones with fragile thin exodermic walls were recovered from this site.

Retrieval by means of froth flotation attested the presence of many genera/species of both carbonized and desiccated plant remains (cf. table no.8). The majority of these are food-plants. They consist of fruit foods such as *Grewia tenax* (Forsk.) Fiori. (Ar. *guddeim*) ,*Ziziphus spina-christi* (L.) Willd. and grain foods such as *Pennisetum* sp. (Ar. *dukhun*) , and *Sorghum* sp. (Ar. *adar*) and *Panicum turgidum* Forsk. (Ar. *tumam*) . The *Sorghum* sp. was identified as being morphologically (shape, size) similar to the wild species, but it was indeed impossible to determine whether the recovered evidence of carbonized grains of *Pennisetum* sp. (Brum.) Stapf. and Hubbard (Ar. *dukhun*) were primitive domesticates[37] or truly wild species. I previously voiced my uncertainty about the form of these grains (Magid, A., 1984: 28). However, it has been stated that 'probable domestic millet' was found at this site between a depth of 2.1 m and 2.45 m (Marks, Á.E., et al, 1985: 275). This statement offers a more nearly certain identification in that, the specimens recovered are more similar to "domestic millet" than the specimens that underlie the results of my identification. One of the main problems of the specimens I am dealing with is that despite their being fairly well preserved, they are not sufficiently well preserved to show clearly the diagnostic features[38] and other essential portions[39] of the species, e.g. the part of the inflorescence subtending the spikelets, and hence to facilitate the task of identifying the form(s) of these specimens. Even with well preserved speçimens, it is often difficult to distinguish between the wild and the domesticated species of *Pennisetum* (Purseglove, J.W., 1985: 205); and this task is even more difficult if the distinction is made only on the basis of individual parts of the plants, e.g. the grains. Thus, the available evidence from carbonized specimens does not permit for any

---

37 Primitive domesticates of cereals, e.g. *Sorghum bicolor* (L.) Moench and probably those of *Pennisetum* sp., may generally look similar in shape and/or size (but not in the diagnostic features) to the wild species (Clark, J.D., and A.B.L. Stemler, 1975: 588).

38 Identifying the form (wild or domesticated) of cereal grain depends on determining the structure of the rachis. If the seed ends in rough rachis fracture, the suggestion is that it represents a domesticated. 'If the base of the rachis fragment is smooth, the suggestion is that it represent wild stock' (Stemler, A.B.L., and R.H. Falk, 1981: 196). As regards the identification of inflorescences, the abscission zone is the diagnostic feature in almost all cereals. That is to say 'in wild cereals, one finds a smooth abscission zone; in domesticates, abscission is repressed and the inflorescence breaks in a series of rough fractures when threshed' (ibid).

39 See note 37.

interpretation beyond the conclusion that the occupants of the Shaqadud cave-site were utilizing a *Pennisetum* sp. whose form is - so far - difficult to identify with certainty due to the distortion of the diagnostic features. In order to solve this problem, more data (i.e. carbonized/desiccated specimens) with better preserved diagnostic features is needed. Species such as *Solanum dubium* L. (Ar. *gubbein*) (cf.pl.no.4) , *Sida alba* L. (Ar. *umm shadeida*), *Crotalaria* sp. (Ar. *soferia*) were also recovered (cf. table no.8). It is indeed, difficult to develop an interpretation with regard to the presence and/or use of these plant species on the site; however, fruits such as those of *S. dubium* are used at present by the nomads of the Butana as fermenting catalysts that are added to boiled milk[40].

Examination of plant impressions in potsherds (recovered from 2 squares[41]) from this site yielded impressions of four seeds (cf. table no.7). Two of these were identified at the genus level. One looks very similar to *Setaria* sp. , and the other does not look morphologically different from the wild *Pennisetum* sp. (*P. americanum* (L.) K.Schum.) (cf.pl.no.5) . These results confirm the contemporaneity of the carbonized/ desiccated seeds of *Pennisetum* sp. and *Setaria* sp. with the period of occupation of the site. However, direct comparison of the carbonized specimens with the impression of the grain of the *Pennisetum* sp. is totally invalid, primarily due to the different proportions of the impressions (Renfrew, J.M., 1973: 18) and secondarily due to the distorted diagnostic[42] features of the carbonized specimens. Since the area of Shaqadud is considered as being within the eastern part of the natural habitat of wild pearl millets (Harlan, J.R., 1977: 376), the finding of these specimens in this area is not unexpected. But as I stressed in the foregoing more and better preserved data is needed in order to provide more comprehensive information.

The recovery of desiccated seeds of *Setaria* sp. under an overturned concave lower grinder from this site (at a depth of 185-190 cm BCD[43]), provides the first evidence from the prehistoric Central Sudan of direct archaeological association of grinders and grain foods.

Generally speaking, the macrofossil plant remains and plant impressions from the Shaqadud cave-site (tables no.8 and 7) indicate that two distinct but complementary adaptations are practised with regard to the exploitation of food-plants. One consisted in seasonal activities of gathering wild fruit foods that consisted of two plant species, *Grewia tenax* (Forsk.) Fiori. and *Ziziphus spina-christi* (L.) Willd. The recovery of only

---

40 Evidence of domestic animals was also recovered from the cave-site at Shaqadud (Peters,J.,1984: pers.comm.).

41 The present writer was given access to examine potsherds from squares F8 and G8, each was 0.5x0.5x3.5 m.

42 This type of situation would probably be manageable if there is a reference collection which consists of fresh and carbonized grains and casts of their impressions in pottery (both of wild and domesticated species).

43 BCD are initials denoting: Below Cave Datum.

Plate no.4 Carbonized seed of <u>Solanum dubium</u>.

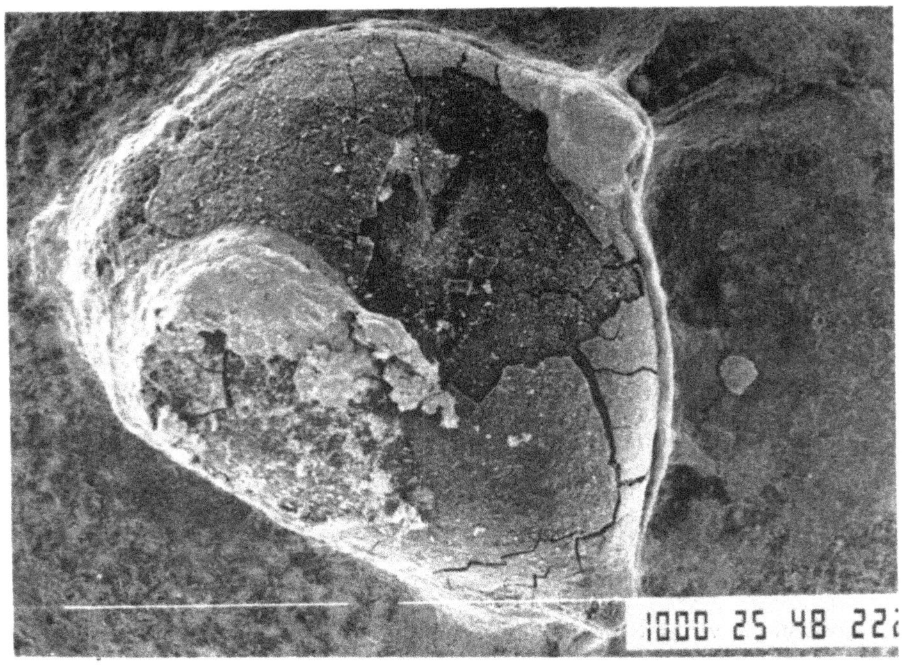

Plate no.5 Positive cast of an impression of a seed of Pennestium sp.

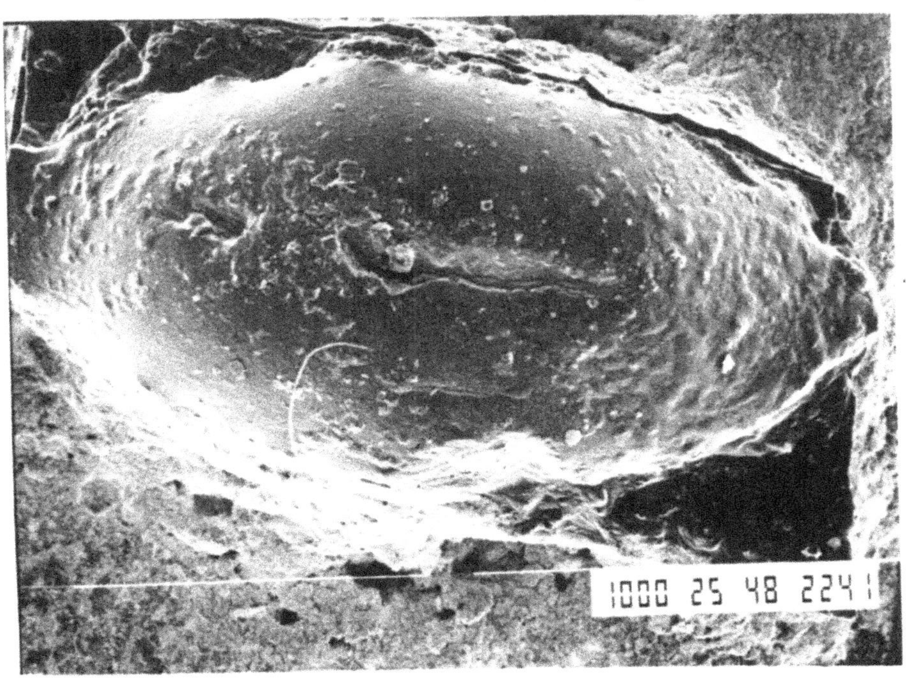

these two species does not necessarily exclude the possibility that other wild fruit foods were exploited, however, excavation of more data is needed to confirm this assumption. The other adaptive practise could be postulated as purposeful planting and harvesting of grain foods, such as *Pennisetum* sp.

4.4.1.1.3.3 The sites in the Khartoum area.

4.4.1.1.3.3.1 The sites on the east bank of the Nile.

As I mentioned in the foregoing, neither systematic nor random flotation of the debris was done at any of the sites on the east bank of the Nile north of Khartoum (Haaland, R., 1987a: 181) for retrieval of macrofossil plant remains. However, seeds of *Celtis integrifolia* Lam. were hand-picked while excavating all of these sites except the site of Kadero II (ibid).

Impressions of plants in pottery were found at all of the sites, Um Direwia, Kadero and Zakiab (cf. fig. no.1). The results of identifying the positive casts of these impressions attested the presence of a few grain impressions of wild *Sorghum* sp.(cf.pl.no.6) and seeds of other food-plants (cf.tables no.9, 10 and 11) . Grain impressions in pottery similar to these were found by R. Haaland and identified by A.B.L. Stemler as being wild sorghum, subspecies *S. verticilliflorum* (Steud.) Stapf. In addition, impressions of what looked very similar to *Setaria* sp. were also found in potsherds from the Um Direwia and the Zakiab sites (cf. tables no.9 and 11). No evidence of *Pennisetum* sp. was found in the collection[44] of potsherds I examined from these sites, but one impression of what has been identified as *Pennisetum violaceum* (Lam.) L. Rich., (wild species) was previously found in a potsherd from the site of Zakiab (ibid). From the fruit foods, an impression of an unripe seed of *Celtis integrifolia*[45] Lam. was recovered from a potsherd from the site of Kadero (ibid: 182). This evidence confirms the association of the desiccated seeds of the same species recovered with the prehistoric (neolithic) settlement of the sites on the east bank of the Nile. In the light of these results from studying the plant impressions in pottery (and remains of material culture), the economic adaptation related to plant exploitation has been interpreted as one based on cultivation of wild cereals, i.e. *Sorghum* sp. and *Pennisetum* sp. supplemented by gathering of immediately eaten wild food-plants, e.g. *Celtis integrifolia* Lam. (ibid). As regards the climatic conditions, it has been concluded that the mean annual rainfall was at least 500 mm; 3 times wetter than the 160 mm of annual rainfall and the prevailing arid conditions in the area at present. The interpretation was based on the presence of both wild *Sorghum* sp. and *C. integrifolia* Lam. which require a

---

44 Dr. R. Haaland has given me permission to examine plant impressions in pottery both from the sites on the east bank and the Shaheinab site on the west bank.

45 The current usage relates all domesticated pearl millets and their close spontaneous relatives to *Pennisetum americanum* (L.) K. Schum. (Harlan, J.R., et al, 1976: 434 footnote).

PLate no.6 Positive casts of impressions of seeds of wild Sorghum sp.

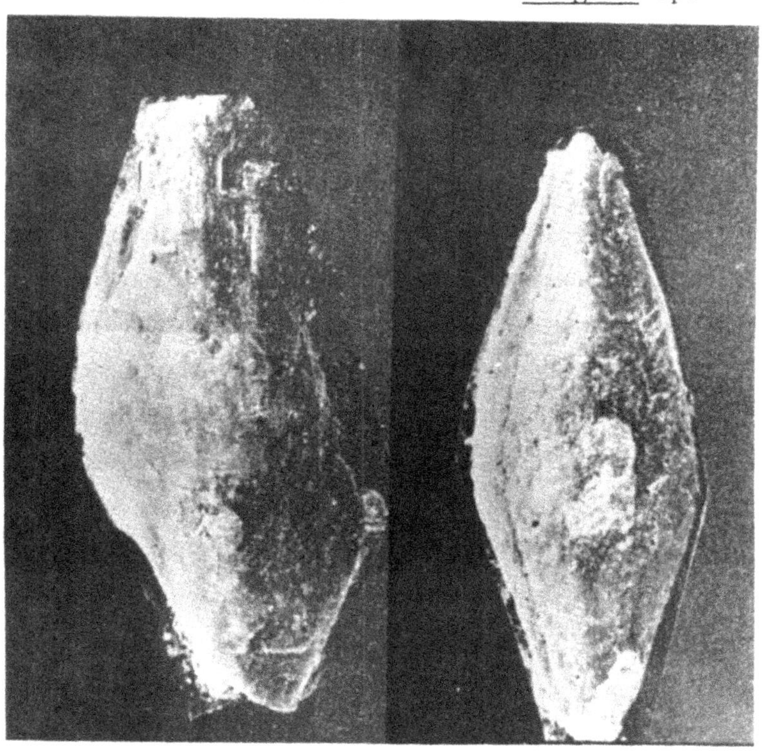

minimum of 500 mm of annual rainfall to grow (ibid). The annual rainfall was probably higher in this area than in the Atbara area but there was a general decline in both areas over time (cf. chapter IX).

4.4.1.1.3.3.2 The sites on the west bank.

The results of the identification of the macrofossil plant remains (both carbonized and desiccated) from the site of Nofalab (cf. fig. no.1) attested the presence of a number of species/genera (cf. table no.12). However, none of these specimens was contemporary with the prehistoric occupation of the site except for the desiccated and carbonized seeds of *Celtis integrifolia* Lam. These are presented here (as those from Abu-Darbein, table no.5) as examples of the expected distribution of such "intrusives" within the cultural debris of prehistoric sites in arid and semi-arid regions. This does not necessarily mean that all macrofossil specimens found in the top levels of the cultural debris are often intrusives. It is advisable in principle to consider such possibility until analysis of the data proves otherwise.

A few desiccated fragments of seed(s) of *Celtis integrifolia* Lam. were the only macrofossil plant remains retrieved from the site of Islang island (cf. fig. no.1); and no evidence of any contemporary carbonized and/or desiccated macrofossil plant remains was recovered from the prehistoric settlement site at the Shaheinab[46] (cf. fig. no.1).

As regards the plant impressions in pottery, only two impressions of seeds were found in the potsherds examined from the site of the Shaheinab (table no.13) and no such evidence was found in the potsherds examined from the sites of Nofalab and the Islang island. One of these impressions of seeds is very similar to that of *Celtis integrifolia* Lam., and the other is morphologically similar to wild sorghum. When compared with grain impressions of wild sorghum from the sites on the east bank, it proved identical to those identified as *Sorghum verticilliflorum* (Steud.) Stapf. Since the sites on the west bank are contemporary to those on the east bank and climatic indicators such as *Celtis* sp. and wild *Sorghum* sp. were also recovered; it is justified to conclude that the climatic conditions were much wetter (c. 500 mm of annual rainfall) than the present arid conditions (c. 160 mm annual rainfall). But the exploitation of plants seems to have been based mainly on gathering wild fruit foods and very little or no cultivation activities despite of the occupants were capable of carrying them out. The physical environment and topography (cf. chapter III in this book ; Magid, A., 1982: 9-16) of the west bank 'were two of the reasons and remained as fundamental obstacle in the way of cultivation activities until recently' (ibid: 105-107).

---

[46] The macrofossil plant remains and plant impressions in pottery were studied previously by the present writer (Magid, A., 1982: 87-93) and the results of that study are included in the present work.

4.4.1.1.3.3 The sites in the southern area.

As previously mentioned (chapter III: 3.4), two sites are excavated in this area: Jebel Tomat and Rabak (fig. no.1).

No evidence of macrofossil plant remains was found from the lower levels (between 20-60 cm) of the cultural layer of the site of Jebel Tomat (appendix no.1). But a few macrofossil plant specimens were recovered in the upper levels (between 0-20 cm). The well preserved and almost fresh shape and size of these specimens cast doubts and uncertainty upon their contemporaneity with the cultural debris; and hence they were excluded from the present work. However, carbonized grains and whose inflorescences of domesticated sorghum, *S. bicolor* (L.) Moench dated to A.D. 245±60, *Ziziphus* sp. (probably *Z. spina-christi* (L.) Willd.), *Acacia* sp., *Salvadora* and *Ficus* sp. were found during an earlier excavation of the same site (Clark, J.D., and A.B.L. Stemler, 1975: 588-591)[47].

Grain impressions of sorghum which is morphologically similar to the wild form were found in potsherds excavated from levels that were radiocarbon dated to 4140±90 B.P. (cf. table no.14, appendix no.1), but it was not possible to identify these to the species level . In addition, impressions of seeds which look similar to *Panicum* sp. and *Setaria* sp. were also found (table no.14) . The recovery of a seed impression that has been identified as *Sesamum* sp.(cf.pl.no.7) marks the first evidence of its type - so far - from the Central Sudan. Since the domesticated *Sesamum indicum* L., (cultivated at present in the Central and West Sudan) still retains the character that its seeds shatter and semi-shatter off the inflorescence at maturity, it is difficult to identify the form (wild or domesticated) of the archaeological evidence recovered from the site. It is also highly unexpected to retrieve any carbonized and/or desiccated seeds of sesame due to their perishable nature.

The macrofossil plant remains (and their impressions in pottery) recovered from this site indicate that both gathering wild fruit foods and cultivation of sorghum were practised. Jebel Tomat is the only site in the Central Sudan at which both wild and domesticated forms of sorghum have been found - so far. Thus it provides a good basis for studying the transition from gathering to cultivation of wild and then domesticated plants as I shall discuss in chapter VIII. In the light of the macrofossil plant remains recovered and their impressions in pottery and the radiocarbon dates obtained from this site (appendix no.1), it appears that the climate (mainly annual rainfall) during the early period of occupation of this site (c. 4400 B.P. and older) was wetter than during the later period (1700 B.P. and younger). The later period seems to have been as arid as it is at present (Clark, J.D., and A.B.L. Stemler, 1975: 588).

---

47 The age of these specimens was not specified, but a series of radiocarbon dates were obtained by the same author (ibid: 589) and quoted in this work in appendix no.1. It was, however, pointed out that these specimens have more or less the same age as that of the domesticated sorghum (ibid: 588).

Plate no.7 Positive cast of an impression of a seed of *Sesamum* sp.

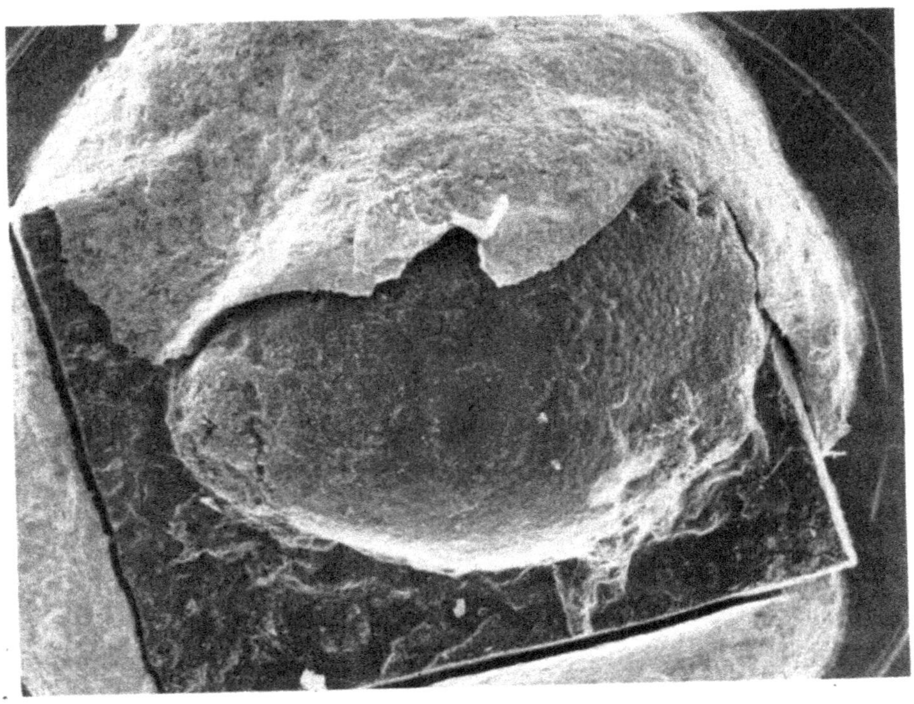

From the site of Rabak further south of Jebel Tomat (cf. fig. no.1), no macrofossil plant remains were retrieved from the debris of this site. The reasons that no specimens (recent or contemporary) were recovered may be related to different factors. With regard to "intrusive" remains; it may be due to the absence of adequate vegetation cover in the vicinity of the site, that no deposition and burial of fallen-off parts of plants took place. In addition, even if there had been an earlier deposition of "intrusives" in the upper levels of the site when there was denser vegetation cover in the vicinity (and less urbanization), this would have probably been swept away due to the erosive action of run-off water from the relatively hilly site area to the low-lying plains during the rainy season. However, a higher degree of confidence about this might be gained by obtaining more samples for flotation and studying more closely the drainage system in this area. As regards the absence of macrofossil remains contemporaneous with the cultural debris, one of the factors is probably the unfavourable preservation conditions discussed earlier in this chapter (cf.4.4.1.1.2).

A few plant impressions in pottery have been found (cf. table no.15). Among these are grain impressions that are morphologically similar to wild sorghum though it is not possible to identify any of them to the species level. The positive cast of one of these wild grains of sorghum (found in square 100x/100y, depth 50-60 cm with radiocarbon date from level 30-40 cm of 4140±90 B.P.) is bigger (measures 6.0x2.5 mm) than the others; and its shape is more like primitively domesticated sorghum. The size and shape of this grain could, in the absence of diagnostic features of its form and comparison with casts of extant wild and domesticated species taken from different part of inflorescences, have led to a totally wrong identification and interpretation. Specimens of such a nature may frequently be found, and many reasons may be put forward to explain them. One of these is that the grain may be derived from the apex of the inflorescence where the grains are bigger than those at the lower or distal part. This grain could also have been produced during a season with higher rainfall than normal, or the whole plant may have been growing in a wetter area with less dense stands than others growing in drier area with dense stands in the same field. Lastly, all of these factors may have cooperated in producing such size and shape. Other identifiable impressions of plants consist of what most likely were seeds of *Curcubitaceae* species, but it is not possible to determine their genus and species levels .

The plant remains recovered from the site of Rabak - mainly impressions in potsherds - indicate that wild sorghum was most likely the main source of food of vegetal origin. It was probably cultivated to secure adequate supplies. More data on macrofossil plant remains and plant impressions in pottery is - indeed - needed in order to tackle the problems raised earlier and to obtain a better picture of the economic adaptation related to the exploitation of food-plants.

CHAPTER V

Archaeological Artefacts Associated with the Exploitation of Food-Plants

5.1 Introduction.

The main emphasis of this chapter is on the archaeological artefacts which are assumed to have been associated with the exploitation of food-plant resources in the Central Sudan in the period between c. 9000-1700 B.P. The recovered biological evidence that consisted of plant remains, both carbonized-desiccated macrofossils and their impressions in pottery (cf. chapter IV) indicates that food-plant exploitation, from mere gathering to cultivation of wild food-plants and to their domestication, developed during the period under study. Accordingly, one would also expect to find artefacts that were functionally related to the exploitation of food-plants, their preparation, storage, etc.

The main problem in the interpretation of the cultural remains seems to be the scarcity of evidence which can be directly related to the exploitation of food-plants, especially to tools which were used in cultivation. Therefore, when trying to interpret the function of the cultural remains I will draw on ethnographic observations which I made among the Fur and the Zaghawa in west Sudan (cf. fig. no.20). It is also important in this context to discuss artefacts made of perishable raw-materials such as wood. Such implements could have been used during the period being studied here. The problem is that we have not found such remains of these. However, ethnographic materials show that these types of tools are still used in the traditional cultivation activities. The archaeological material discussed in this context consists of pottery and lithic material.

5.2 Pottery.

The earliest evidence of pottery in the Central Sudan is dated to 9339[1]±110 B.P. and 9370±110 B.P. (Khabir, A.M., 1987: 378). These dates are considered among the

[1] The oldest known radiocarbon date for the cultural tradition, Early Khartoum, with which the earliest evidence of pottery is associated is 10060±150 B.P. (Caneva, I, 1983: 149). However, this date is doubtful though not yet been completely discarded.

Fig. no. 20 Some of the present regions, towns and tribes in the Sudan cited in the text.

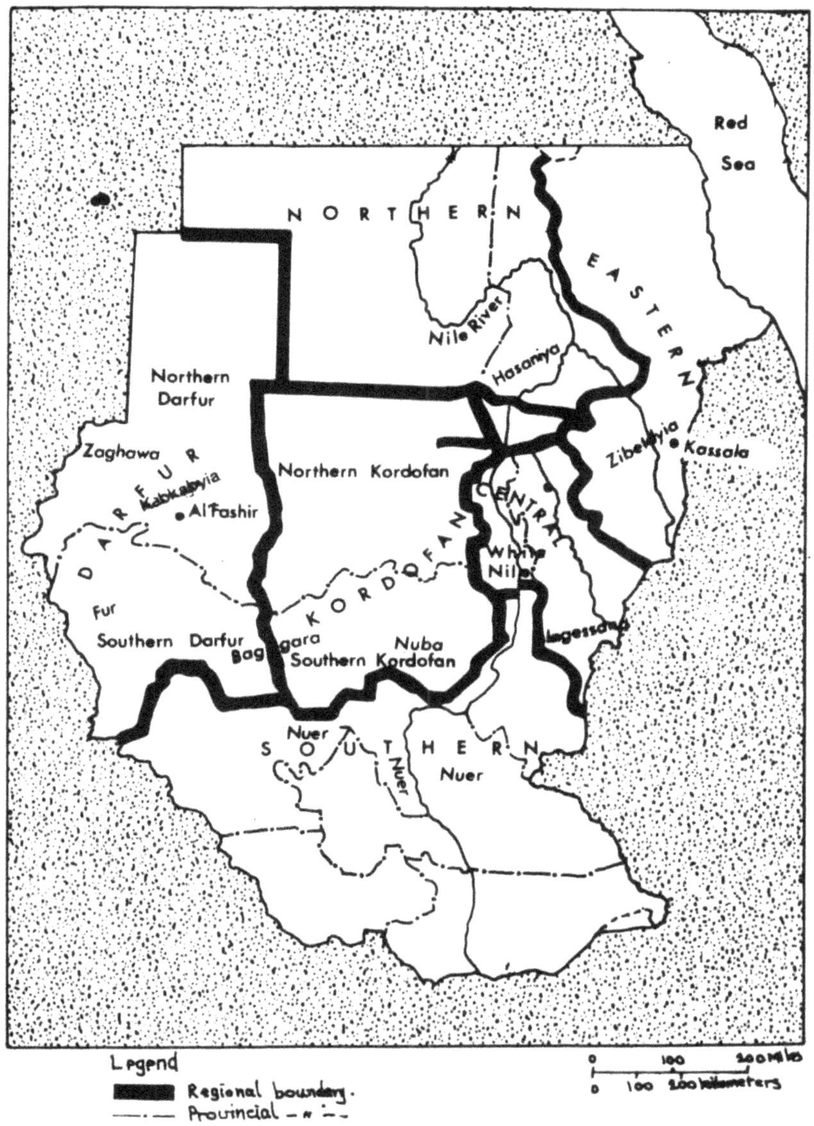

A combination of figures after Nelson, H.D., 1982: p.xx and Rinehart, R., 1982: 88

oldest dates for pottery in Africa. Other, almost similarly early dates from Africa, 9130±70 B.P. are derived from the site of Ti-n-Tor-ha-Ost, Acacus in southwest Libya (Barich, B.R., 1978: 223). The focus of the present research is on the activities of exploitation of food-plants with an emphasis on the development from gathering to cultivation of wild and then domesticated species. Accordingly, the main emphasis of studying pottery are as follows:

1) the use of pottery in food preparation and storing, and
2) the economic and adaptive implications of its introduction and use with regard to the transition from gathering of food-plants to their domestication.

In order to pursue these objectives, I found it necessary to classify and relate the data to the general culture history of the Central Sudan between c. 9000-1700 B.P. With regard to the classification of the pottery, I employed the following criteria:

1) the shape and size of the pottery as estimated from measurements of the diameters and thicknesses of the walls of rimsherds,
2) the technique of pottery production,
3) temper,
4) surface treatment, and
5) decoration patterns.

On the basis of the criteria mentioned the pottery material recovered can be classified into three cultural phases:

1) Pottery associated with the Early Khartoum tradition (c. 9000-6000 B.P.).
2) Pottery associated with the Shaheinab tradition (c. 6000-5000 B.P.).
3) Pottery associated with the post-Shaheinab traditions (c. 5000-1700 B.P.).
   a) the Shaqadud cave-type of pottery.
   b) the Jebel Moya type of pottery.

5.2:1 <u>The pottery of the Early Khartoum tradition</u>
(cf. fig.no 21and table no.16).

The early Khartoum type of pottery was recovered from the sites of Aneibis, Abu-Darbein and the bottom levels of the Islang (cf. fig. no.1). As the radiocarbon dates show (appendix no.1), the Early Khartoum type of pottery is the oldest type in the Central Sudan.

I suggest that the Early Khartoum pottery was used for storing, cooking and perhaps serving food of vegetal origin. Sutton stated that before the invention of pottery, vessels which were capable of both tolerating heat and retaining liquids were almost non-existent and that roasting over an open fire or in ashes was the only means of cooking (Sutton, J.E.G., 1974: 530). He suggested that the first ceramic vessels were used for cooking soup, porridge, stews of meat or fish, etc.; and hence he considered these first

Fig.no.21 Wavy-line pottery (Early Khartoum)

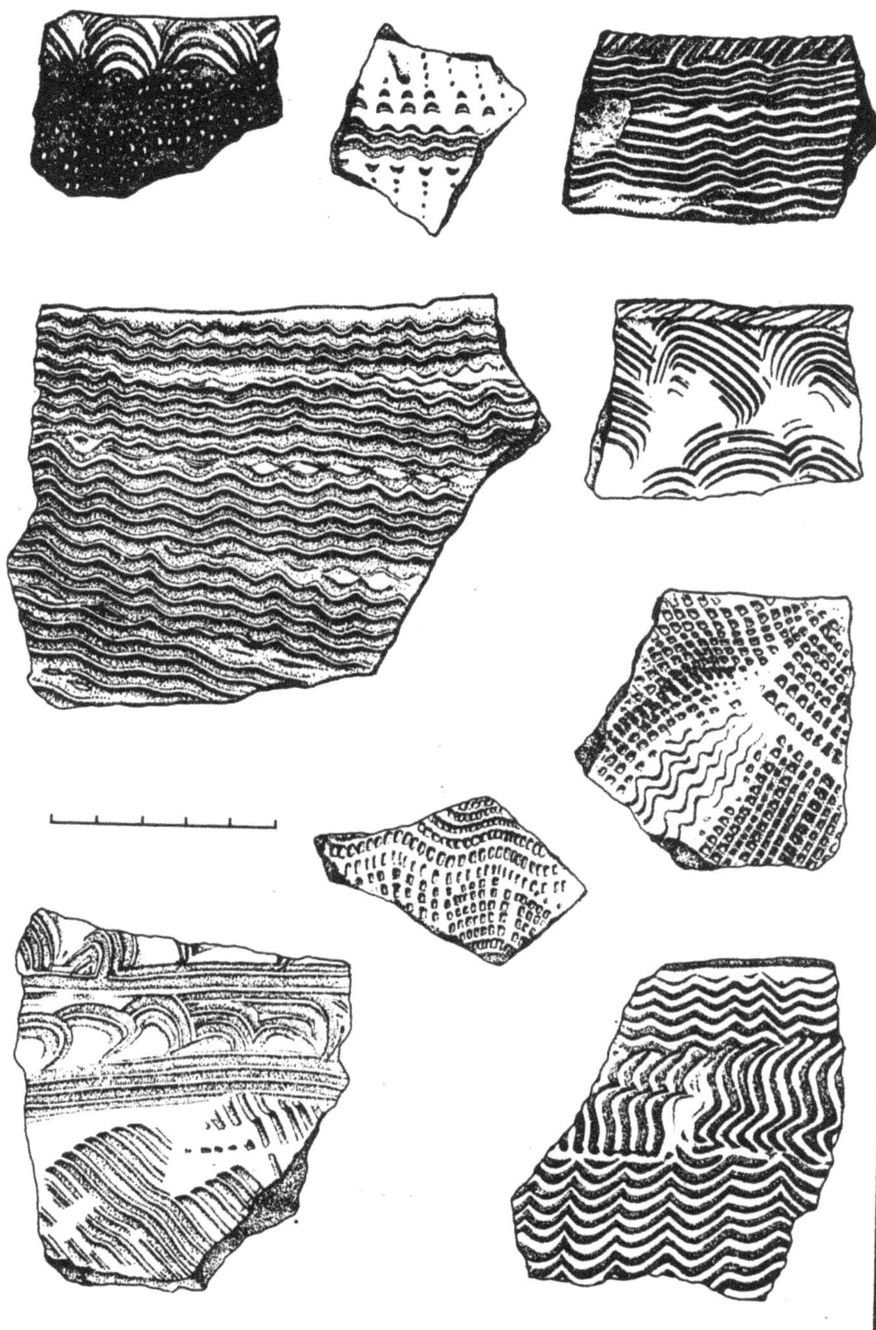

After Caneva, I., 1983: 182

Table no. 16

Disribution of wavy-line and dotted wavy-line(Early Khartoum) pottery (and other types) at Islang site.

Square no.	Level	Wavy-lines	Dotted wavy-lines	Impressed pattern	Incised pattern	Plain ware	Total	Unidentified sherds
1	Surface	-	-	14	2	8	24	3
	1	-	3	27	4	21	55	5
	2	-	5	44	2	10	61	7
	3	5	3	4	-	3	15	1
	4	6	-	1	-	1	8	-
	5	4	-	-	-	-	4	1
	Total	15	11	90	8	43	167	17
1323	Surface	-	-	11	-	1	12	1
	1	-	-	21	1	3	25	2
	2	-	8	16	2	8	34	5
	3	9	2	8	-	4	23	1
	4	3	-	2	-	3	8	-
	Total	12	10	58	3	19	102	9
1628	1	-	-	13	-	6	19	5
	2	-	-	35	4	11	50	7
	3	-	4	23	2	4	33	4
	4	7	2	8	1	5	23	2
	5	5	1	2	-	3	11	2
	Total	12	7	81	7	29	126	20
3180	1	-	-	33	-	12	45	23
	2	-	-	47	2	19	68	11
	3	-	-	104	4	10	118	14
	4	-	13	79	-	9	101	14
	5	2	7	49	-	3	61	7
	6	4	3	11	-	-	18	4
	7	15	3	-	2	2	22	2
	8	6	1	-	-	-	7	3
	9	2	-	2	-	5	9	5
	Total	29	27	325	8	60	449	83

After Magid,A.,1982: 72

vessels as the origins of the home pots (ibid). Sutton's interpretation clearly emphasizes the importance of pottery as a new means of cooking.

Although no complete pots or reconstructable fragments of broken pots were found, the recovered rim, base and sidesherds indicate that the pottery was of a bowl shape (Arkell, A.J., 1949: 85). Estimations of the diameters of some of these bowls were made by measuring the rimsherds[2]. The results show that the diameters of the pots range from very small, e.g. 10 cm, to very large ones >50 cm (Magid,A.,1982: 71). This variation in diameter may be interpreted to mean that the pots with the small diameters were used for cooking and those with wide diameters were used for storage. It has been suggested that the coil technique was used in the production of this pottery (Arkell, A.J., 1949: 86); Arkell, A.J., and P.J. Ucko, 1965: 148; Caneva, I., 1983: 159).

Another characteristic feature of Early Khartoum pottery is that it is unburnished. The degree of burnishing is considered one of the basic criteria for the interpretation of the function(s) of prehistoric pottery (Haaland, R., 1987a: 149-151). Since this pottery is unburnished, the criteria of burnishing can not be used in interpreting its function(s).

The material used as temper for the clay was fine and coarse angular grains of quartz-sand (Arkell, A.J., 1949: 81; Caneva, I., 1983: 161) and grains of mica. In addition, the pottery is well fired (Arkell, A.J., and P.J. Ucko, 1965: 148). Thus the end product is a very hard ware. A few rimsherds and sidesherds with holes were found (Magid,A.,1982: 71). Arkell suggested that these holes were used to attach handles, e.g. of leather, to the pots or as repair holes for cracked or broken ones. It is to be assumed that the repaired pots could not be used for cooking or holding and storing liquids. They, might, however, have been used for storing solids, e.g. fruits, roots, etc.

The wavy lines and the dotted wavy lines are the characteristic decorative features of the Early Khartoum pottery, though, dots, vees and zizag impressions (cf.fig.no.21) together with plain ware are also common (Arkell, A.J., 1949: 81, 87; Khabir, A.M., 1987: 278).

5.2.2 The pottery of the Shaheinab tradition
(cf.fig.no.22 and tables no.17 to 19).

The Shaheinab type of pottery which will be discussed in this context has been recovered from the sites in the Khartoum area and the site of Rabak (cf. fig. no.1).

The comparison of the pottery of the Shaheinab tradition with that of the Early Khartoum one shows the following basic differences:

1) Surface treatment: while all the pottery of the Early Khartoum is unburnished, the characteristic feature of the Shaheinab pottery is burnishing.

---

2 The small quantity of the rimsherds recovered (with length of 3 cm or more) does not provide a representative or informative figure that would allow one to make measurements and estimations of the wall-thicknesses of the bowls.

Fig.no.22

Some of the decoration patterns of the pottery of Shaheinab type

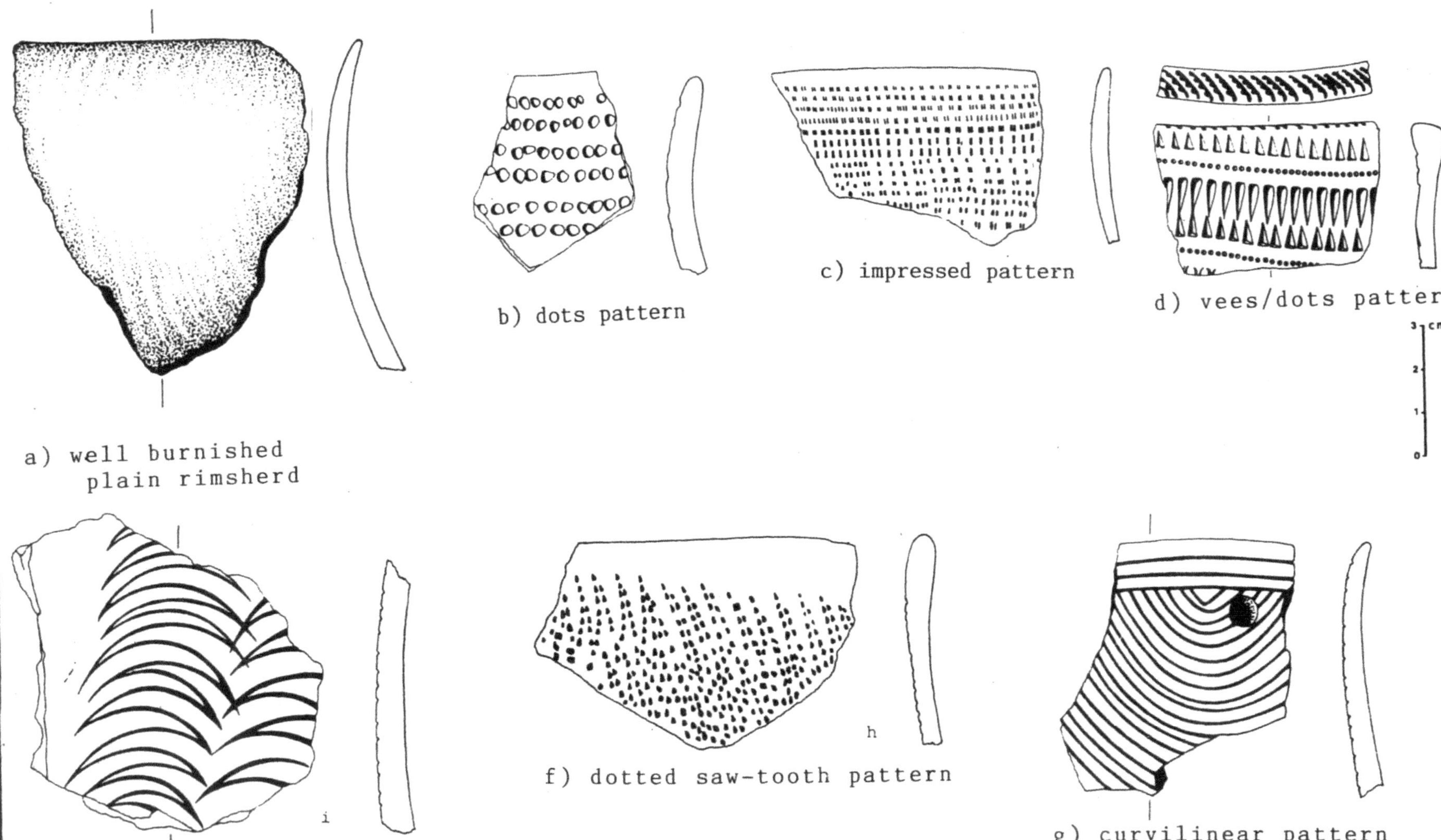

After Haaland,R.,1981: 166-170

Table no.17 Decoration patterns of pottery of Shaheinab type excavated from 2 trenches ($8m^2$) at Kadero site

Decorated sherds in gram	C - 8	C - 23	SUM	%
Vees/dots	20	55	75	1,7
Vees	130	10	140	3,1
Dots	90	430	520	11,7
Catfish-spine decoration	20	15	35	0,8
Dotted incised lines	20	15	35	0,8
Incised lines	5	10	15	0,3
Dotted saw tooth	170	70	240	5,4
Saw tooth	50	15	65	1,5
Combed	40	9	49	1,1
Undecorated	500	200	700	15,7
Bottoms	420	290	710	16,0
?	-	15	15	0,3
Fragments	1000	750	1750	39,2
Others	60	50	110	2,5
Total gram	2525	1934	4459	100,1

After Haaland,R.,1981: 161

Table no. 18 Distribution of postherds of Shaheinab type according to their decoration patterns at the Nofalab site

Decoration patterns / Square no.	100e	51h	52h	100j	1225L	51n	Total
1. Dots	18	120	261	161	207	119	886
2. Vees	9	54	144	66	92	39	404
3. Dots/vees	10	30	111	20	87	36	294
4. Dotted straight lines	3	22	48	21	62	22	178
5. Fish-net	1	-	11	2	4	-	18
6. Finger nail impression	1	-	9	-	5	-	15
7. Dotted saw-tooth	-	-	13	3	6	4	26
8. Saw-tooth	-	1	-	-	1	-	2
9. Incised straight lines	5	73	155	92	160	68	553
10. Curvilinear pattern	1	4	35	4	18	2	64
11. Combed	5	2	19	3	4	2	35
12. Plain	19	98	330	125	208	87	867
Total	72	404	1136	497	854	379	3342
13. Unidentified sherds	9	8	24	18	19	13	91

After Magid, A., 1982: 59

Table no.19

Decoration patterns of pottery of Shaheinab type excavated from square 107x/100y at Rabak site .

Type of decoration	Levels 14-4		Level 3		Level 2		Level 1	
	Nos.	%	Nos.	%	Nos.	%	Nos.	%
1 Dots	36	9,7	27	29,0	17	38,6	14	22,2
2 Vees/dots	2	0,5						
3 Catfish spine	9	2,4	5	5,4	3	6,8	1	1,6
4 Dotted saw tooth	29	7,8	2	2,2	1	2,3	3	4,7
5 Saw tooth			1	1,1	2	4,5	1	1,6
6 Undecorated (wiped)	269	72,3	41	44,0	15	34	19	30
7 Decoration non Khartoum type	13	3,5	3	3,2				
8 Jebel Moya type of decoration			10	10,8	4	9,1	20	31,7
9 Finger nail decoration					2	4,5	3	4,7
10 Others			1	1,1			2	3,1
11 Base	14	3,8	3	3,2				
Sum	372	100,0	93	100,0	44	99,8	63	99,8

(1) The size of the square is 4 square metres.Levels 14-4 are classified together since there is no difference in their ceramics and their C 14 dates indicate their contemporaneity.

(2) The study of the archaeological remains from the site is in progress.

2) The material used as temper in the clay for the pottery of the Early Khartoum tradition is basically of mineral origin, e.g. quartz and mica. The temper of the Shaheinab pottery is partly quartz; but organic materials, mainly plants, were also used in addition. Although both the Early Khartoum and the Shaheinab pottery is well fired, the latter is not as hard as the former. This is most likely due to the use of the "soft" organic temper. It would be of interest for further research to see if there is any association between the beginning of cultivation and the introduction of organic temper, mainly plant materials in pottery production.
3) The wavy lines and dotted wavy line pattern of decoration characteristics of the Early Khartoum pottery disappeared during the Shaheinab tradition.

The pottery of the Shaheinab tradition consists of open-mouthed globular pots with rounded bottoms (Arkell, A.J., 1953; 69-71). The degree of burnishing has been correlated with the estimated size of the pots based on the measurement of rimsherds and their thicknesses (cf. tables no.20 and 22). This has shown that the pots had different functions (Haaland, R., 1987b: 48, 59; Magid, A.: the present work).These functions can be summarized as follows:

1) Cooking:	using for this purpose, highly burnished small pots (<30 cm in diameter) with thin walls (mostly <5 mm).
2) Containers for liquids and solid substances:	using mainly poorly/slightly burnished large pots and bowls (> 30 cm in diameter) with relatively thick walls (>5 mm).
3) Food[3]-serving:	using mainly exotic pots, highly burnished and carefully decorated pots. These pots are mostly small in size.

This shows that the degree of burnishing, the size and to some extent the thickness of the rimsherds are criteria indicative of the functions of the pots.

Thin-section analysis and X-ray photographs have shown that the coiling technique was not used. It is most likely that the paddle and anvil technique was used for the manufacture of this type of pottery. This suggestion is based on ethnographic observations of similar pots made among the Fur and the Zaghawa (cf. fig. no.20).

The material used as temper for the clay consists predominantly of fine-grained quartz; but organic materials, e.g. of vegetal origin, are now introduced as an additional type of temper. A fabric analysis of some sherds typical of the Shaheinab pottery reveals that it is well fired at temperatures between 700-800°C (Haaland, R., 1987a: 150)

[3] The function of food serving is also interpreted on the basis of I. Hodder's ethnographic study among the Nuba of west Sudan which he presented in a public lecture at the Historisk Museum, University of Bergen in October, 1981.

Table no.20 Diameters and wall-thickness of pottery of Shaheinab type

Diameters in cm \ Thickness in mm	<5	>5 to <10	>10 to <15	>15	Total
10 to 20	12	3	-	-	15
20 to 30	4	10	1	-	15
30 to 40	1	12	2	-	15
40 to 50	-	4	10	1	15
50 to 60	-	-	-	2	2
Total	17	29	13	3	62

After Magid,A.,1982: 59.

Table no.21 Diameters of pottery of the Shaheinab type as estimated from measurements of rimsherds recovered from square 100 x/ 100 y at Um Direwia site*

Type of decoration	0-10 cm	10-20 cm	20-30 cm	30-40 cm	40-50 cm	50-60 cm	Fragments	Sum
Vees/dots			1	3	10	12	59	85
Vees	1		1	1	4		13	20
Dots				2	2	2	25	31
Catfish spine	1			2	4	2	25	34
Dotted impressed lines			2	1	3	1	5	12
Incised lines			1	1			14	16
Dotted saw tooth	1		1			1	6	9
Saw tooth				1	1	2	1	5
Combed				1		1	3	5
Undecorated	1		3	2	2	4	52	64
Red burnished ware		9	5				12	26
Black burnished ware	1	6	1				5	13
Curvilinear	2	4					7	13
Special sherds				2			1	3
?		1	2			1	20	24
SUM	7	20	17	16	26	26	249	360
%	1,9	5,6	4,7	4,4	7,2	7,2	68,9	99,9

* See foot note on tables no. 17 and 22.

After Maaland ,R.,1981: 185.

Table no.22

Shaheinab type of pottery with holes from Zakiab site.

(A) Rimsherds(estimated diameters in cms.are based measurements of rimsherds)

Type of decoration	Sum	<10 cm	10-20	20-30	30-40	40-50	50-60	small fragments
vees/dots	5						2	3
vees	1				1			
catfish spine	1							1
dotted incised lines	3		1	1	1			
dotted saw tooth	3				1	1		1
undecorated	2							2
red burnished	4	1			1			2
?	2				1			1
Sum of sherds	21	1	1	1	5	1	2	10

(B) Sidesherds.

Decoration		hole repaired both faces	repair from one face	?
vees/dots	6	5 (1 sherd with two repair holes)	1	
dots	4	2	2	
catfish spine	3	2 (1 sherd with two repair holes)	1	
dotted saw tooth	2		1	1
saw tooth	1		1	
undecorated	2	1		1
burnished red ware	1	1		
combed	1	1		
?	2	2		
Sum	22	14	6	2

After Haaland,R.,1981: 192-193.

The pottery of the Shaheinab tradition is characterized by a relatively wide variety of decoration[4] patterns as compared with that of Early Khartoum. The impressed decoration is the most common among the sherds recovered and it includes the zigzag, vees and dots patterns (cf. tables no.17 to 19 and fig.no.22). The wavy and dotted wavy lines have disappeared.

In general, the pottery of the Shaheinab tradition is similar to that of Early Khartoum in the following ways:

1) The pottery[5] was perhaps used for the same basic purposes; namely food preparation and storage. It seems, however, that the Shaheinab pottery is more specialized; well burnished pots for cooking, poorly burnished pots with large diameters for storage.
2) The shape of the pottery of the Early Khartoum tradition (open globular pots) remained the same during the Shaheinab period.
3) Some Early Khartoum decoration patterns on pottery, mainly vees and dots, continued to exist in the Shaheinab pottery.

5.2.3 <u>The pottery of the post-Shaheinab traditions</u>.

The post-Shaheinab types of pottery which are included in this study are:

I) The pottery from the Shaqadud cave-site.
II) The Jebel Moya type of pottery from the sites of Jebel Tomat and Rabak.

Analysis of these types of pottery has not yet been completed in its final form. Thus, I shall depend mainly on the published information as regards these two types.

The comparison of the data from the Shaheinab and the post-Shaheinab periods, shows that the pottery of the former continued to be produced and used but that new types were also introduced; namely, the Shaqadud cave-pottery and the Jebel Moya type. This is evident in the following:

1) The open-mouth globular pots of the Shaheinab tradition are also present in the post-Shaheinab period, but a variety of other forms (shapes) with different sizes is introduced. Paddle-and anvil was the main technique used in pottery production both during the Shaheinab and the post-Shaheinab traditions.
2) The surface treatment of pottery of the Shaheinab tradition, viz burnishing continued during the post-Shaheinab; but wiping, as evident from the pottery of the cave-site of Shaqadud, seems to have been a newly introduced technique of surface treatment.

---

[4] More information on decoration and decoration technique is discussed by Haaland (Haaland, R., 1981: 162-185; 1987a: 145-150) and Magid (Magid, A., 1982: 59-64; Mohammed-Ali, A.S., 1982: 75).

[5] Rimsherds and sidesherds with holes (cf. table no. 24) were also recovered from the sites of Shaheinab type. As in the Early Khartoum pottery holes made for repairing cracked pots indicate that these pots were not used for cooking or holding or storing liquids such as vegetable stew.

3) The temper materials in the clay are basically the same, i.e. a mixture of organic and mineral elements, both during the Shaheinab and post-Shaheinab traditions. It seems, however, that organic temper was increasingly used during the post-Shaheinab.
4) The vees/dots and zigzag decoration patterns survived from the Early Khartoum into the Shaheinab tradition and continued to exist during the post-Shaheinab period but new patterns were also introduced (cf.figs.no 23 and 24).

I) Pottery from the Shaqadud cave-site:

It has been stated that the types of pottery recognized from the cave-site of the Shaqadud are not known from elsewhere in the Central Sudan (Marks, A.E., et al, 1985: 273). A summary of these types is presented on fig no.25; but no information is available at present as regards their size, shape, function, etc.

II) The Jebel Moya type of pottery (cf.figs.no.23 and 24).

The Jebel Moya type of pottery was excavated from the uppermost levels of the sites of Jebel Tomat and Rabak and dated to 4490±100 B.P. (Haaland, R., 1984: 43). No complete pots or reconstructable pots of broken sherds were found.

With regard to the function of the pottery, the well established specialized production of pottery, e.g. cups, pots and jars to be used for specific purposes, e.g. drinking, cooking and storing of the Jebel Moya tradition had most likely started already during the Shaheinab period when cooking pots were well burnished; and poorly burnished pots were used for, e.g., storage. Furthermore, the wide variety of forms and sizes of the Jebel Moya pottery as compared with pots from the earlier phase indicates that each product was used to perform one specific function and that multi-function use was less important.

5.2.4 Association of pottery with exploitation of food-plants.

The association of pottery with the exploitation of food-plants is summarized in the following points:

1) The introduction of pottery probably demarcated the beginning of a new adaptation in which already known, potential food-plants were now exploited, for instance the beginning of utilizing seeds and grains of cereals. These types of food-plants require preparation and hence a certain level of technological knowledge before they can be processed and eaten. In this context, pottery might have provided the basic requirement for cooking these seeds and grains before serving them as food. Thus, its introduction played a significant role because
   a) It facilitated the preparation and consumption of food-plants which otherwise would have been unpalatable.

Fig.no.23

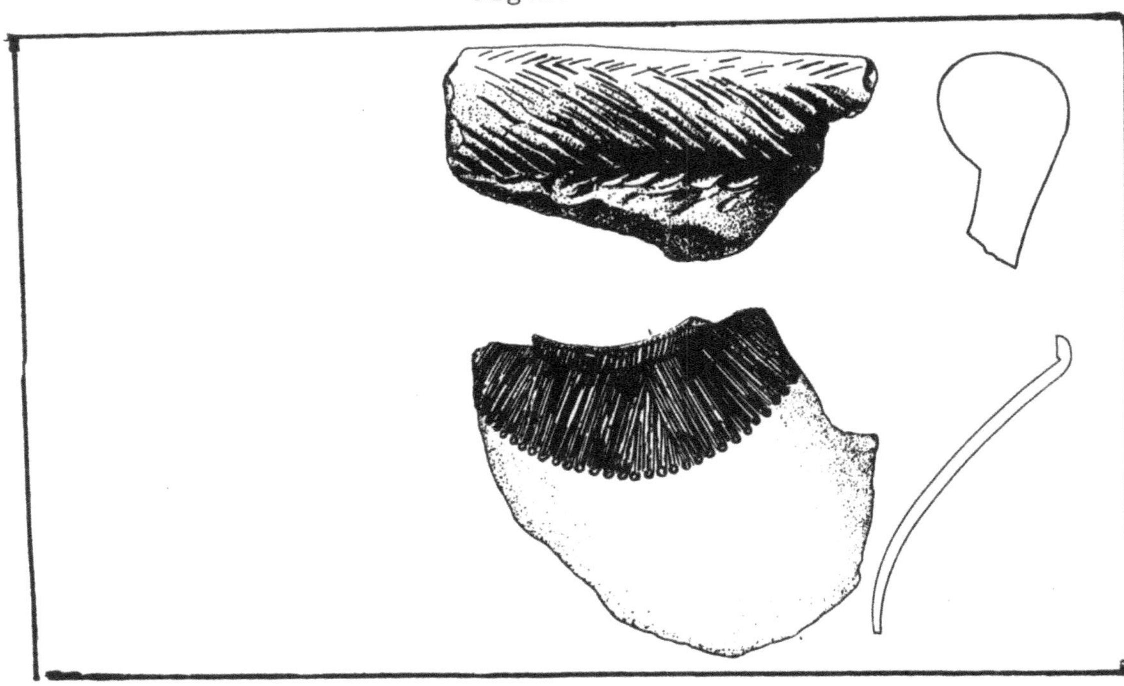

Some illustrations of the pottery of
Jebel Maya type excavated from the
site of Rabak (top 3 levels)

(After Haaland,R.,1984: 45)

Fig.no.24 Examples for the variability of shapes of the Jebel Maya pottery

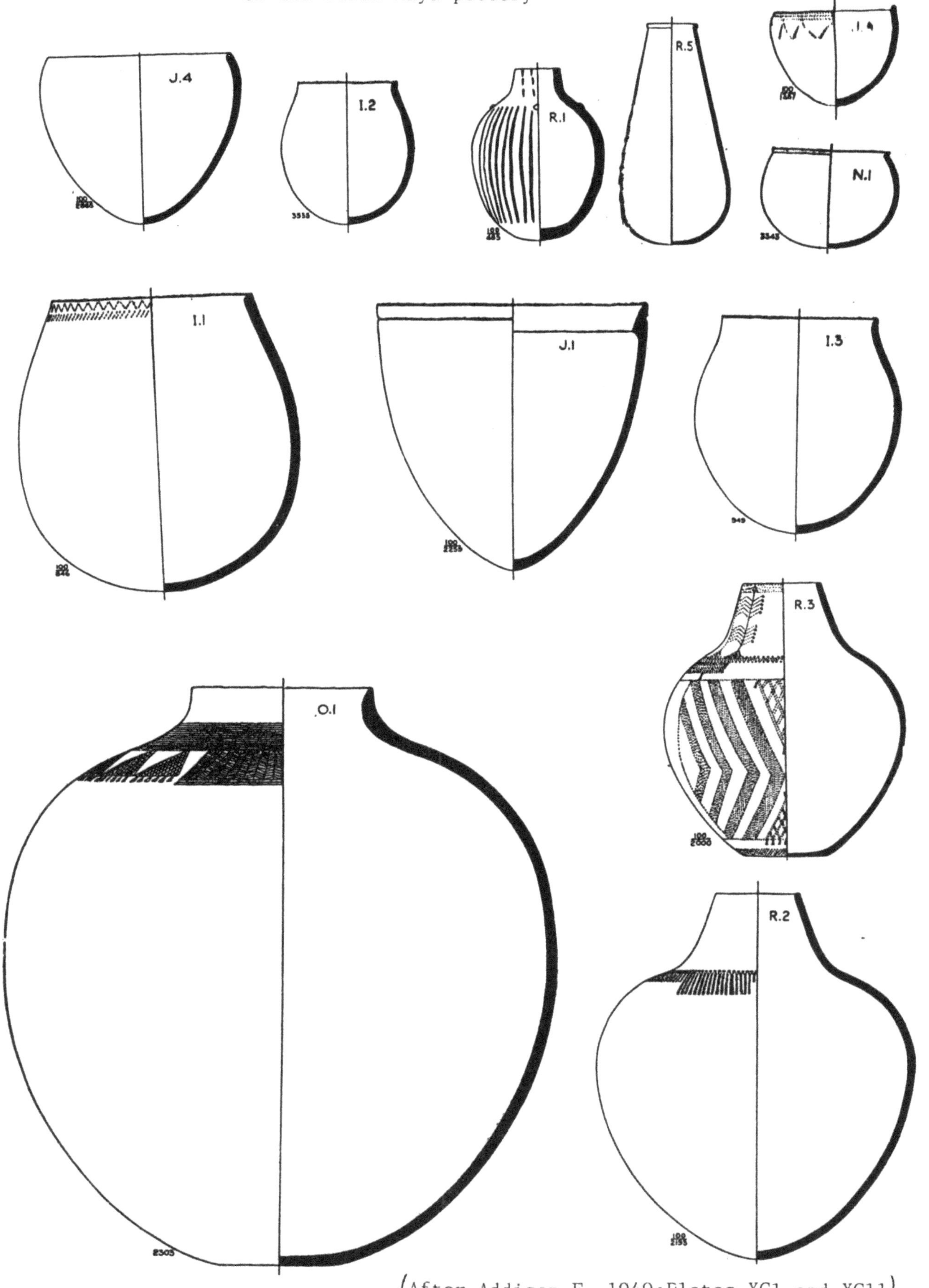

(After Addison, F., 1949: Plates XC1 and XC11)

Fig no. 25 Seriation of cave ceramics by provisional types and grouped levels. A, Thick Black Slipped; B/C, all Thin Black Slipped; D/E, Thin Impressed; F/G/H, Red Slipped; I, Fiber Tempered; J, Fingernail Impressed; K/L, Wiped and Burnished and Streaky Burnished; M, Grooved Unslipped. These provisional types are undergoing additional study. The figure is based on the cave ceramics only from the 1981/82 field season and, thus, is merely suggestive of the final pattern.

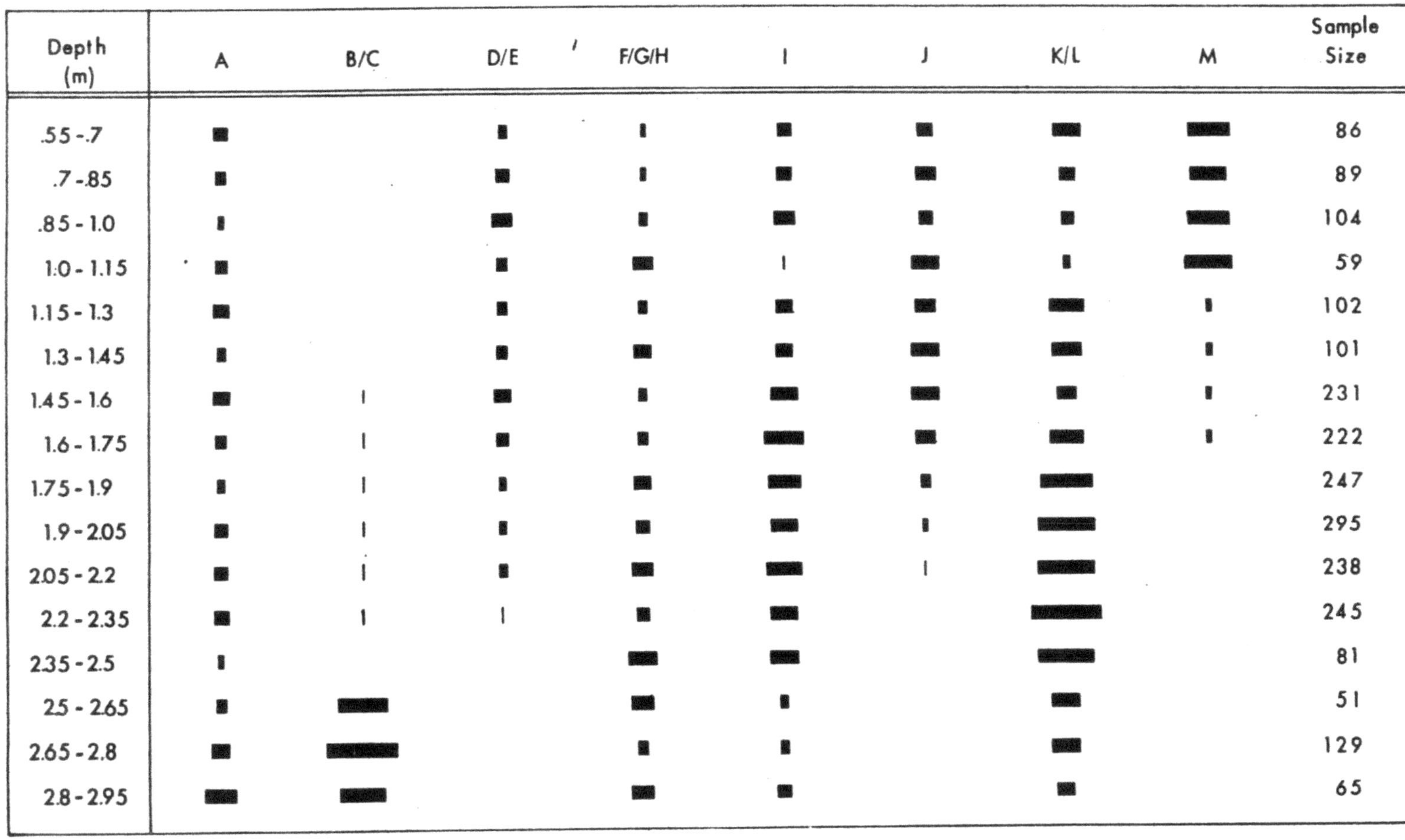

Marks, A.E., et al, 1985: 274

b) It probably provided a new means or better alternative for preparing food-plants which were previously prepared differently; for instance, both ground and unground seeds/grains can be prepared, e.g. as porridge or boiled grains, by using pots.

c) It provided a new or better means for food-serving facilities.

2) Pottery containers would also provide means of storage for the durable food-plants, e.g. seeds, berries, fruits and nuts to be used during periods of need or when they were not available in nature. In this connection, I observed (in 1985) in the area between El-Fashir and Kabkabyia in Darfur (cf. fig. no.20) in the Sudan that the inhabitants are still using huge clay jars (Fig.no 26) locally called *dabanga* for storage of the harvest both of wild and domesticated grain-foods. The same type of clay jar was used in north Sudan until recently for storing of yields of cultivated crops (Herzog, R., 1957: PL.no.26).

Since pottery is generally too heavy and fragile to be used by nomadic gatherers, its presence in archaeological sites indicates some degree of sedentism among these food gatherers.

The foregoing discussion emphasized the role of pottery in promoting the gathering of food-plants during the Early Khartoum tradition. It then became possible to cook food-plants and equally importantly to improve storage facilities. The same types of activities seem to have continued and pottery was used for more or less the same purposes during the following Shaheinab phase. Gathering immediately eaten food-plants was still practised but there is also botanical evidence for grains of several wild grasses, e.g. *Sorghum* sp., *Pennisetum* sp., *Setaria* sp., etc. (cf. chapter IV). These finds indicate that the cultivation of wild cereals had already started (cf. chapter VIII). Although pottery continued to be used for the same purposes, it seems likely that it was during the Shaheinab period that pottery was used for more specialized functions such as cooking, storing and serving food.

With regard to the post-Shaheinab period, it seems that the role of pottery remained unchanged during its early phase; that is the period of occupation of the Shaqadud cave-site when gathering and cultivation of wild food-plants was still practised. Thus, pottery was perhaps used for the same purposes as those it was used for during the previous Shaheinab period. It was only during the latter part of the post-Shaheinab that the earliest evidence of domesticated plants, namely *Sorghum bilocor* (L.) Moench, occurs. As I shall show in chapter IX, the process of domestication was initiated in the Central Sudan most likely by a pottery-using nomadic population about 1700 years ago. The question that immediately arises is: how could that be explained in the light of my previous statement that pottery is too heavy and fragile to be used by nomadic people? There is no contradiction in this argument if we consider the fact that nomadic pastoralists are

Fig.no.26

Huge clay-jars for storage
of grain-foods
(Darfur,W.Sudan)

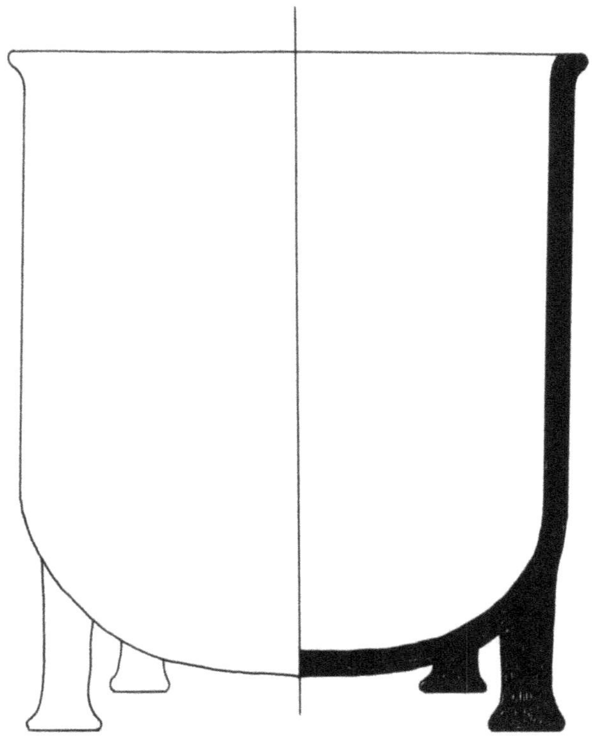

Lid of the clay-jar

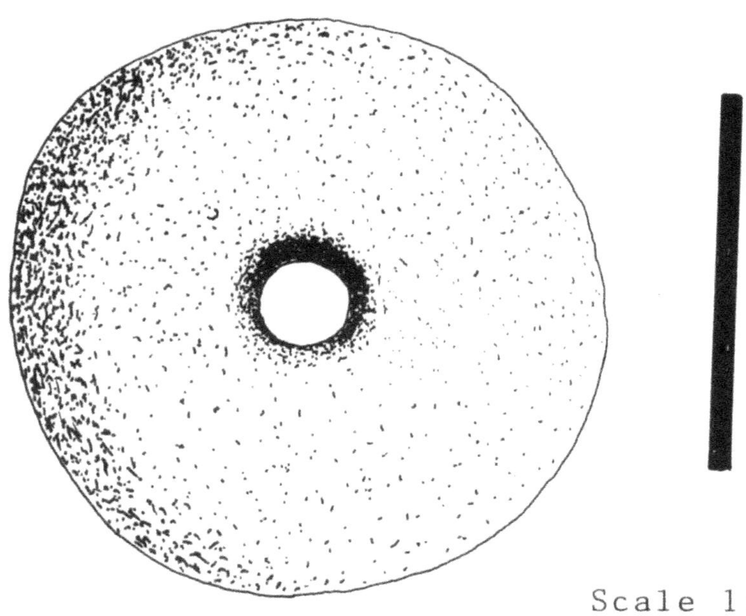

lid-thickness

Scale 1:20cm.

different from the nomadic food gatherers in that the former could use their animals, e.g. cattle, as a means of transport. Thus they have the advantage of transporting their pots, while nomadic gatherers would have to carry these bulky and heavy pots themselves. Ethnographic observations show that pottery is used by nomadic pastoral groups in the Sudan. Examples are the Baggara cattle nomads of western Sudan, the Hasaniya sheep-and-goat nomads of the Central Sudan and the Beja and the Zibeidyia camel, sheep-and-goat nomads in eastern Sudan (cf. fig. no.20). These peoples cultivate their crops during the rainy season and use pottery for cooking and serving food, etc. Thus, the prehistoric nomadic pastoralists of the Central Sudan were probably also using basic pottery equipment needed in their daily life in addition to other necessary equipment they made from hides, wood, etc.

In short, the information available from the Central Sudan generally indicates that pottery may have played an important role in promoting gathering activities and cultivation of wild food-plants by providing new and better means of food preparation and storage but it is unlikely that it had any direct role in initiating the process of their domestication.

5.3 Lithic Artefacts.

The main emphasis in this section will be on six types of lithic tools which have been seen by various archaeologists as being functionally related to the exploitation of food-plants. Examples of the different types of artefacts are shown in tables from 23 to 27, but the detailed study of their classification, technology, etc. is beyond the scope of the present work[6]. However, these points will be referred to when seen relevant to the discussion. The following six tool-types will be discussed:

1) Lunates.
2) Knife (?sickle)-blades.
3) Grinders.
4) Rubbers.
5) Stone rings.
6) Gouges.

These lithic artefacts will be discussed within the context of the major activities of gathering and cultivating wild and domesticated plants in the Central Sudan as represented by the sites of the Early Khartoum, Shaheinab and post-Shaheinab types.

---

[6] For detailed information on the general characteristics of the lithic industry in the Central Sudan in the period between c. 9000 to 1700 years ago, see Caneva, I., 1983: 203-233; Haaland, R., 1981: 60-154; 1987a: 63-82; Magid, A., 1982: 27-54 and Zarattini, A., 1983: 234-242.

Table no.23 Cross tabulation of lithic tools and raw-materials from Abu-Darbein(Early Khartoum type of site).

Tools	Chert	Quarty	Others	Sum	%
Boorers	47			47	6,96
Groovers	52			52	7,70
Micropoicons	10			10	1,48
Sum engraving tools	109			109	16,14
	218			218	32,29
Lunates	64			65	9,62
Backed blades	39			39	5,77
Backed tools	25			25	3,70
Trapesoide points	3			3	0,44
Triangle points	3			3	0,44
Truncations	15			15	2,22
Obliqucly backed blades	9			9	1,33
Crescent scraper	1			1	0,15
Backed knife	1			1	0,15
Sum backed tools	160			161	23,85
Tanged arrow heads	3			3	0,44
Scrapers	86	2		88	13,03
Scrapers concave	13			13	1,92
Notched tools	89			89	13,18
Denticulates	15			15	2,22
Multiplefunction tools	83		1	84	12,44
?		3	1	4	0,66
Sum tools	667	6	2	675	99,50
Tools irregularly retouched	322	2	1	325	
Microblades	16			16	
Blades	46			46	
Platformcores	5			5	

Table no. 24 Cross tabulation of lithic tools and raw-materials from Islang(late Early Khartoum-early Shaheinab type of site)

Tools / Raw-material	Quarts	Nile-pebbles	Fossil-wood	Rhyolite	Basalt	Volcanic tuff	Total	%
1. Engraving tools (groovers & borers)	2	-	-	4	1	-	7	7.7
2. Engraving tools (?)	-	-	-	-	-	-	-	-
3. Scrapers	5	2	-	11	1	1	20	22
4. Scrapers (?)	2	-	-	3	1	-	6	6.6
5. Lunates	9	1	-	11	-	-	21	23
6. Backed tools	8	1		7		-	16	17.6
7. Multiple-function tools	1	-	-	2	-	-	3	3.3
8. Notches	7	1	1	2	1	-	12	13.2
9. Denticulates	-	-	-	-	-	-	-	-
10. Tanged arrow-heads	-	-	-	-	-	-	-	-
11. Scales-pieces	-	-	-	-	-	-	-	-
12. Beaked tools	1	-	-	3	-	-	4	4.4
13. Gouges	-	-	-	-	-	-	-	-
14. Polished fragments	-	-	-	1	-	-	1	1.1
15. Burnishers	1	-	-	-	-	-	1	1.1
16. ?	-	-	-	-	-	-	-	-
Total	36	5	1	44	4	1	91	100.1
%	39	5.5	1.1	48.4	4.4	1.1		100
17. Irregularly retouched tools	30	4	1	43	7	2	87	

2, 5, 11, 12 and 13 are presented here for comparison between this site and Nofalab in terms of tools and raw-materials.

After Magid, A., 1982: 42.

Table no.25

Cross tabulation of lithic tools and raw-materials from Nofalab(Shaheinab type of site).

Tools \ Raw-material	Quartz	Nile-pebbles	Fossil-wood	Rhyolite	Basalt	Volcanic tuff	Total	%
1. Engraving tools (groovers & borers)	7	1	-	4	-	-	12	4.8
2. Engraving tools (?)	6	1	-	1	2	-	10	4
3. Scrapers	11	3	2	7	2	1	26	10.4
4. Scrapers (?)	4	-	1	-	1	-	6	2.4
5. Lunates	5	1	-	2	-	-	8	3.2
6. Backed tools	6	1	-	3	1	-	11	4.4
7. Multiple-function tools	10	1	-	6	-	1	18	7.3
8. Notches	29	2	-	8	1	1	41	16.5
9. Denticulates	1	-	-	-	-	-	1	0.4
10. Tanged arrow-heads	3	-	-	-	-	-	3	1.2
11. Scaled pieces	2	-	-	-	-	-	2	0.8
12. Beaked tools	66	7	-	16	1	-	90	36.1
13. Gouges	-	-	-	4	-	-	4	1.7
14. Polished pieces	-	-	-	5	2	-	7	2.8
15. Burnishers *	-	-	-	-	-	-	1	0.4
16. ?	5	2	-	1	-	1	9	3.6
Total	156	19	3	57	10	4	249	100
%	62.7	7.6	1.2	22.9	4	1.6	-	100
17. Irregularly retouched tools	93	31	4	42	9	5	184	-

* It was not possible to identify the raw-material.

After Magid, A., 1982: 31.

Table no.26

Cross tabulation of lithic tools and raw-materials from Kadero(Shaheinab type of site)

	Quartz	Nile Pebble	Fossile Wood	Rhyolite	Basalt	Volcanic	SUM	%
Borers		2					2	1,1
Groovers	10	6					16	9,1
Beaked tools	6	2		4			12	6,8
Groovers one side retouched	10	5		1	2	1	19	10,8
Convex Scrapers	1	8		6	2		17	9,7
Scrapers?	1	3		2	1		7	4,0
Concave Scrapers	1						1	0,6
Notch	2	1		5			8	4,5
Dentics		1					1	0,6
Lunates	13	10					23	13,1
Backed tools	16	7		2	1		26	14,8
Gouges polished one side				20			20	11,4
Gouges unpolished			1	5			7	4,0
Gouges polished both sides				3			3	1,7
Fragments Gouges				8			8	4,5
Gouges, edges completely polished				6			6	3,4
SUM	60	45	1	62	6	1	176	100
%	34,3	25,5	0,6	35,2	4,0	0,6		99,9
Tools irregularly retouched	210	76	1	66	8	8	469	

After Haaland,R.,1981: 103-104

Table no.27 Cross tabulation of lithic tools and raw-materials from Rabak(1)(Shaheinab type of site)

	Quarts	Quartzite	Rhyolite	Other material	Sum	%
Convex scrapers	29	30	3		62	13,2
Concave scrapers	17	12	1		30	6,4
Scrapers	11	12			23	4,9
Groovers	13	4			17	3,6
Borers	8	3			11	2,3
Backed flakes	22				22	4,7
Lunates	19				19	9,1
Truncated flakes	8				8	1,7
Denticulates		1			1	0,2
Retouched flakes	133	105	20	2	260	55,4
Bipolar flakes	12	3			15	3,2
Polished axe fragments				1	1	0,2
Sum of tools	272	170	24	3	469	99,9

(1) The table includes all the excavated data except squares 107x/100y and 111x/100y

5.3.1 <u>Lunates</u> (cf.fig.27).

Evidence of lunates was recovered in the Central Sudan from sites of Early Khartoum (c. 9000-6000 B.P.), Shaheinab (c. 6000-5000 B.P.) and post-Shaheinab (c. 5000-1700 B.P.) types (Arkell, A.J., 1949: 42-43; 1953: 26; Caneva, I., 1983: 223-224; Haaland, R., 1987a: 73-76; Magid, A., 1982: 34-35).

It has been suggested that hafted lunates (cf.fig.no.28) dated to c. 12000 B.P. were used as sickles (Wendorf, F., 1968: 943; Wendorf, F., and R. Schild, 1976: 276-277). The interpretation of the function and the reconstruction of the hafting of these lunates are based on specimens (found at the site no.890 in Tushka in north Sudan) with traces of lustrous edges (ibid).

The interpretation suggested by Wendorf as to how the lunates were hafted and what function they performed is not applicable in the case of the lunates which were recovered from the Central Sudan for the following reasons:

1) Examination of the lunates under stereo-lighted microscope (x40) did not show any visible traces of sickle-gloss on them that would indicate they were probably used as tools to cut food-plants. The use of the scanning electron microscope, SEM, (cf. pl.no.8) confirmed this conclusion.
2) The tools are too small (cf. diagram no.3) to have been used as sickles if they had been hafted (Magid, A., 1982: 35). The available botanical evidence from the Central Sudan (chapter IV) particularly for the period between c. 6000 and 1700 B.P. shows that *Sorghum* sp. (both wild and domesticated) and wild *Pennisetum* sp. were the main cereal species exploited. The lunates are too small to cut the thick (Haaland, R., 1981: 120) and solid stalks of these plant species.
3) It is evident that there was a noticeable decrease both in the number and size of the lunates (tables no.28 and 29; diagram no.3) from the period of Early Khartoum tradition to those of the Shaheinab and the post-Shaheinab one. The increase over time of the number of grinders is considered an indication of an increasing emphasis on the exploitation of food-plants. Thus if lunates were used for the exploitation of food-plants, they would have also increased in number over time.

Another interpretation of the function of the lunates is that they were 'most commonly used as the cutting parts of composite projectile weapons, in particular arrows' (Clark, J.D., et al, 1975: 323) which were hafted in shafts consisting of two parts, a mainshaft and a foreshaft (ibid: 329, 333). This interpretation is on a picture on a predynastic slate palette (cf.fig.no.29) from Egypt dated to the second half of the 4th millennium B.C. as well as on ethnographic sources (ibid: 323.324). When pointing the similarities between the Egyptian and the Early Khartoum lunates, Clark implicitly suggested that the lunates from Early Khartoum were also used to tip the arrows as their cutting part (ibid: 324). It has also been stated that lunate arrow tips similar to the

Fig.no.27 Lunates from Shaheinab type of sites

After Haaland,R.;1981: 114

Fig.no.28 Suggested method of hafting of of lunate-sickle(from Tushka)

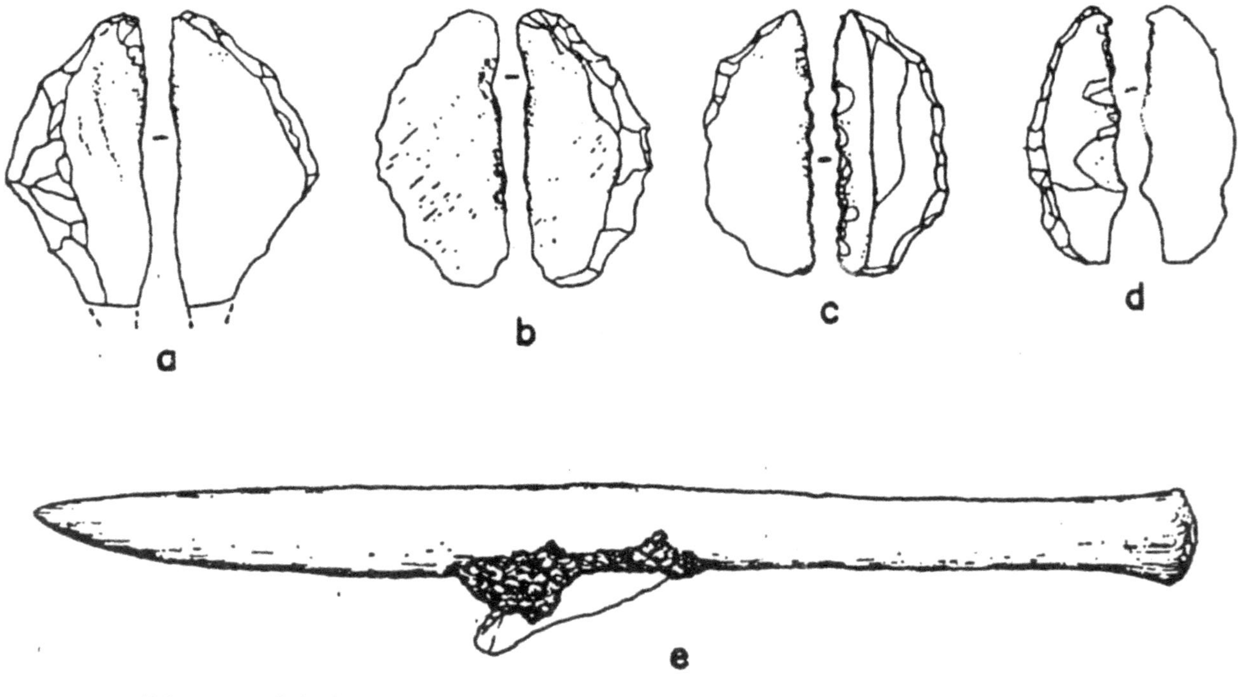

Pieces with lustrous edges from Tushka and reconstruction of probable method of hafting. Stipple indicates area of sheen. Hatching on "b" indicates area covered by traces of adhesive. The hafting technique is reconstructed from an actual flint-edged sickle found near Lucerne, Switzerland,

(After Wendorf,F.,and R.,Schhild,1976:277)

Table no 28

Raw-material and measurements of lunates from Islang
(Late Early Khartoum early Shaheinab type of site)

Square	Level	Raw-material	Maximum Length millimetres	Maximum Width millimetres	Total in each level
1323	2	Quartz "	20 21	8 12.5	2
	3	Quartz Rhyolite "	20 20 25	10 12.5 13	3
	4	Quartz Rhyolite	15 25	10 13	2
1628	2	Nile-pebbles Rhyolite	12.5 17.5	8.5 11	2
	3	Quartz "	17 18	8 11	2
	5	Rhyolite "	24 25	12.5 15	2
	6	Rhyolite	27.5	15	1
3180	3	Quartz	12.5	7.5	1
	4	Quartz Rhyolite	20 20	10 12.5	2
	6	Rhyolite	25	12.5	1
	9	Quartz Rhyolite "	17.5 20 30	9 12.5 15	3
Total number of lunates			-	-	21

After Magid,A.,1982: 44

Table no. 29

Raw-material and measurements of lunates from Nofalab
(Shaheinab type of site)

Square	Level	Raw-material	Maximum length (in millimetres)	Maximum width (in millimetres)	Total in each level
1225L	2	Quarts	12	5	1
100e	1	Rhyolite	15	10	2
		Quartz	17	9	
	2	Rhyolite	20	12.5	1
52h	1	Quartz	12	5	2
		"	10	7	
	4	Nile pebbles	17	10	1
	5	Quartz	10	7	1
Total number of lunates		8	-	-	8

After Magid, A., 1982: 33

Diagram no. 3 Scatter diagram showing tendency of decrease in size and number of lunates over time

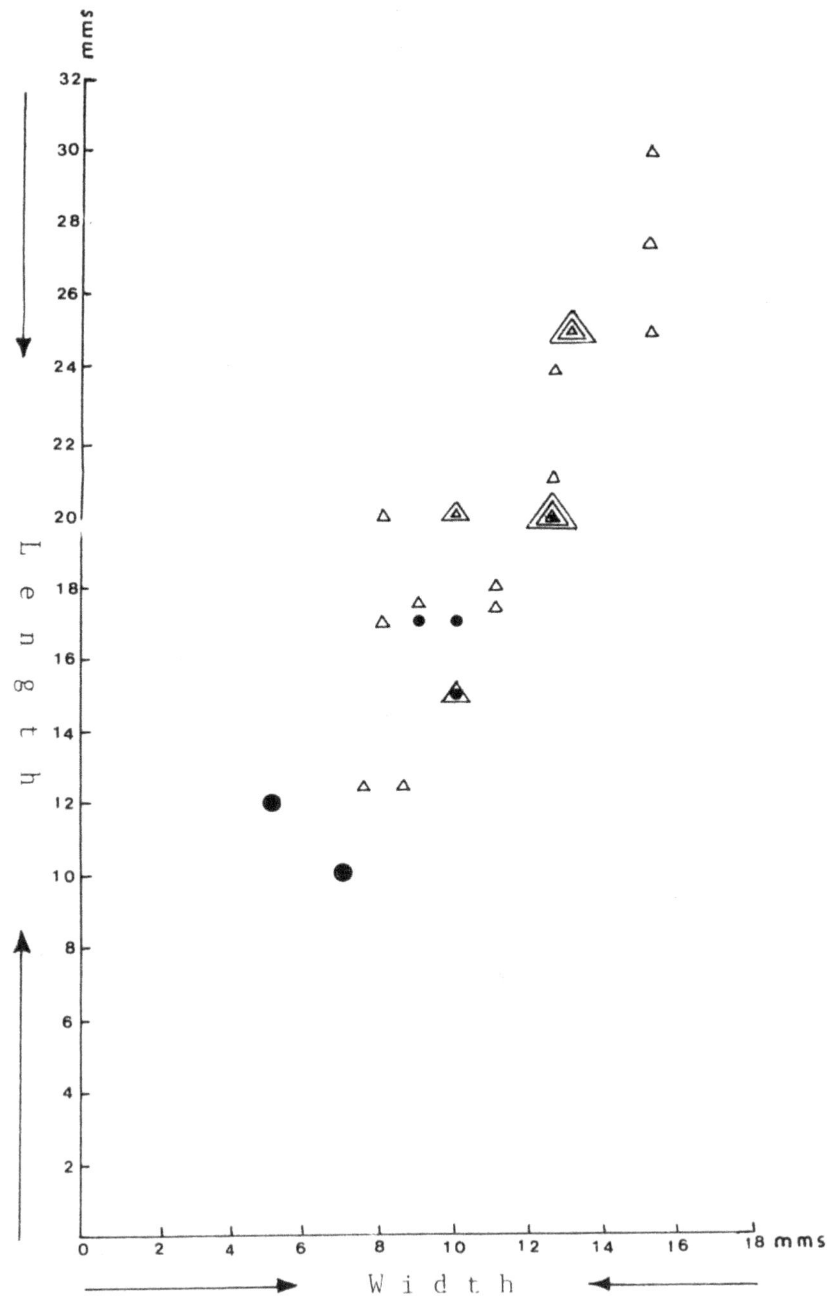

● The Nofalab lunates (Shaheinab type)
△ The Islang lunates (Early Khartoum type)

After Magid, A., 1982: 34 with some modification

Plate no.8 An example for wear pattern on lunates

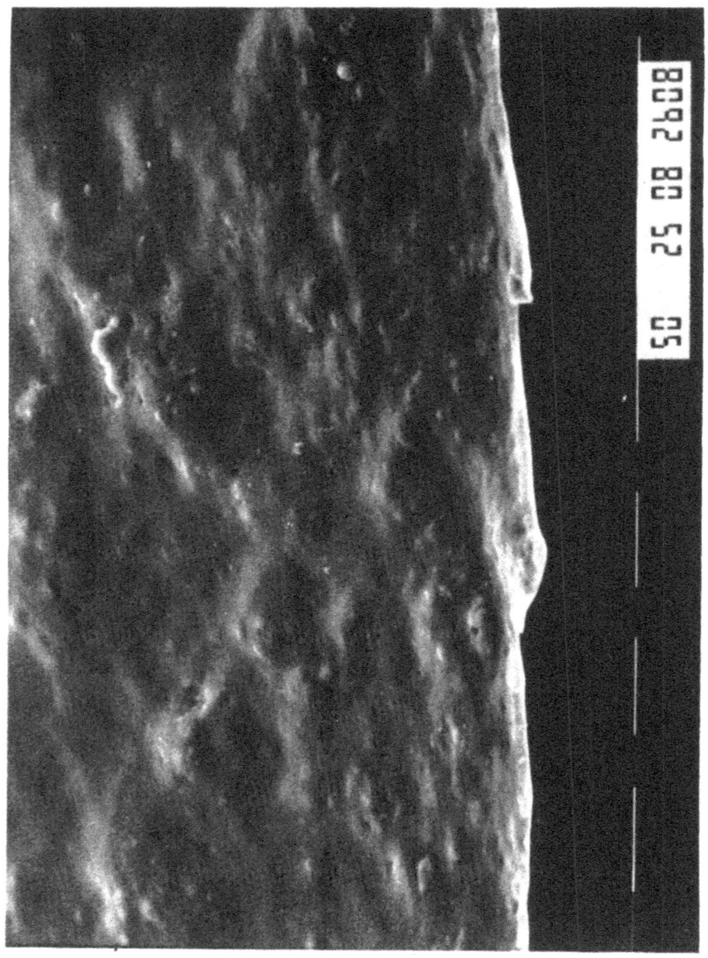

Fig. no.29 Archaeological evidence of lunate-arrow tips

Fragment of a late Predynastic slate palette showing a man from lower Egypt transfixed by a chisel-ended arrow that has passed out through his back

After Clark, J.D., et al, 1975: 325
(with enlargement x 3)

Egyptian ones were also found in the royal tombs of el-Kurru dated to c. 860 to 643 B.C. (Clark, J.D., et al, 1975: 358). Haaland extended Clark's argument and pointed out the 'even closer similarities' between the lunates from Egypt and those from the sites of Shaheinab type as regards their small size and noted that the size of those from the Early Khartoum period are relatively larger (Haaland, R., 1981: 113-115). Haaland further argued that the lunates from the sites of the Shaheinab type were hafted and used (ibid). Thus, the foregoing discussion clearly favours the interpretation that the lunates recovered from the Central Sudan were used as arrow-tips.

Since these putative lunate arrow-heads were presumably used mostly for hunting (ibid: 113), the decrease in their number and size over time in the Central Sudan (diagram no.3) is most likely attributable to a decreasing reliance on hunting and an increasing emphasis on domesticated animals. The decrease in the size of the lunates is also combined with a tendency toward a change in the type of raw-material from which they were made. During the Early Khartoum period when hunting was an important economic activity, lunates were mostly made from raw-materials such as rhyolite, chert and Nile pebbles which are characterized by their good flaking and texture properties. Most of the Shaheinab type of lunates are made from quartz pebbles (cf. table no.31) which are characterized by their bad and uncontrolable flaking properties. Thus, the tendency towards the production of smaller lunates made from raw-materials which consisted mainly of small pebbles and were characterized by the bad flaking properties, reflects "indifference" in the selection of raw-material and the production of lunates and hence also indicates a decreasing emphasis on hunting activities.

5.3.2 Knife (?sickle)-blades.

The characteristic feature of all the excavated sites from the Central Sudan is the complete absence of tools that can be directly associated with cutting/harvesting wild and/or domesticated food-plants, e.g. *Sorghum* sp. and *Pennisetum* sp. However, two artefacts (recovered from the site of Um Direwia) were classified as knives because they have steeply trimmed edges opposite to thin cutting ones (Haaland, R., 1981: 126). On the basis of the shape and size of these artefacts (having lengths of more than 3 cm), it was suggested that they had been used for cutting (ibid), but without specifying the suggested cutting function. On the assumption that they might have been used for harvesting of food-plants, an examination of their microwear use pattern was made with the aid of a SEM (cf.pl. no. 9). Parallel striations[7], on the cutting edge is the type of

[7] Although it is virtually impossible to determine whether the tool was used for cutting wild or domesticated plants (Unger-Hamilton, R., 1985: 121), experimental studies have shown that a striated polish is often produced on a cutting edge due to contact of the tool with the soil (ibid) indicating that the plant was cut from the bottom of the stem. This striated polish can also be produced by repeated use of the artefact to cut grasses and similar substances containing silica (Wendorf, F., and R. Schild, 1976: 277).

Plate no.9   An example of wear pattern on knives

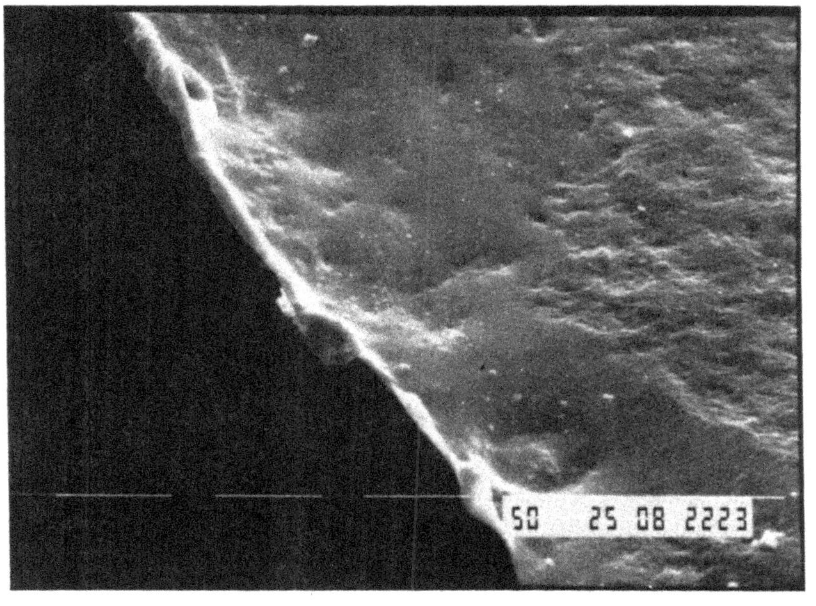

utility wear one would expect if the lunates had been used for harvesting, but the results of the SEM examination did not show any of these.

5.3.3 Grinders.

Grinders from the Central Sudan were recovered from the sites of the Early Khartoum, Shaheinab and post-Shaheinab traditions (cf.figs.no.30 to 33 and pl.no.10). Although there are differences among the grinders from these cultural periods as regards, e.g., their size and shape, they can all be classified as upper or lower grinders. Most of them are made from sandstone. The upper grinders are generally smaller than the lower ones, and show polished flat surfaces due to the use of both faces (Magid, A., 1982: 39); and their shape is generally oval or rounded. The majority of the lower grinders are bigger generally with one polished concave face (the working surface); while the reverse face is very seldom worked on.

Most of the grinders (both upper and lower) were found broken, and the sizes of the fragments varies from large pieces to small unidentifiable ones. Big fragments with traces of fire were found. These were interpreted as having been used to make hearths (Haaland, R., 1981: 131; Magid, A., 1982: 40).

There are slight differences between the grinders of the Early Khartoum and the Shaheinab types in that the latter are more standardized in shape (Zarattini, A., 1983: 234) and are generally larger than the former. In addition, the grinders from the Shaheinab phase are smaller than those from the post-Shaheinab phase (from, e.g., the sites of the Shaqadud cave and Jebel Tomat). Thus there seems to have been a tendency to increase the size of grinders over time. There was also a tendency for the saddle quern type to become more frequent. The diversity and increased numbers of grinders over time probably indicates an increased emphasis on foods of vegetal origin, regardless of whether they were cultivated or not.

Several suggestions have been proposed for the function(s) of the grinding tools, but the main emphasis in our discussion is on the interpretation which relates these tools to the exploitation of food-plants. The grinders recovered from the sites of the Early Khartoum culture were interpreted by Arkell as having been used only to grind ochre (Arkell, A.J., 1949: 51, 53). This is, however, considered as only one function of these grinders (Haaland, R., 1987a: 200; Magid, A., 1982: 40-41; Zarattini, A., 1983: 234). The small size of the grinders of this culture has been attributed to a need for portable daily equipment (ibid) used for crushing nuts, stones of fruits of *Ziziphus* sp. to extract the edible seeds, etc. However, no direct association between these grinders and the processing of grain/seed food has been established. The recovery of grinders correlated with the introduction of pottery in the Central Sudan during the Early Khartoum period should be considered as an indication of the beginning of exploiting grain/seed foods.

Fig.no.30

Lower grinder of the Early Khartoum type

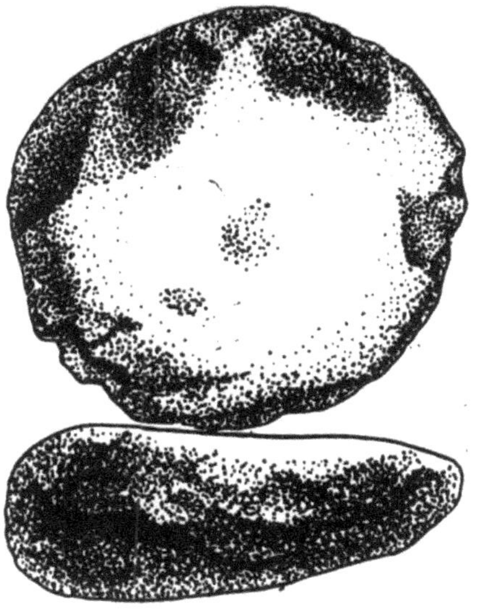

Scale 1:1

After Arkell,A.J.,1949:Plate 29

Fig.no.31

Lower grinder of the Shaheinab type

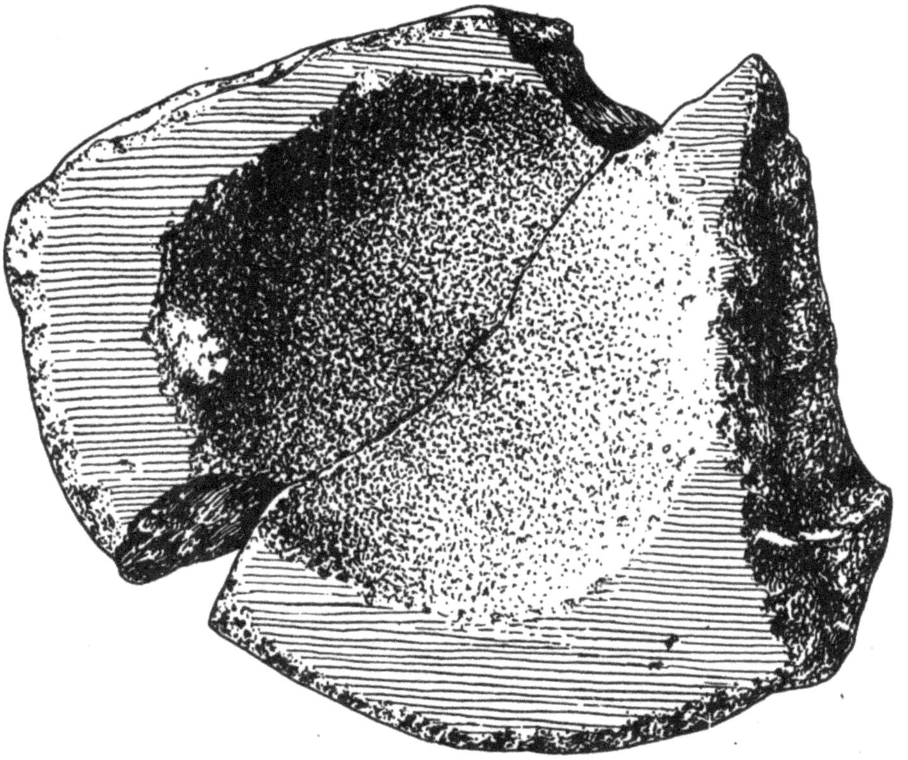

Scale 1:1

After Arkell,A.J.,1953:Plate 22

Fig.no.32

Fine-grained lower grinder
(Post Shaheinab:Shaqadud-
cave-site)

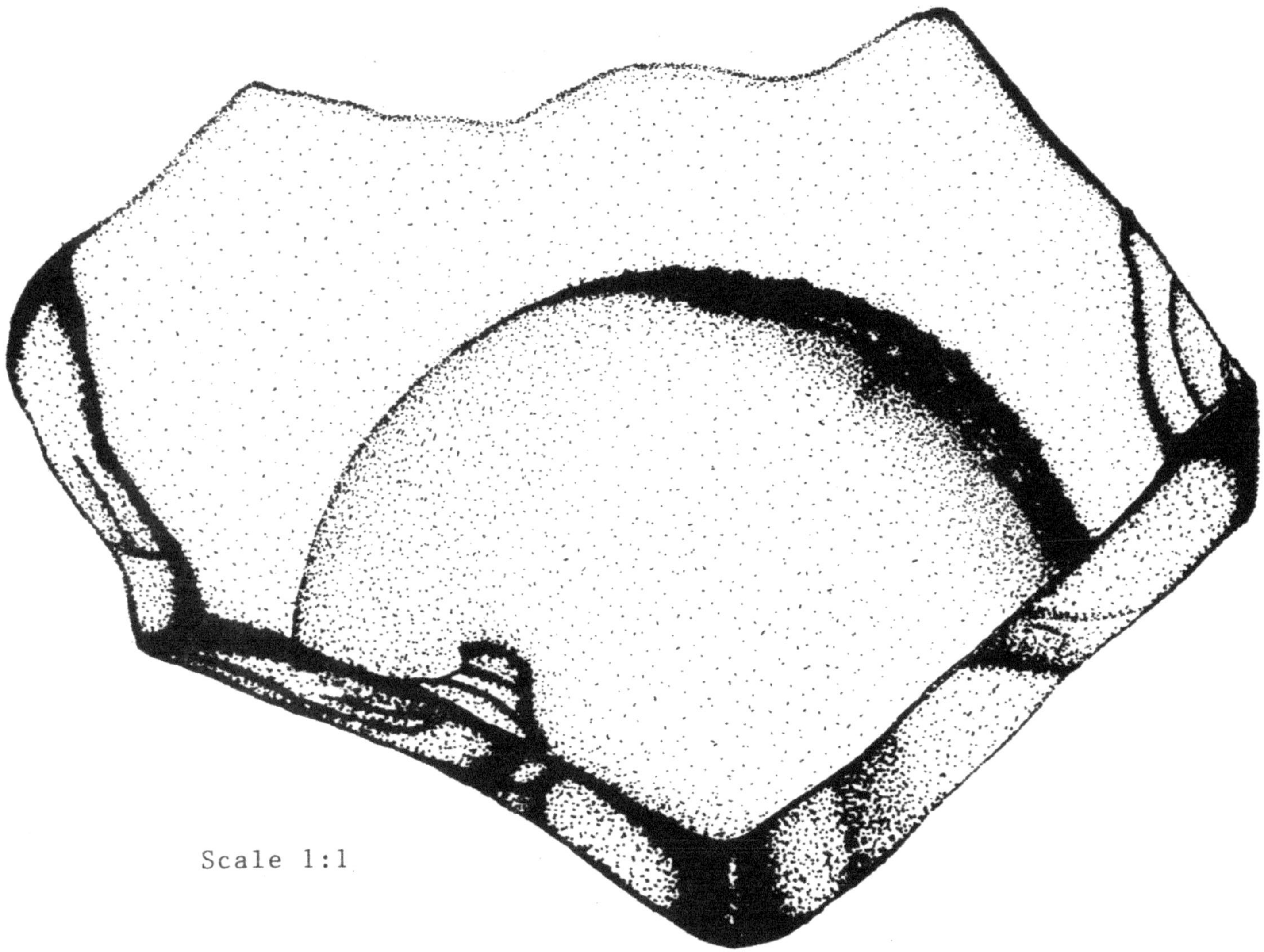

Scale 1:1

Fig.no.33 Coarse-grained lower grinder (Post-Shaeinab:Shagadud cave site)

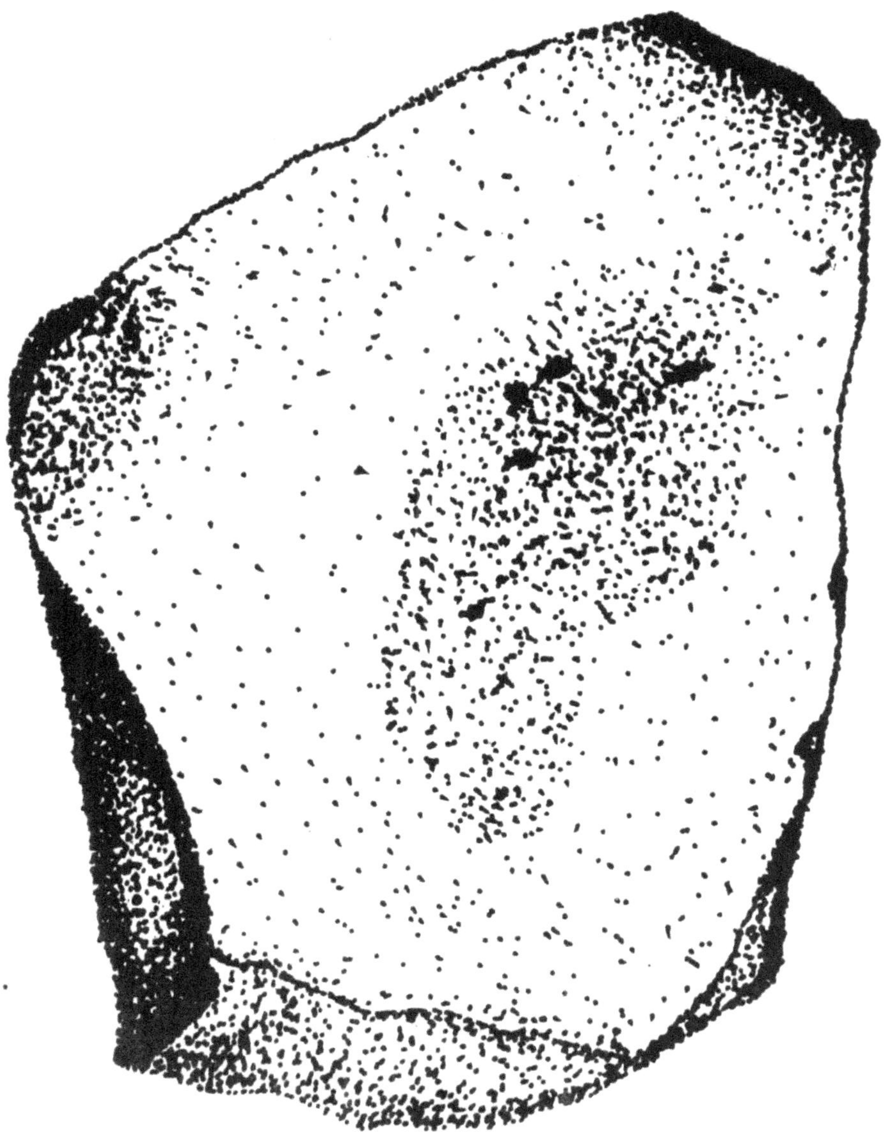

Scale 1:1.

Plateno. 10

QUERNS

After Addison, F.,1949:pl.LXxIV

Food-plants of this type, e.g. grains-seeds, certainly required processing and preparation, e.g. threshing the husks and/or grinding etc. Pottery and grinders are obviously the only basic equipment to perform such tasks. Thus, it is worth the effort to include in future research plans provision for verification of this suggestion.

With regard to the grinding stones from the sites of the Shaheinab tradition, Arkell suggested that since 'there seem to have been no cereals but only ochre to grind' (Arkell, A.J., 1953: 54) grinding stones were still used for grinding ochre and for crushing quartz and clay for pottery production (ibid: 44, 47, 49, 54). However, during the last decade new evidence, such as impressions of cereal grains, were recovered from several sites of the Shaheinab type (Haaland, R., 1987a: 181-183); Magid, A., 1982: 89-93; Klichowska, M., 1978: 42-43; this text chapter IV). The correlation of this new botanical evidence with the large number of grinders (Haaland, R., 1981: 214; Magid, A., 1982: 40-41; compare tables no. 30 and 31) provides the basis for arguing that the grinding equipment was mainly used for processing grain-foods. They might have also been used for grinding vegetal foods other than grains (cf. chapter VIII).

The extremely numerous grinders found in the sites of the Shaheinab tradition indicate an increased importance of vegetal foods such as sorghum and perhaps the beginning of their cultivation (Haaland, R., 1981: 215). It should, however, be mentioned that the Shaheinab tradition is a complex in which differences occur in the frequency of grinders as well as of other tools when individual sites are compared. For instance, the frequency of grinders is relatively low in the dry season camps such as the Zakiab site as compared with the very high frequency of grinders found at sites such as Um Direwia (cf. tables no.31 and 32).

The recent excavations of sites of post-Shaheinab type, e.g. the Shaqadud cave-site, revealed archaeoethnobotanical evidence[8] that basically agrees with and largely confirms Haaland's view that the technological knowledge of the population had reached a level where they probably had already started the cultivation of wild food-plants beside the gathering activities. The cave-site of Shaqadud which represents the older sites of the post-Shaheinab period (cf. radiocarbon dates in appendix no.1), provided the earliest *in situ* evidence that shows the direct association of grinders with wild grain foods: A lower concave grinder was found with its concave (working) face covering mummified and carbonized grains of *Setaria* sp. This evidence strongly indicates that grinders had been used for processing wild food-plants, e.g. mainly grain-foods. Evidence of grinders was also recovered from the younger sites of the post-Shaheinab tradition such as Jebel Tomat. The earliest evidence of domesticated cereals, namely *Sorghum bicolor* (L.) Moench, from the Central Sudan was found at this site. It is most likely that grinders were used, during this period more than at any time before, for grinding food-plants.

---

[8] More detailed information on the culuivation of wild food-plants is presented in chapter VIII .

Table no. 30 Ground tools from Islang(late Early Khartoum early Shaheinab type of site)

Square no.	Upper grinders		Lower grinders		Unidentified fragments of grinders	Total no. of grinding tools	Hammer stones		Ochre	Total of all
	Complete	Fragment	Complete	Fragment			Complete	Broken		
1	1	1	-	6	6	14	1	-	3	18
1323	2	-	-	5	4	11	2	-	8	21
1628	1	2	1	7	5	16	-	1	2	19
3180	-	3	-	8	11	22	2	-	7*	31
Total	4	6	1	26	26	63	5	1	20	89

* One is polished

After Magid,A.,1982: 47

Table no.31

Ground tools from Um Direwia
(Shaheinab type of site)

	90x 100y	92x 100y	94x 100y	96x 100y	98x 100y	100x 100y	102x 100y	104x 100y	110x 100y	122x 100y	122x 102y	Sum
Lower grindstones	90	120	110	256	250	258	148	82	60	9	4	1378
Upper grindstones	320	460	430	1500	1300	991	720	442	320	51	42	6576
Polished fragments	172	159	215	350	350	750	460	250	300	15	10	
Flakes from grinders	500	560	650	2600	2500	805	710	390	350	30	25	
Rubbers	20	5	3	8	20	38	26	-	14	-	3	
Small grinders with dimples	7	4	7	9	35	20	15	-	10	3	-	110
Hammerstones, sandstones	20	25	38	190	250	56	33	52	14	10	6	
Hammerstones, other material	76	47	82	192	213	100	59	60	50	9	6	
SUM	1205	1380	1535	5105	4918	3018	2171	1276	1118	127	96	

cont	122x 104y	100x 102y	100x 104y	100x 106y	100x 108y	100x 110y	100x 112y	90x 84y	92x 84y	94x 84y	104x 68y	
Lower grindstones	5	190	108	78	50	62	17	5	15	11	2	543
Upper grindstones	15	713	540	344	184	320	90	110	120	122	10	2568
Polished fragments	30	340	180	270	125	150	110	120	10	25	17	
Flakes from grinders	40	630	800	415	420	350	250	216	90	40	23	
Rubbers	1	15	-	2	4	5	-	5	-	-	-	
Small grinders with dimples	1	25	-	4	3	2	-	3	3	3	-	44
Hammerstones, sandstones	10	26	78	44	27	36	4	19	9	13	5	
Hammerstones, other material	11	85	64	84	35	39	31	13	15	33	7	
SUM	113	2024	1770	1241	848	964	502	491	262	247	64	11928

After Haaland,R.,1981: 152-153
with some additions.

Table no. 32

Ground tools from Zakiab(Shaheinab type of sites)

	81x 100y	83x 100y	85x 100y	87x 100y	89x 100y	91x 100y	93x 100y	95x 100y	97x 100y	99x 100y	100x 100y	102x 100y	Sum
Lower grindstones	-	-	-	-	-	-	-	-	-	1	-	13	14
Upper grindstones	1	-	15	10	3	10	-	-	10	15	-	39	103
Polished fragments	7	9	5	11	10	9	-	3	12	-	6	-	
Flakes from grinders	-	-	10	120	-	100	-	50	7	-	47	-	
Hammerstones, sandstone	-	23	-	-	-	-	2	-	-	2	-	-	
Hammerstones other rawmaterial	-	-	-	-	-	3	-	-	-	1	1	2	
SUM	8	32	30	141	13	122	2	53	29	18	54	54	

cont.

	102x 98y	102x 96y	102x 94y	102x 102y	102x 104y	100x 102y	120x 100y	102x 116y	102x 120y	102x 122y	95x 89y	93x 89y	91x 89y	
Lower grindstones	8	8	2	3	2	6	1	1	2	-	3	3	1	40
Upper grindstones	32	29	26	26	3	18	15	19	15	5	13	15	5	221
Polished fragments	9	24	17	6	10	2	3	27	22	8	-	12	-	
Flakes from grinders	-	-	-	10	15	-	-	3	50	-	-	1	-	
Hammerstones, sandstone	1	-	-	-	1	-	40	-	1	-	-	-	-	
Hammerstone other rawmaterial	-	4	-	2	-	6	-	-	2	-	-	1	1	
SUM	50	65	45	47	31	32	60	50	92	13	16	32	7	378

After Haaland,R.,1981: 150-151 with some additions.

An ethnographical study of the grinding equipment (e.g. its production, types, how long it can be used before being resharpened or breaking, etc.) was conducted; and its results were correlated with archaeological material in the work of Haaland (Haaland, R., 1981: 131-137; 1987a: 79-83). Haaland's work as well as other studies on ethnographical materials correlated with archaeological data (Arkell, A.J., 1953: 25,46; Magid, A., 1982: 39-41) indicate that the production and utilization of grinders recovered from the archaeological sites in the Central Sudan were generally practised as follows:

1) The rough shaping of the grinders was done at the quarry sites and the final shaping was completed at the settlement site.
2) Resharpening was probably done whenever the working surface was worn smooth. The period of use before the next resharpening was dependent both on how intensively the grinders were used and on the type(s) of food-plants ground on them.
3) Breakage of the grinders was a product of a prolonged use, with several resharpenings that resulted in their thinning and fragmentation/breakage. These factors may explain why most of the grinders recovered are broken.
4) Hammer stones seem to have been the main tools which were used for shaping, pecking and resharpening the grinders. Thus, they are considered as being indirectly associated with the grinding of food-plants.
5) The recovery of coarse-grained and fine-grained grinders (cf.figs.no.32 and 33) is interpreted as an intentional selection rather than a random collection of raw-material from the quarry sites. These different textures of grinders most likely indicate that each type of grinders, coarse or fine-grained, was used at a specific stage in the grinding process. For instance, the coarse-grained grinders were probably used in the first stage of coarse grinding and the fine-grained grinders used at the last stage to obtain a finer and more thorough grinding of the same substance.

To sum up, grinding tools were present in the Central Sudan throughout the period being studied here (c. 9000-1700 years B.P.). In other words, there is evidence of grinders since the period when gathering was the only mode of exploitation of food-plants and throughout the development of this activity to the cultivation of wild and the domesticated forms. It was probably when grinders came into use and most likely also pottery that the population of the Central Sudan was able to consume grain-foods. However, it is unlikely that grinders played a major role in the process of domestication, but it is likely that they accelerated the pace in the development towards the time when the cultivation of wild plants started and their domestication was initiated.

5.3.4 Rubbers.

Rubber blades (cf.fig. no.34) are generally defined as 'flattish [mostly] elongated [or oval] tools, of which the two surfaces are smooth and more or less convex, with one or both long edges more or less sharp or keeled. The tip is usually rounded' (Arkell, A.J., 1953: 51). It has been suggested that the straight edges and the curved part approaching the tip rather than the tip itself was the working part (ibid). The dimensions of these rubbers, however, are not standardized and vary greatly. Sandstone is the main raw-material from which they were made.

It has been suggested that rubbers could have been used for polishing stone tools such as gouges, stone axes, chisels, etc. (Semenov, S.A., 1964: 68). But it has also been suggested that they were used for polishing wood and bone in the same manner as sand-papers are used today (Arkell, A.J., 1949: 64; Clark, J.D., et al, 1975: 333). It is also likely that rubbers were used for polishing the handles and the working edges of wooden tools associated with cultivation activities, e.g. wooden hoes and sowing tools. Another possibility is that sandstone rubbers might have been used for polishing the interior walls of the dry fruits of *Strychnos spinosa* Lam. (Ar. *bukhusa*) which were probably used as containers for food-plants as well as other substances. This interpretation is based on archaeological and ethnographic evidence. The archaeological evidence consists of sandstone rubbers. Although no evidence of *Strychnos spinosa* Lam. has been recovered from the Central Sudan, with the savanna type of environment with patches of forest grasses and shrubs that prevailed, the growth of this species would not be unexpected. The ethnographic material from the nomads of the Shaqadud area shows that they use rounded or oval sandstone tools and coarse-grained sand to polish the interiors of the fruits of *Strychnos spinosa* Lam. which they use as a house utensil. In short, the material discussed indicates that sandstone rubbers probably can be indirectly related to activities of food-plant exploitation.

5.3.5 Sandstone-rings (cf.fig.no.35).

Sandstone-rings of different sizes were recovered from the sites both of the Early Khartoum and the Shaheinab cultures (Arkell, A.J., 1949: 41; 1953: 50; Haaland, R., 1987a: 132, 134-135; Zarattini, A., 1983: 241) including the site we excavated recently in the Central Sudan. Sandstone is the predominant raw-material from which these stone-rings are made, but other types of raw-material, e.g. sod-granite are also used (ibid).

Arkell suggested that the stone-rings with a hole diameter[9] > 3 cm were probably mounted as weights for sticks (Arkell, A.J., 1949: 63, 64). Since most of the artefact recovered are broken, Arkell also suggested that they had been intentionally broken in

[9] The present writer is not aware of any published study on the function of the small stone-rings or those with small hole diameters (< 3 cm).

Fig.no.34

Sandstone Rubbers

(1) From Early Khartoum site

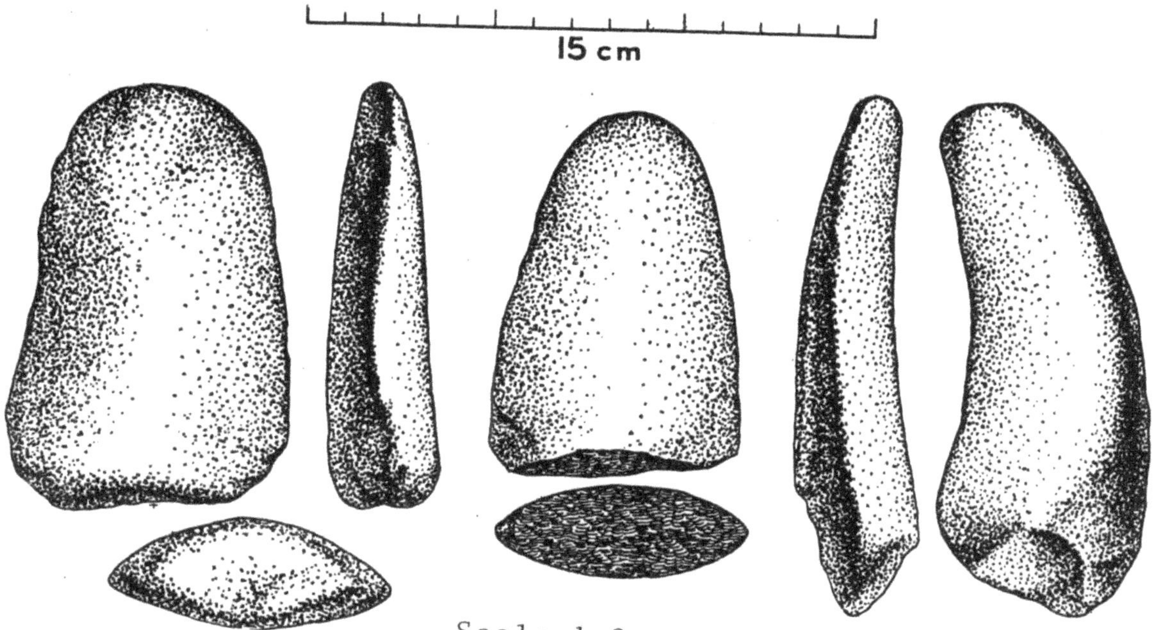

Scale 1:2

After Arkell,A.,J.,1949:Plate 37

(2) From Shaheinab type of sites

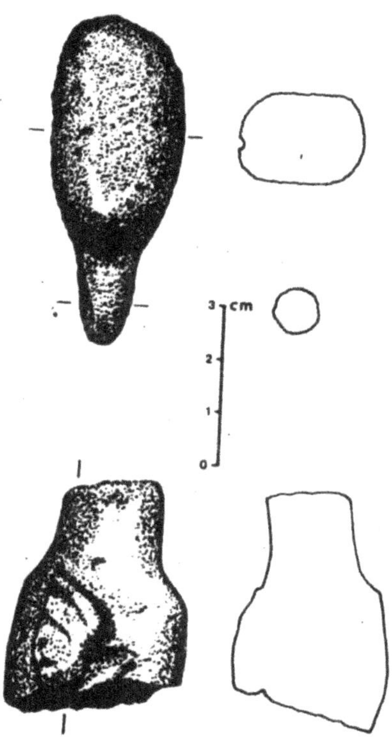

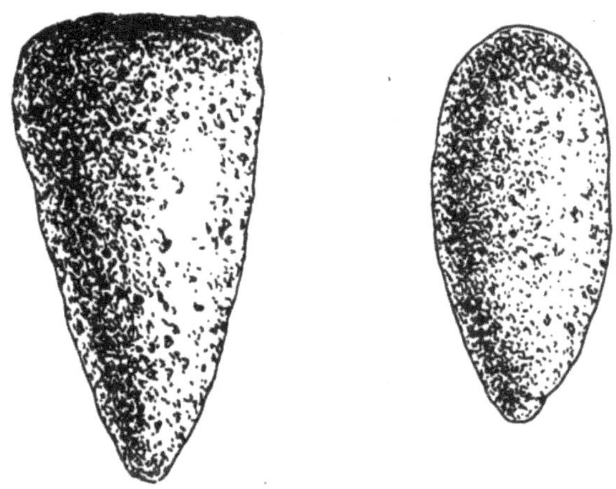

Scale 1:2

After Arkell,A.,J.,1953: .22 and
Haaland,R.,1981: .141.

Fig.no.35 Sandstone-Rings

(1)From Early Khartoum sites.

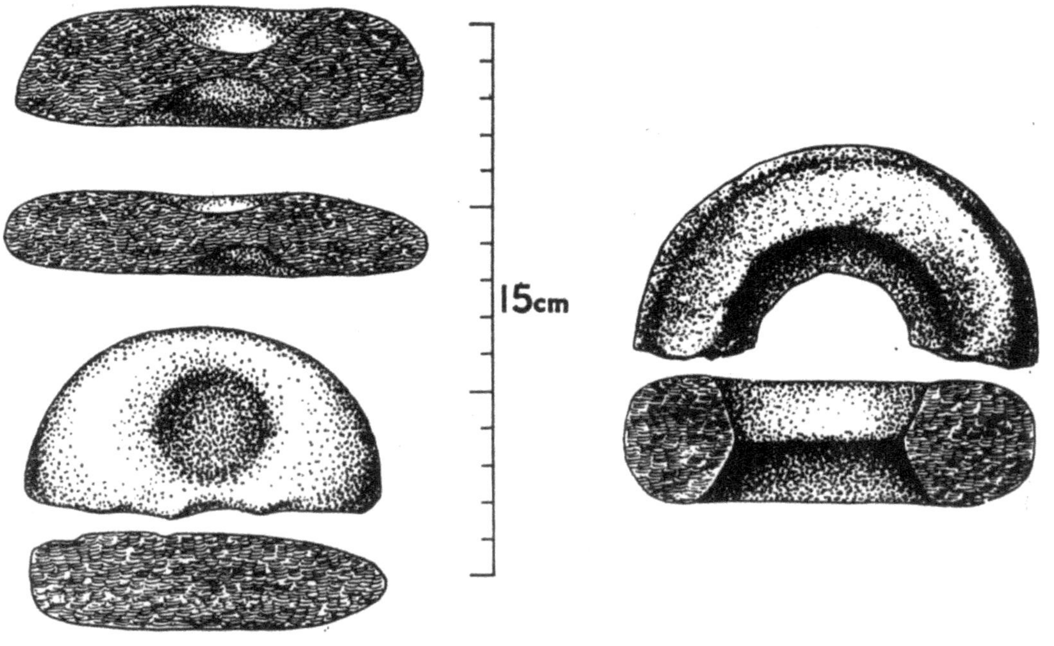

After Arkell,A.J.,1949:Plate 36

(2) From Shaheinab type of sites

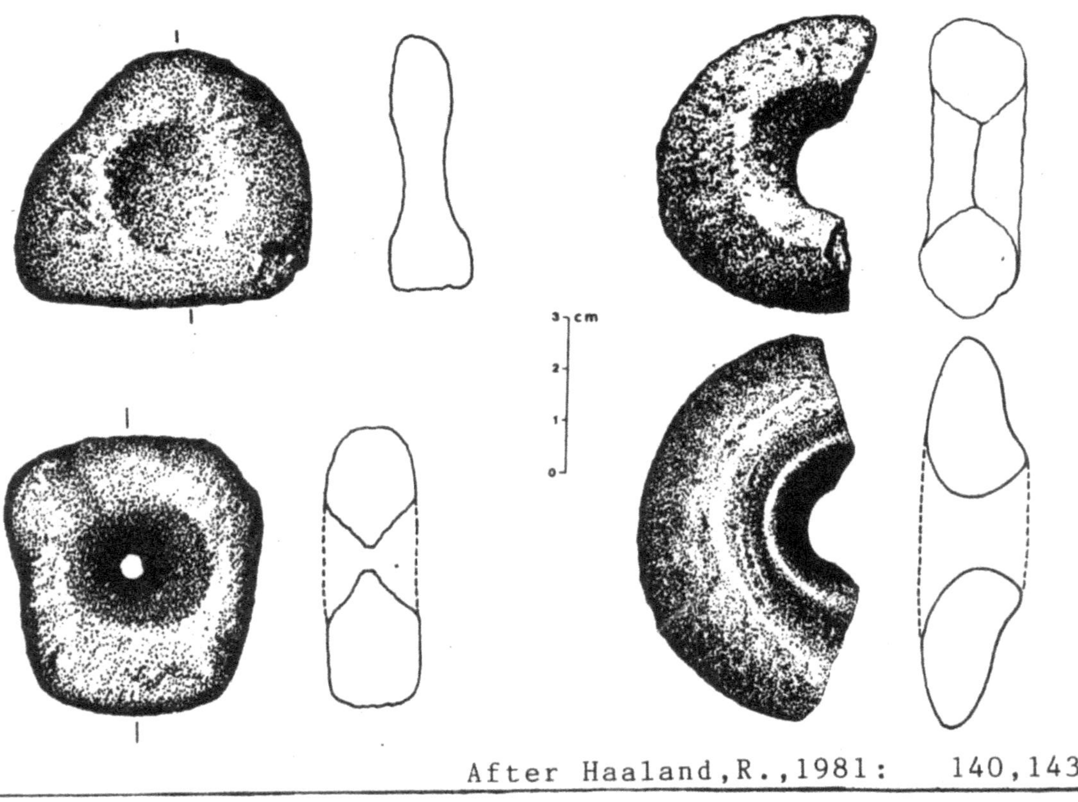

After Haaland,R.,1981: 140,143

order to be used as rubbers for shaping shafts of wood and bone (Arkell, A.J., 1949: 63; 1953: 50, 51), but most of them retained their hour-glass perforation which indicates they were not used as rubbers.

Another interpretation suggests that the big[10] stone-rings were used as weights for digging sticks like those the Bushmen of the Kalahari desert are still using (Haaland, R., 1981: 141). Or 'they could have been used as weights on hoes for breaking up ground for cultivation' (ibid) in the same manner as they were used until a few decades ago in the Harar plateau in Ethiopia (Clark, J.D., 1944: 32).

In short, most of the emphasis of the interpretations of the function of the stone-rings from the prehistoric Central Sudan is on their association with one activity or another of the exploitation of food-plants, e.g. gathering or cultivation. However, no suggestions have been offered - so far - as regards the function of the small stone-rings or why most of the recovered evidence consists of broken or fragmentary stone-rings, etc. In the future research more attention should be paid to these issues as well as to further testing of the standing interpretations.

5.3.6 Gouges.

'Gouge' is a conventional term given to tools found in the Central Sudan and elsewhere which are characterized by a conical 'slightly elongated' outline with a thin, oval-pointed cross-section. Their curved, hollow working edges are made by oblique flaking and their rounded butt-ends are carefully flaked and trimmed (Arkell, A.J., 1953: 31; Caton-Thompson, G., and E.W. Gardener, 1934: 20). The gouges, which are always made on rhyolite, are considered the characteristic tool type found on the sites of the Shaheinab culture (Arkell, A.J., 1953: 25, 31). Archaeological research shows that the geographical distribution of these gouges is localized between the sixth cataract, the Sabaloka cataract, and Jebel Aulia[11] (cf. fig. no.1) in the Khartoum area (Haaland, R., 1987a: 208, 221).

Since polishing is often considered an indicative criterion of the function of tools (Arkell, A.J., 1953: 31; Semenov, S.A., 1970: 68-69), a random sample of one hundred gouges was picked for the study of their surfaces[12]. In addition measurement of their dimensions was also made (appendix 4, cf. fig. no.36).

---

[10] See footnote no.9.

[11] The present writer observed the presence of similar artefacts to the gouges as regards their general outline at the site no.7 in Kashm el-Girba area but they are smaller in size.

[12] It has been observed during this study that all the gouges are bifacial tools and, hence, that it is difficult to distinguish whether they are core of flake tools. Thus, for conventional reasons, the wider and thicker end with oblique flaking will be called distal end; and the opposite smaller carefully trimmed end will be called the proximal end. Furthermore, the side on which the hollow oblique flaking is done (at the distal end) will be called the upper face, and the opposite one will be called the lower face (cf. fig. no.36).

Fig.no.36 Sketch illustrating different parts of a gouge*

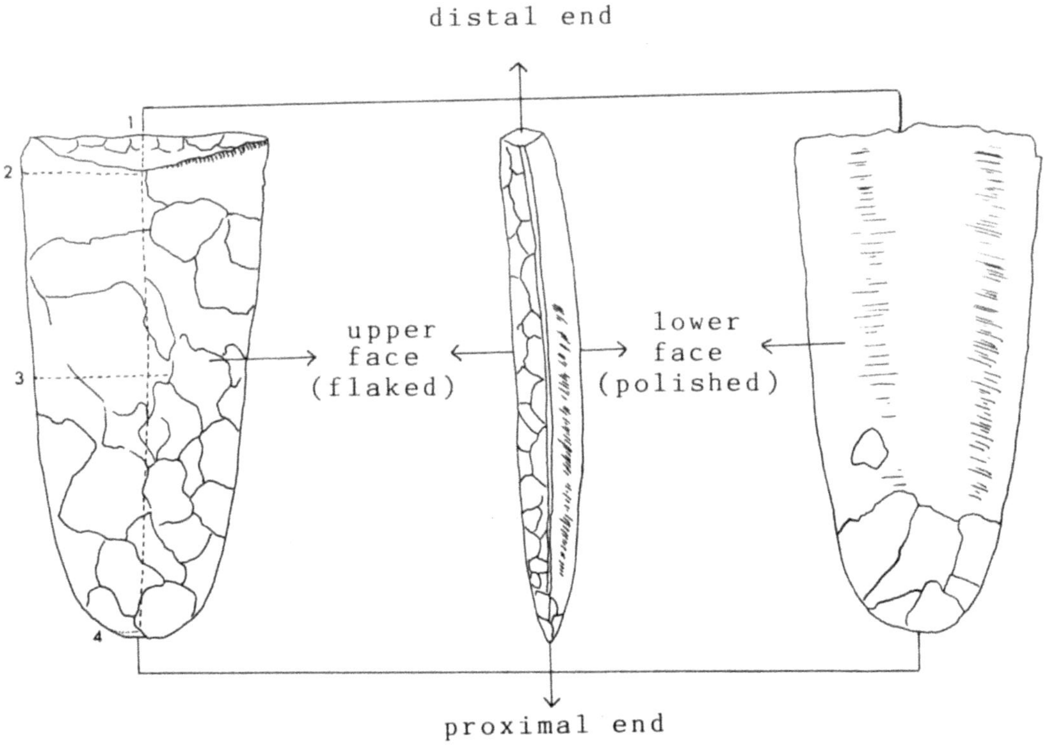

Legend:

- 1,2,3 and 4 are stations where measurements of maximum length and maximum breadth and thickness at distal, middle and proximal parts were taken (in their respective order).

*
It was impossible to apply an objective typological criteria to identify, e.g. dorsal and ventral faces , proximal and distral ends. Thus functional criteria related to the working edge and the polishing of the surfaces and the manner of hafting was applied to distinguish the different parts illustrated above.

Five types of gouges are distinguished as follows:

Type I: gouges flaked on both faces (fig. no.37).

Type II: gouges polished on both sides (fig. no.38).

Type III: gouges flaked on the upper face and polished on the lower one (fig. no.39).

Type IV: gouges flaked on the upper face and partly flaked, partly polished on the lower one (fig. no.40).

Type V: gouges partly flaked partly polished on both faces (fig. no. 41).

The majority (57 specimens) of the population studied is flaked on the upper face and partly flaked, partly polished on the lower one (type IV). The partial polishing on these and type V is performed mainly on the half of the side closer to the distal (working) end (cf. figs. no.38 to 41).

The study of the gouges also included their measurements at different points (cf. fig. no.36, appendix no.4). These criteria were used in the interpretation of the possible function(s) of gouges and to assess how long they were used before they became too short to perform their function(s) properly. The measurements of the gouges show the following general trends (histograms from 4 to 6):

1) The maximum length among the five types of gouges ranges between 7 and 8 cm represented by ten gouges from the collection studied. It has been reported that the longest gouge found - so far - in the Central Sudan is 11 cm long (Haaland, R., 1981: 121-122). Lengths that range between 6 to 7 cm are represented by 26 gouges, but the majority of the sample has a length that ranges between 2 and 6 cm. Despite their relatively short lengths (2 to 6 cm), most of this latter group of gouges retains both its distal and proximal ends which means the gouges were not broken but rather reduced due to repetitive sharpening whenever the working edges became blunt. The broken gouges are recognized by their often missing proximal end and infrequently missing distal and proximal ends (cf. fig. no.42). Breakage of these gouges, mainly towards the proximal end, is attributed to the fact that this part of the gouge is generally thinner and hence weaker than the distal end (cf. appendix no.4). The average lengths of each type of these gouges are shown on histogram 7. These averages indicate that the minimum length of a functionally efficient gouge is between 5 and 6 cm.
2) The majority of the specimens have a breadth that ranges between 3 and less than 5 cm at the distal (working) end and 1 to 2 cm at the proximal end. These measurements indicate that the working edge (distal end) of the gouges was efficiently used when it was in a range of 3 to 5 cm. It was also observed that the breadths in range between 3 to 5 cm are positively proportional to the lengths in the range between 6 and 8 cm (cf. histogram no.7)

Fig.no.37 Gouges type I

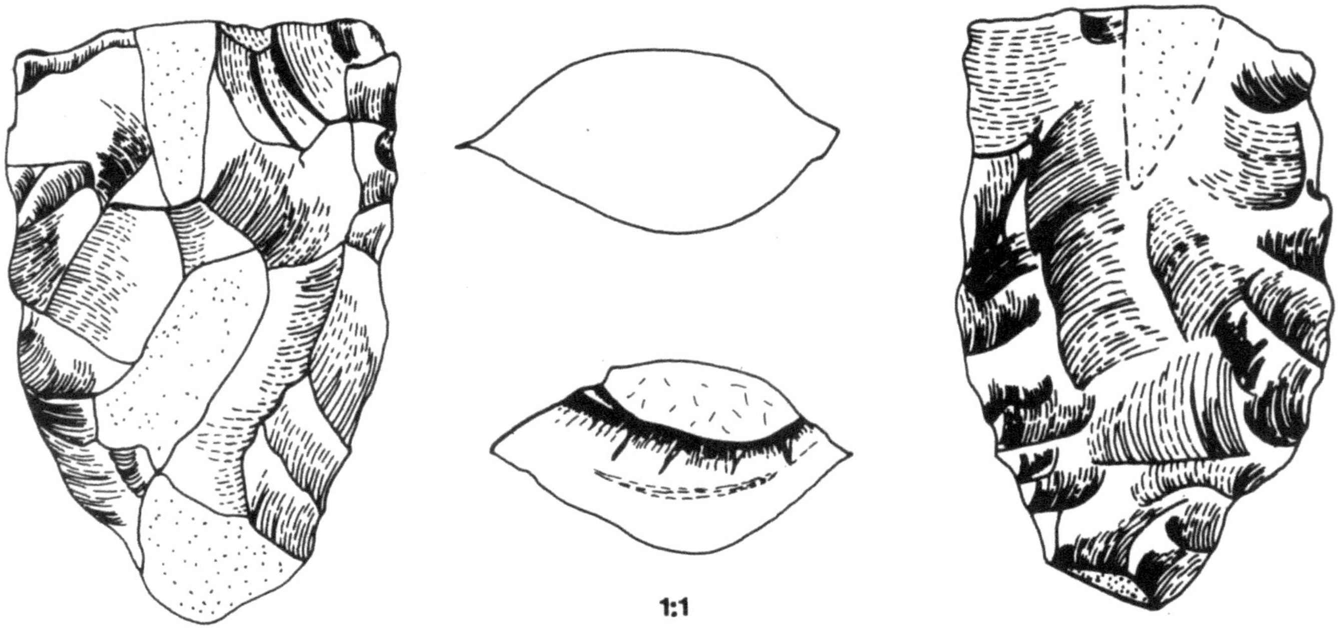

Fig.no.38 Gouges type II

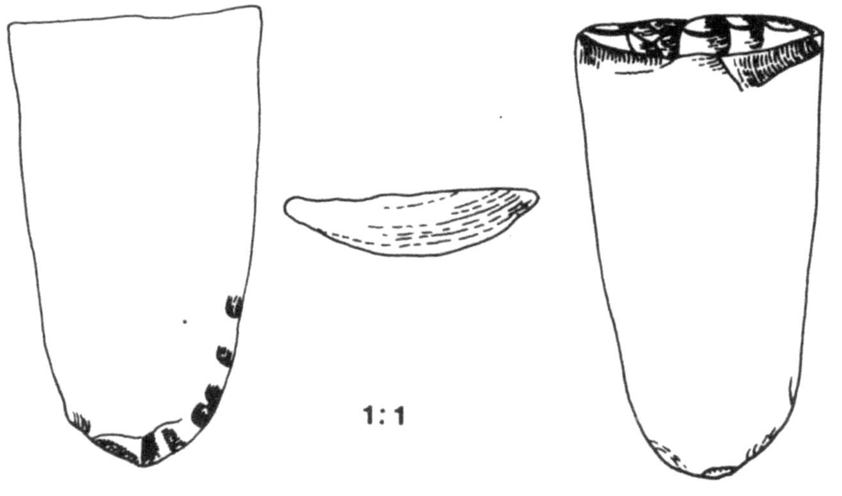

Fig.no.39 Gouges type III

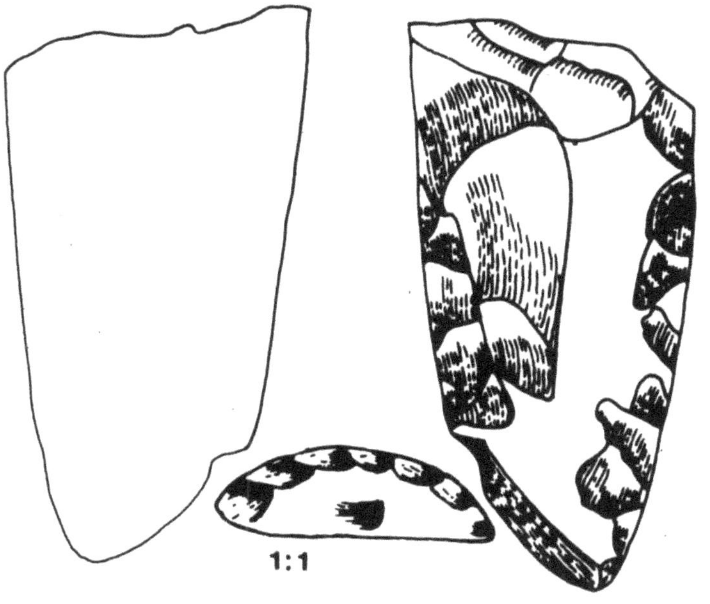

Fig.no.40 Gouges type IV

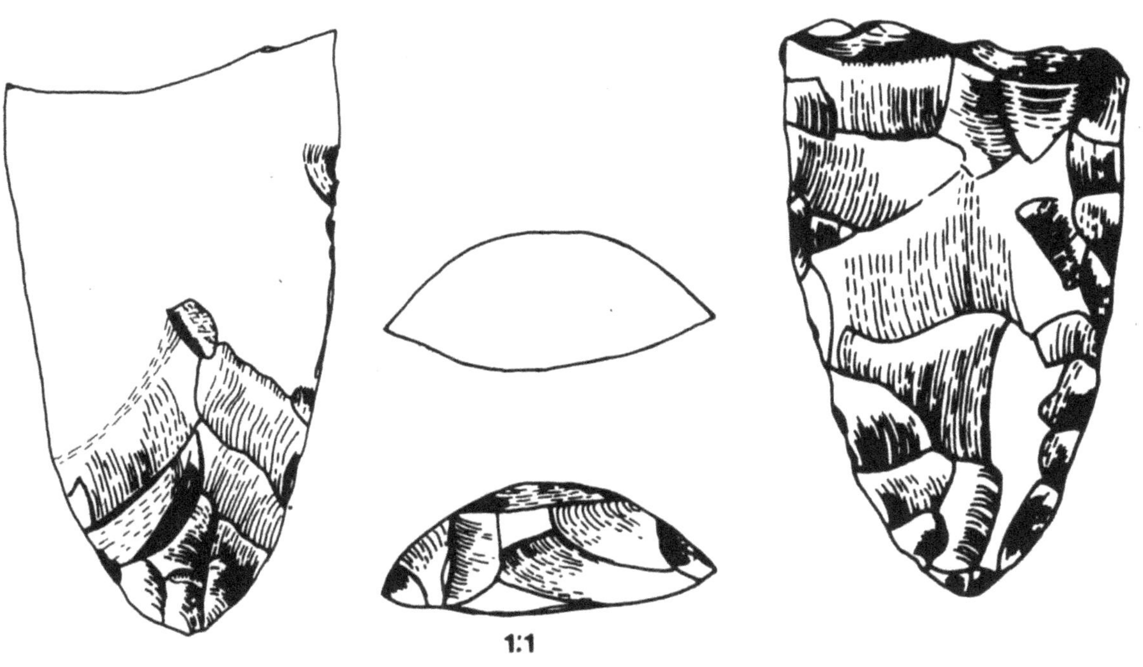

Fig.no.41

Gouges type V

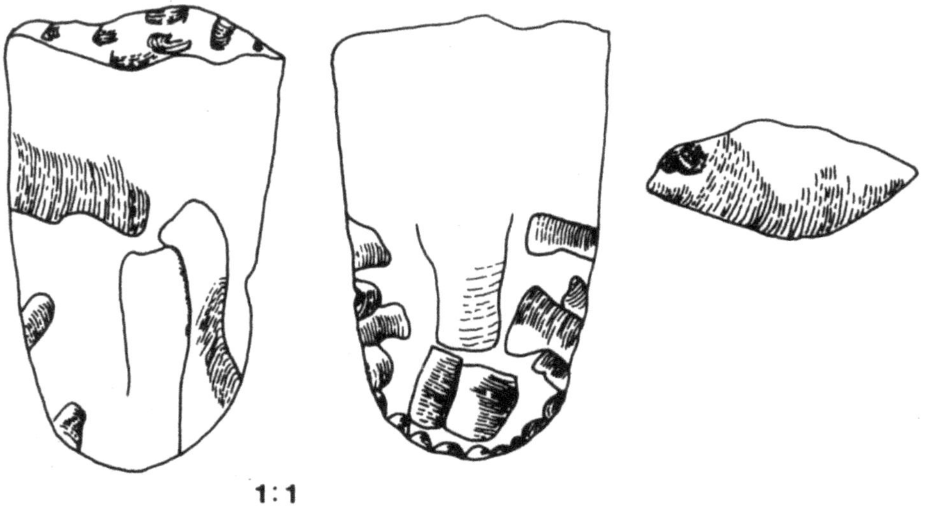

1:1

Histogram no4

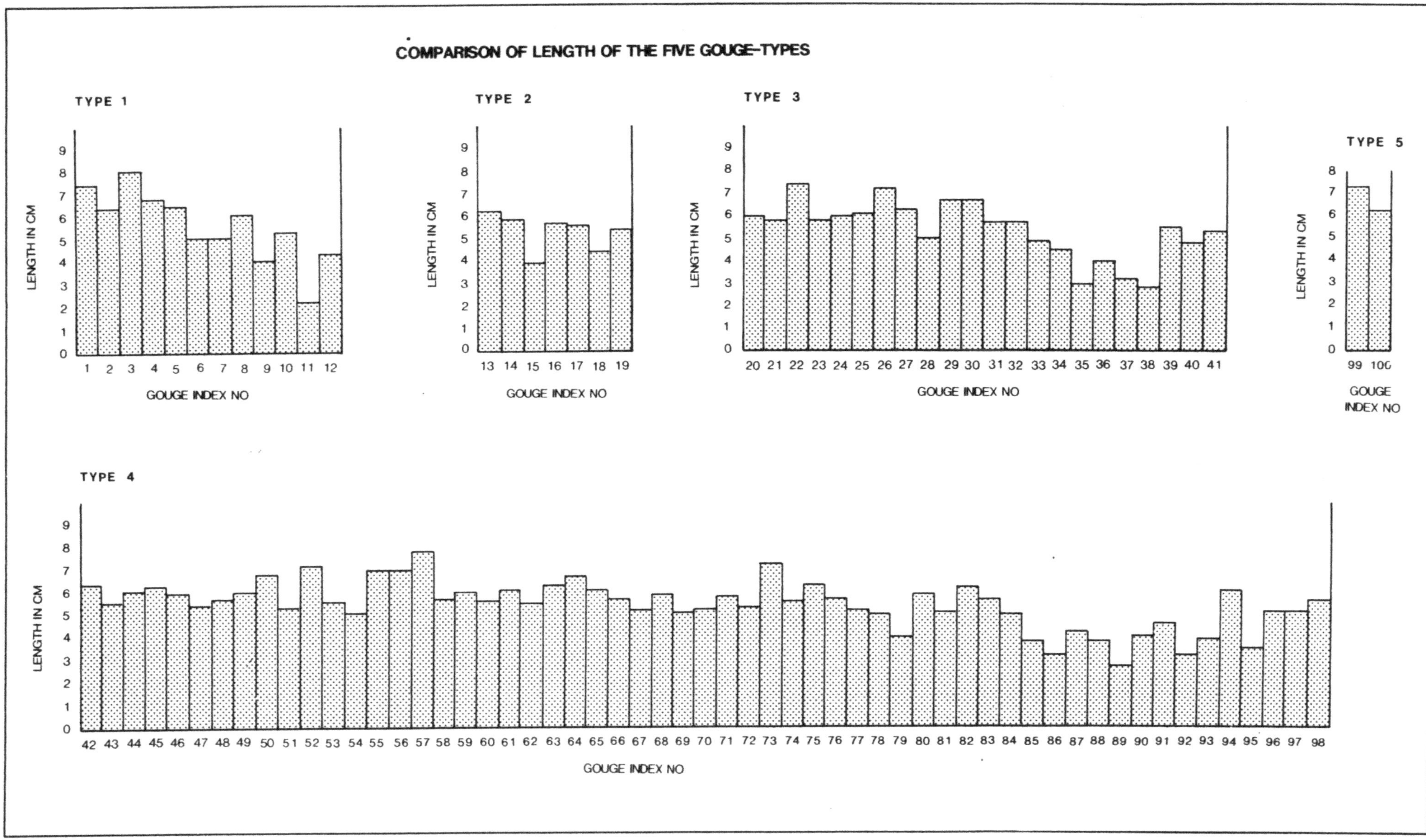

Histogram no5

COMPARISON OF BREADTH OF THE PROXIMAL AND DISTAL ENDS OF THE FIVE GOUGE-TYPES

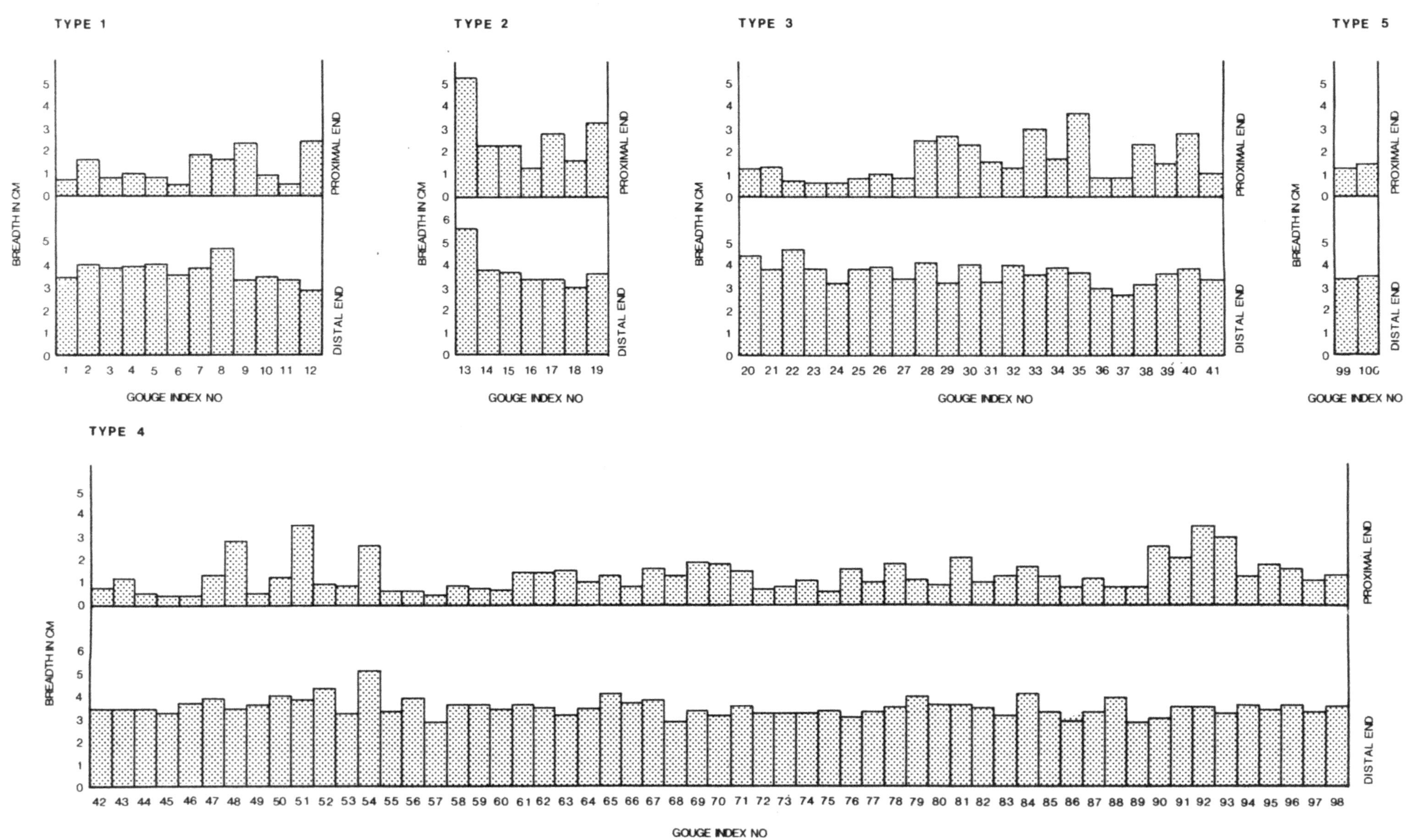

COMPARISON OF THICKNESS OF THE PROXIMAL AND DISTAL ENDS OF THE FIVE GOUGE-TYPES

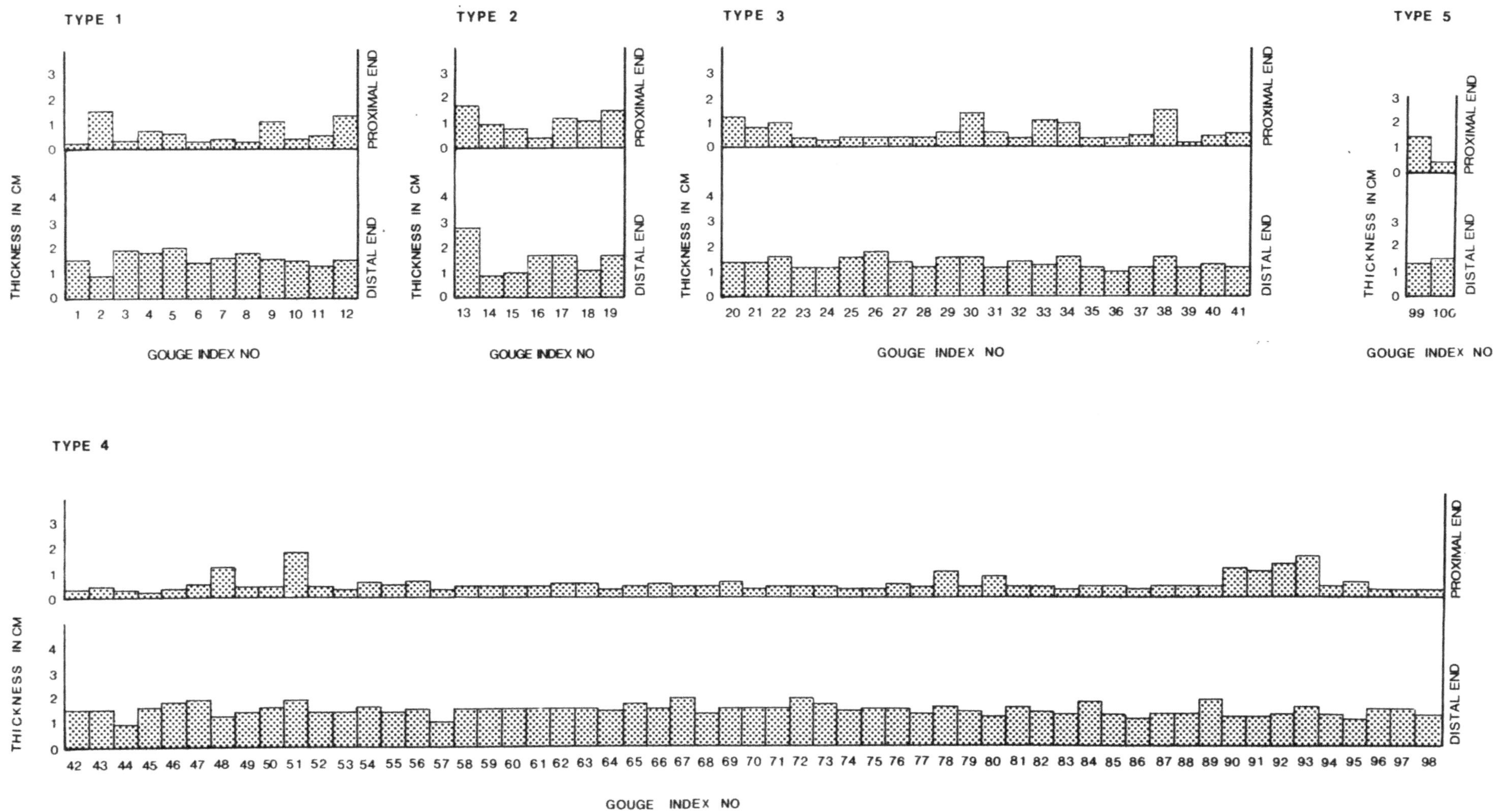

Fig.no.42 A gouge with missing distal and proximal ends

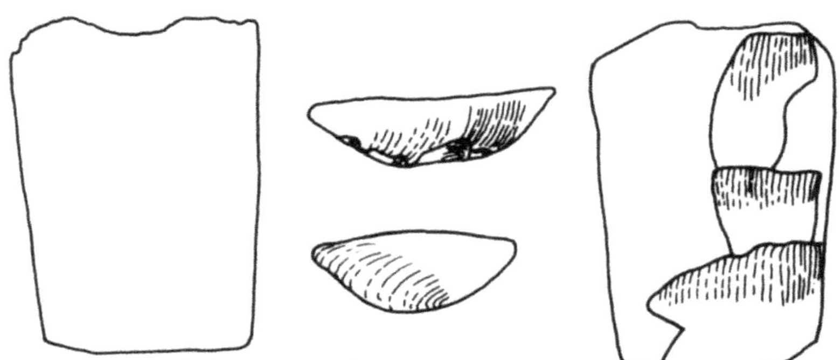

1:1

Histogram no7

COMPARISON OF AVERAGE LENGTH OF THE FIVE GOUGE-TYPES

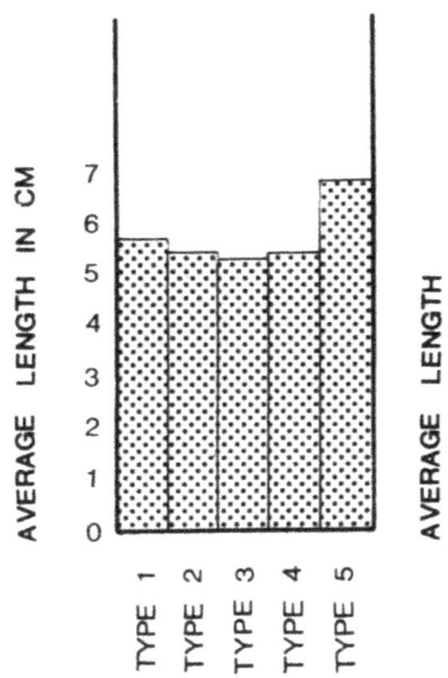

COMPARISON OF AVERAGE THICKNESS AND BREADTH AT THE PROXIMAL AND DISTAL ENDS OF THE FIVE GOUGE-TYPES

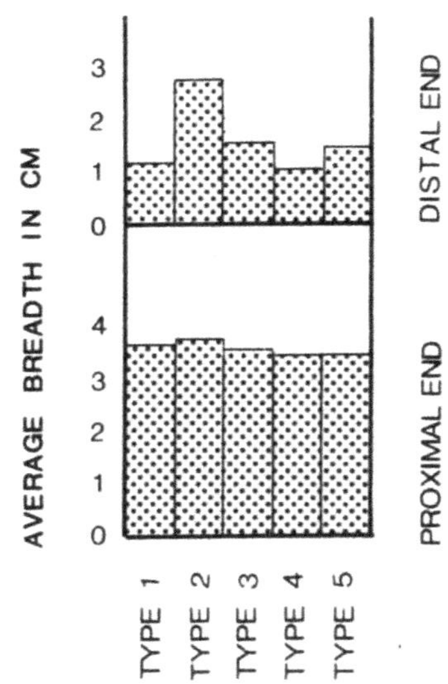

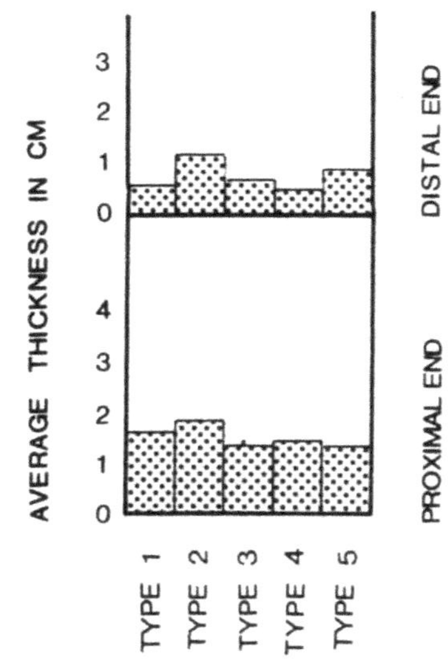

3) The thickness of the gouges varies mostly between 1 and 2 cm at the distal end (excluding the flaked edge) and one or less at the proximal end (cf. fig. no.36 and histogram no.7).

One interpretation of the function of the gouges suggests that they were probably also used as hoes for tilling the soil (Haaland, R., 1981: 121-122; 1987a 76-77, 221). The earliest evidence of the gouges goes back to c. 6000 B.P. when, according to our assumption, the cultivation of wild plants was first practised. In a previous study (Magid, A., 1982: 38), I suggested that since there is no evidence from the area of any tools which can be interpreted as being used as hoes, Haaland's interpretation can be considered as starting point for further study because these gouges occurred during the period (c. 6000-5000 B.P.) when cultivation of wild cereals is assumed to have started. The objectives of that study, e.g. to examine the possibilities and efficiency of gouges as tilling tools of the soil, should emphasize the following (ibid):

1) Whether these gouges were hafted and in what manner they were hafted?
2) What wear pattern would be expected on the gouges if they were used as hoes for tilling the soil?

The gouges must have been hafted if they were used as hoes. This assumption is based on the fact that the length of the gouges (histogram no.4) is too short to provide both hand grip and a working end long enough for tilling the soil. Furthermore, these gouges would have had to be mounted as a prolongation of a short part of an inflected or curved handle (cf. fig. no.43) because a perpendicular attachment to the handle would produce a short working-end which would be inefficient in use (ibid). Analysis of the wear pattern on the gouges is indispensable for a more comprehensive interpretation of the function of these tools. In addition, the only way to reconstruct the manner of hafting is to examine the wear pattern and to determine the function. It is generally known that there is a high degree of wear on digging tools, 'almost as much as being subjected to abrasive agents' (Semenov, S.A., 1970: 129). The characteristic traces of wear pattern would cover a good part of the surface of the lithic tool, and its edge would be very blunted due to its blows in the soil (ibid: 133). In addition, it is expected that at least the working part of the tool would be polished on both sides, and all the hollows and facets on its surface (which were not removed by earlier deliberate grinding) would be rubbed out and disappear 'because practically almost every point on the tool surface would be subjected to friction against the soil' (ibid). Microscopic traces of the wear pattern consist of irregular striations which run in all directions (ibid). In the light of this information, a study of the general shape of the gouges and microscopic examination of their wear pattern (both by conventional lighted stereo-microscope and SEM) were conducted and yielded the following results:

Fig.no.43 Illustration of the suggested method of hafting and use of gouges[(1)]

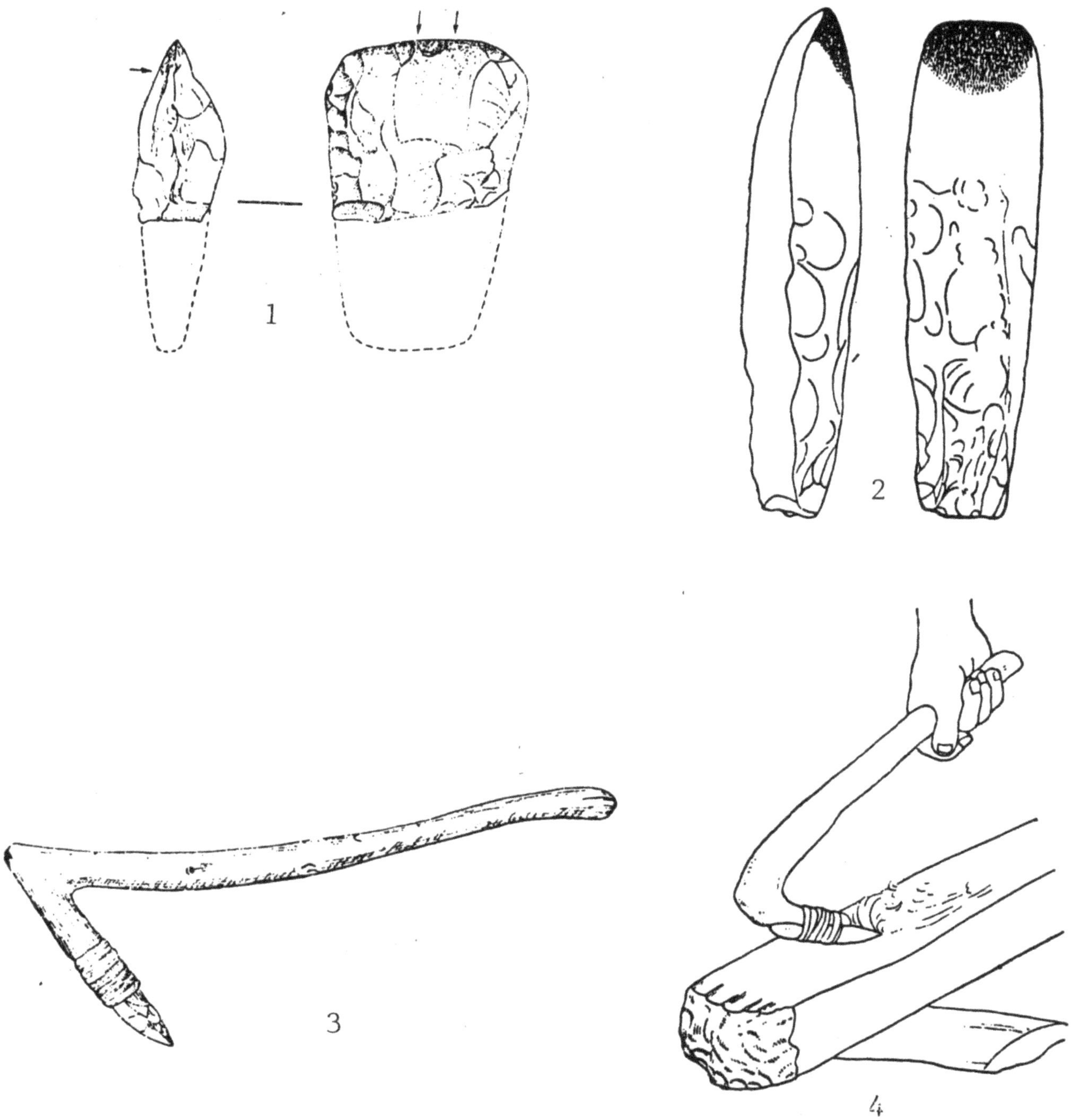

1 the hafted part of the gouge (dotted line).
1&2 the working edge(pointed with arrows in no. 1 and shaded in no.2).
3 inflected handle and method of hafting
4 hafted gouge at work.

After Semenov,S.A.,1964: 127,135(with modified presentation).

1) As regards the general shape
   a) The gouges which are flaked only (type I) and those which are flaked on one (upper) side (type III) are obviously not used to till the soil; otherwise they would have been polished and/or partly polished (on the working part) of both sides.
   b) The working edge (blades) of most of the gouges did not show any indications of bluntness; in fact the majority of the specimens studied were characterized by relatively sharp edges. Furthermore, many of the gouges (including some of the polished and partly polished ones on both sides) had an obliquely flaked working blade without any traces of polish. If they had been used to till the soil, they would have been highly polished.
2) The wear patterns on the gouges which are polished or partly polished on both faces (types II and V) do not show any indications that they were used as hoes. Irregular intersecting wear striations should be in evidence if they had been used to till the soil (Semenov, S.A., 1970: 133). The striations on these gouges are defined in a pattern of lines parallel to the axis of the tools (cf.pl.no.11).

The results obtained indicate that it is unlikely that the gouges in the studied collection were used for tilling the soil. In addition, these results show that the combined use of gouges as hoes and adzes is impossible. This is because the degree of abrasion of the working edge resulting from the tilling of the soil would have been so great that the tool could not have been restored for use as an adze merely by sharpening (ibid: 129). Hence it is most likely that the recovered gouges from the Khartoum area were not used in any cultivation activities. There is another consideration which does not seem to have been accounted for namely, the nature of the soil in the cultivable land in the Khartoum area. That is to say, the alluvial plains (cf. fig. no.5) where the crops, e.g. sorghum, were cultivated, can be tilled by, e.g. merely a wooden hoe (cf. fig.44a and b) without needing to use lithic ones. In addition, short, peg-like wooden tools or *seluka* (cf. fig. no.44c and d) could also have been used for sowing. This suggestion is based mainly on the fact that wooden tools are still used in traditional rainfed and irrigated cultivation activities in the Sudan. For instance wooden hoes (cf. fig. no.44a and b) as well as other wooden tools are used in rainfed cultivation in the Nuba mountains in west Sudan (cf. fig. no.45). The *seluka* (cf. fig. no.44c) is one among many wooden tools (cf. fig. no.46a and b) still used in the traditional irrigated cultivation practices along the Central Nile Valley in the Sudan. Arkell interpreted the gouges as carpentry tools used for 'hollowing tree trunks to make dug out canoes' (Arkell, A.J., 1953: 31). These canoes were assumed to have been an essential means for hunting water animals, collecting water molluscs, fishing and perhaps transporting lithic raw-materials which are not available in the area of the sites, e.g. transporting rhyolite from the sixth cataract (Sabaloka) area (ibid). This interpretation indicates that gouges were used in a manner similar to the use

Plate no.11   An example of wear pattern on gouges

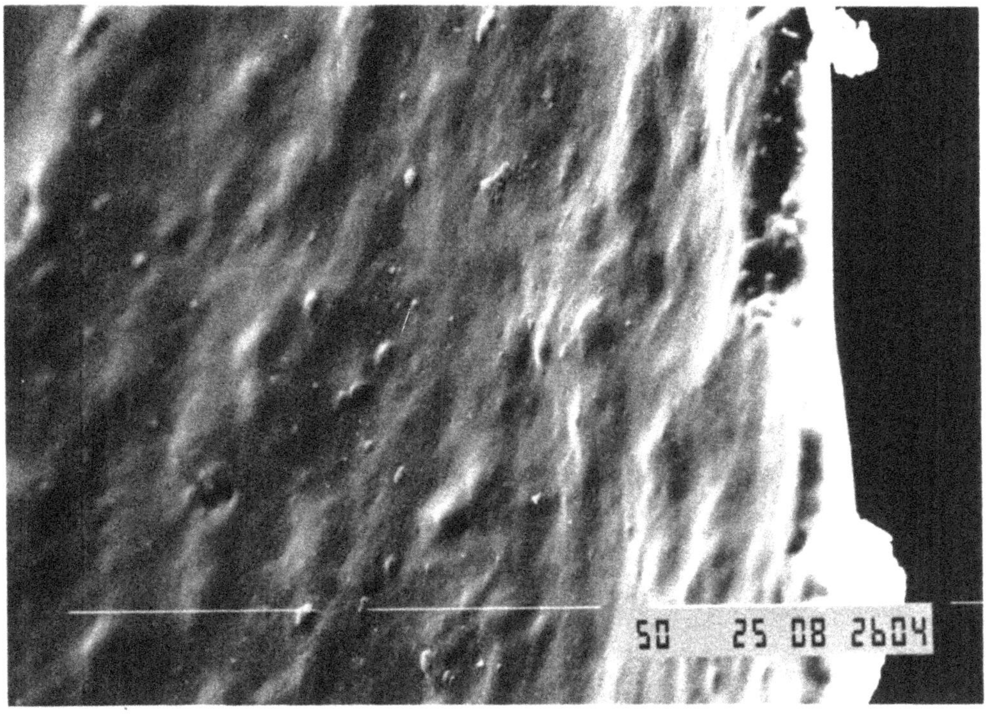

Fig.no.44 Cultivation wooden tools

[a,b,hoes
c,d,sowing tools]

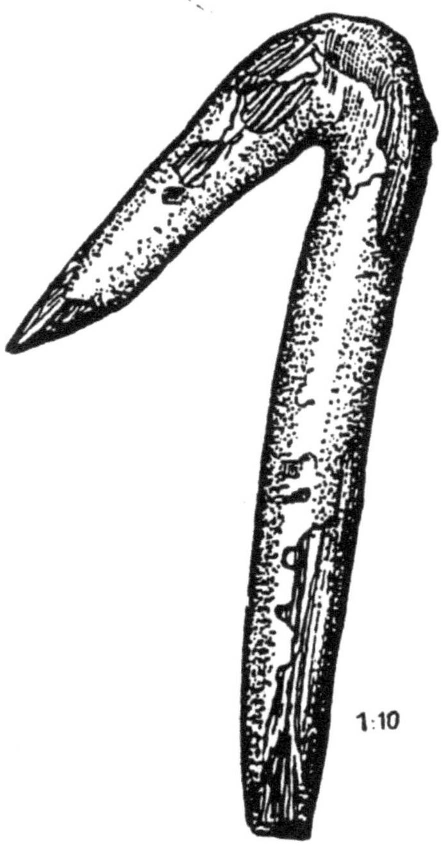

1:10

a

1:10

c

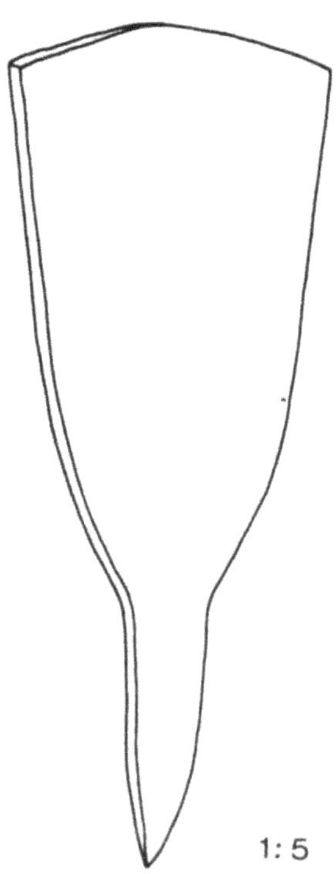

1:5

b

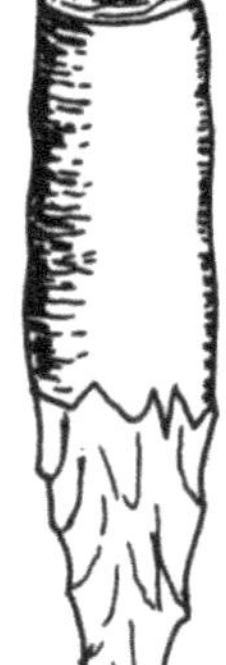

1:10

d

Fig.no.45 Cultivation wooden tools

[a,b,weeding sticks
c,d,e,threshing sticks]

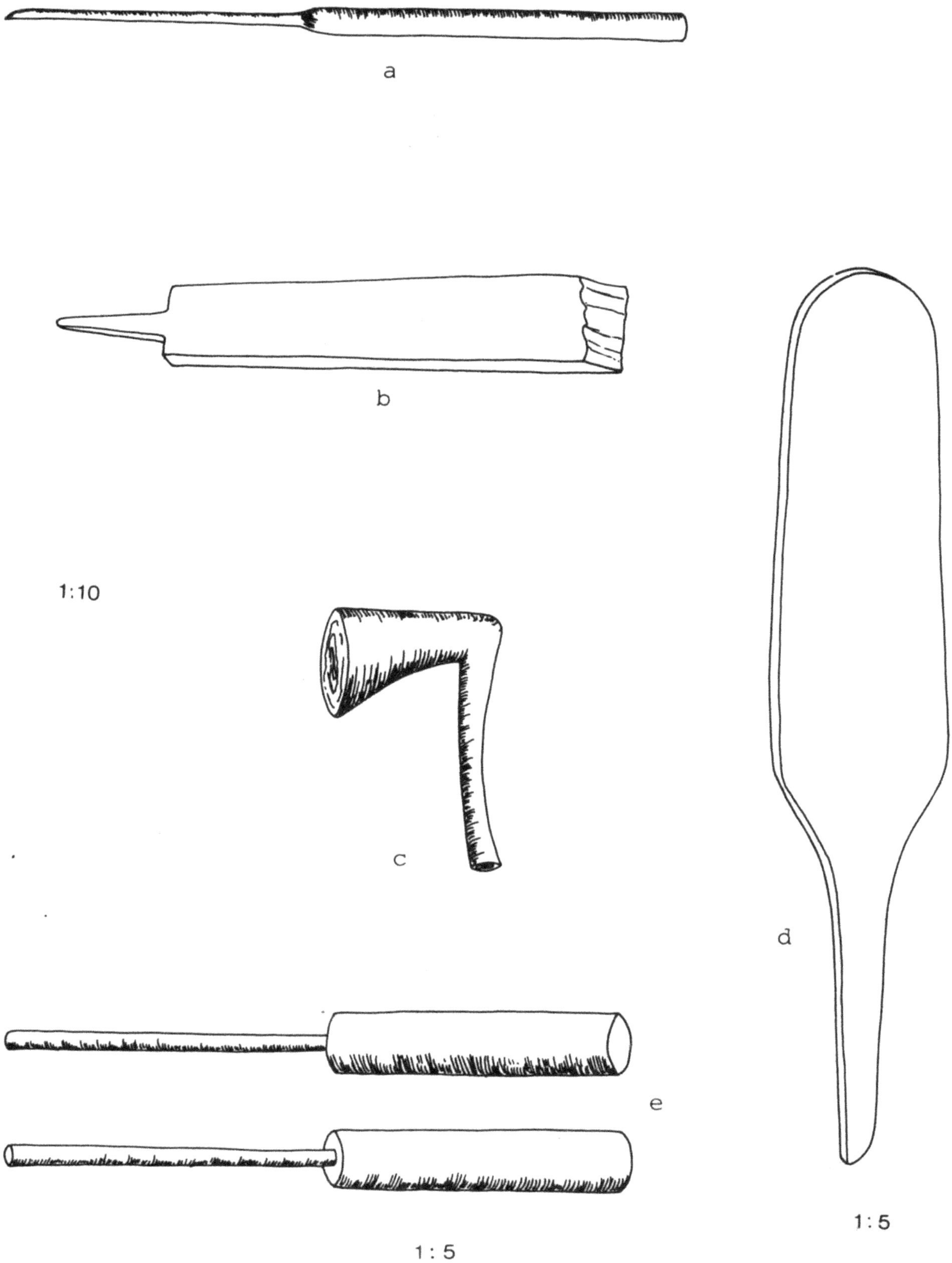

Fig.no.46-A Cultivation wooden tools: rakes

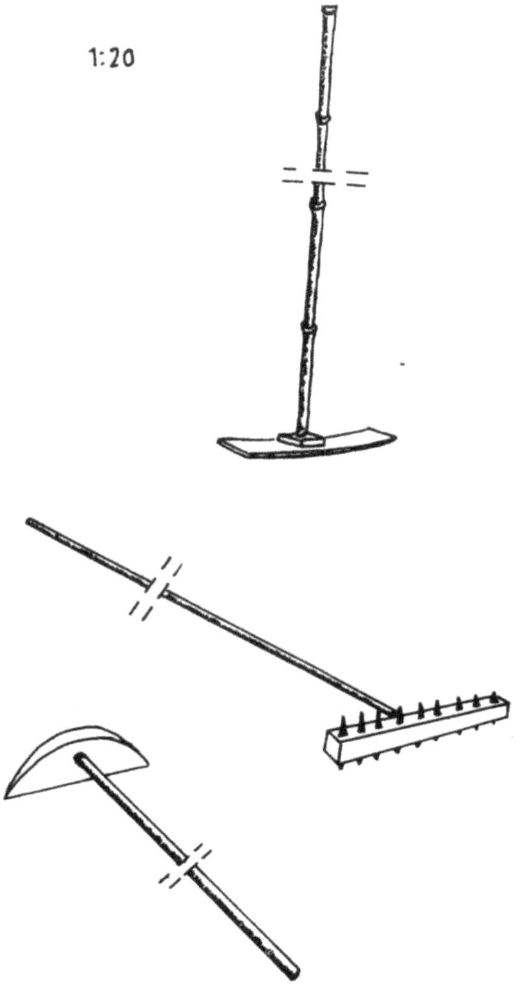

Fig.no.46-B Earth - Levelling - Board (Ar.wasuq)

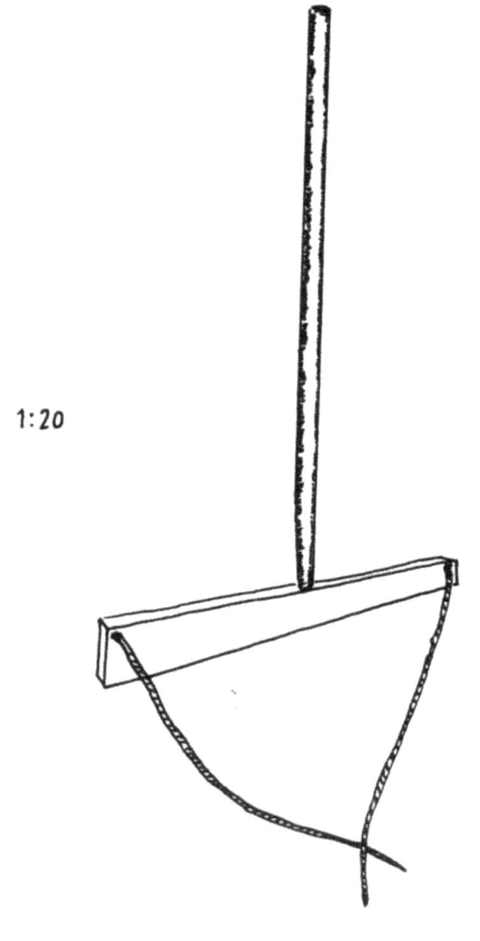

of adzes in the Sudan at present (ibid). Accordingly, the gouges must have been hafted to a handle. In order to avoid the resistance of the wood and trunks, which can be very high if the tool falls at a right (90°) angle (Semenov, S.A., 1970: 131), the gouges must have been mounted at an angle (< 90°) to the handle (Arkell, A.J., 1953: 31) as suggested in fig. no.43. Polishing is considered as one of the basic criteria for lithic tools which were used on wood-working because it reduces the resistance of the tool and facilitates its penetration into the wood (Semenov, S.A., 1970: 68). This means only flaked tools only have high resistance and hence are inefficient as wood-working tools (ibid). It is important that part or all of the lower face of the gouge is polished because it is the side which would meet most, if not all, of the resistance from the material (wood) struck; the upper face does not necessarily need to have been polished (ibid). But polishing hard stone, e.g. rhyolite, by grinding is a very tiring process which requires knowledge and demands persistence and time (ibid). However, as mentioned in the foregoing, it is essential to polish at least the working part "blade" of the tool in order to function properly (Arkell, A.J., 1953: 31). Rubbers made from river pebbles, petrified wood and sandstone are believed to have been used for polishing the tools (Caton-Thompson, G., and E.W. Gardener, 1934: 27; Semenov, S.A., 1970: 69). Taking into consideration the sample studied and all the gouges which were recovered from the sites of the Khartoum area, it is certainly evident that the majority of them (c. 88%) are polished to a certain degree. Although most of these are partly polished (cf. histograms from 4 to 6; appendix no.4) it was consistently observed that evidence of polishing is always present on the part of the surface closer to the cutting 'blade' edge (distal end) of the gouges. This phenomenon corresponds with Semenov's interpretation of polishing of the working part (Semenov, S.A., 1970: 68). Thus, those gouges which are only shaped and flaked, e.g. type I, are unlikely to have been used for wood-working.

The wear pattern on the lithic carpenting tools appears under magnification as parallel groves which are relatively thick and dense at the working edge but narrowing at their upper margins (on the surface of the tool). These groves are perpendicular to the working edge (Semenov, S.A., 1970: 129-131). The SEM examination of some gouges in our collection shows the same pattern as expected in lithic tools used as wood-working adzes (cf. pl. no.11).

If these gouges were used for wood-working, they must have been resharpened (and probably polished) whenever they become dull. The result of this repetitive process would be the reduction of their length (cf. fig. no.47) and also the breadth of their working end until they would have been too short to be hafted, e.g. less than 6 cm, and too narrow to function properly, e.g. less than 3 cm (cf. histograms no. 4 and 5). Thus, polishing, wear pattern and the variably reduced sized of the gouges all indicate that they

Fig.no.47 Resharpening and size reduction of gouges

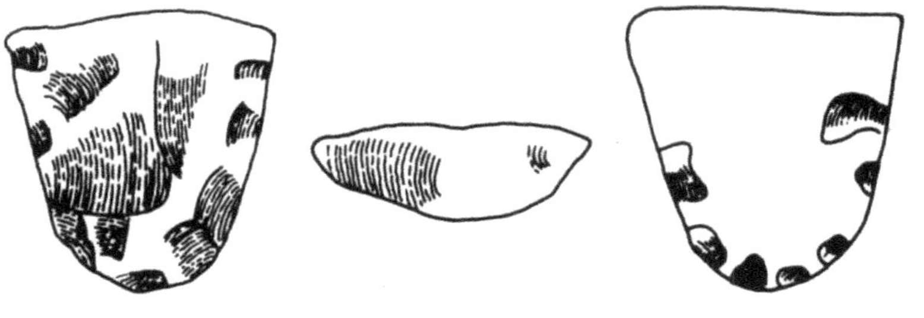

1:1

were used as a carpentry tool, e.g. an adze. As I mentioned in the foregoing, the few broken gouges are different from the others in that either the proximal end or infrequently the distal end or both are missing (fig. no.42). In the light of this new information, my previous assumption (Magid, A., 1982: 37) as well as Haaland's (Haaland, R., 1981: 121-122) that the gouges recovered from the Khartoum area were broken is now shown to be invalid.

In short, the results of the study of the gouges agree with the interpretation that gouges were used as wood-working tools, e.g. adzes, and falsifies the assumption that they were used (as a hoe) for tilling the soil. The question as to why the geographical distribution of these gouges is restricted to the Khartoum area in the period between 6000 and 5000 years B.P., remains to be answered by future research.

5.4 Summary.

The archaeological remains recovered from the Central Sudan during the period being studied here exhibit evidence of artefacts which can generally be associated with the exploitation of food-plants. The evidence recovered is seen as being related to the processing and preparation of food-plants, but no evidence was found of any tool which can be associated or directly related to the cultivation activities of these food-plants, e.g. land preparation, sowing, harvesting, etc. The only tools which might be indirectly related to the cultivation activities are the sandstone rubbers which are believed to have been used for shaping and polishing wooden and bone artefacts. Therefore, in the absence of any tools or artefacts which can be associated with or related to cultivation activities, I will argue that wooden tools were most likely the main tools which were used in these activities. Such wooden tools probably ranged from forked branches and wooden staffs which were picked up at the site of activity, e.g. the cultivation field, to proper and specialized tools such as those wooden tools (cf. figs. from 44 to 46) used in the traditional cultivation practices in the past and the present Sudan.

CHAPTER VI

Ethnographic and Ethnobotanic Study on Gathering of Food-Plants: Case study W.Sudan

The focus of this chapter will be on ethnographic and ethnobotanic observations primarily related to the gathering of food plants in the western Sudan (North Darfur) (cf. fig. no.20). However, relevant examples from other parts of the world will also be discussed. I shall try through these observations to establish a general relationship between the ecological variables (of the area) and the adaptive responses of the people. The implications of these observations will be used in the interpretation of the archaeological remains.

6.1 Gathering Wild Food-Plants: "General Remarks".

Generally, gathering of food-plants is one type of subsistence and economic activity performed by people in their quest for food of vegetative origin. Foraging and collecting are the two main inseparable forms of gathering. The former denotes the consumption of the gathered food-plants on the spot as they are gathered; the latter is conventionally used in this context to denote the gathering of food-plants to be consumed later on, mainly at the settlement area. As we shall discuss later on, collecting and bringing home food-plants could be for three reasons:

1) Division of labour necessitates that the food gathered be brought back to the settlement to be shared by the members of e.g. the family.
2) Spatial reasons: the food gathered may not be available in the immediate vicinity of the settlement area, and gatherers may have to walk long distances to collect it.
3) Temporal reasons: Since some food-plants (especially the preferable ones) are not available the year round, or during drought periods, gatherers have to bring home food-plants and store them for future use. This is, however, a late development in the history of gathering (Sahlins, M., 1972: 17).

Gatherers always seem to forage while collecting. Thus these activities are simultaneously and inseparably performed. It is, therefore, rather difficult to distinguish, in practice, between foraging and collecting, though, theoretically one can distinguish between them on the basis of the time and place of the consumption of the food-plants gathered.

Both forms share two ways of gathering the edible parts of the plants (e.g. fruits, bulbs, etc.). These are: Picking and Digging-out. The main criteria for distinguishing between these types are:

(a) The position of the edible part on the plant; whether it is on the aerial part of the plant or on the sub-surface part.

(b) The energy in-put and labour.

(c) The time required.

In addition, the technique and equipment used are also included, but as a secondary criterion.

1. Picking: denotes removal of edible seeds, fruits, nuts, leaves and branches or brushing and/or bringing together those which have fallen on the ground.

The energy and labour devoted to this type of gathering is not great. In addition, the time spent on this activity is fairly short; several ethnographic and historic accounts of different gatherers' communities show that an average length of time per person per day spent on food-gathering and preparation is between four and five hours (Sahlins, M., 1972: 11). "Moreover they do not work continuously", and the food quest is highly intermittent (ibid). For instance among the !Kung "a woman gathers on one day enough food to feed her family for three days" (Lee, R., 1968: 37). Thus although food-collecting in this case was the primary productive activity "the majority of the people's time ..... is spent in other pursuits" (Lee, R., 1969: 74). But sometimes extra time is devoted during the dry season when the food resources are scarce and long distances have to be walked in search of them; this is particularly true when the dry season advances after an unsuccessful rainy period. Picking may be performed by one of two means:

(a) Without tools (e.g. hands) when the edible parts of the plant are within the reach, e.g. from lower parts of the plant, grasses, and from the ground.

(b) With tools such as sharp-edged tools, long hooked poles e.g. made of wood, or forked tools, branches and brooms. Tools with a sharp working edge are used to detach edible parts which are difficult to pick by hand (even if they are within reach), e.g. juicy branches, fruits, etc. The hooked poles are used to bring down the edible parts of the plants which are otherwise difficult to pick by hand. Forked tools, branches and brooms are normally used to sweep grains and seeds from the ground.

2. Digging-out: which is the act of unearthing the edible root-vegetables[1]. Digging-out root-vegetables is an intensive, time-consuming and labour-demanding activity as compared with picking. It also requires hard physical effort to dig out root-vegetables

---

[1] Root-vegetable is used in this context and afterwards to designate all the edible underground parts of the plants (tubers, bulbs, roots, underground stems), otherwise specifications will be given.

especially if the top-soil and sub-soil are compact and the root-vegetables are deeply buried. Most often extra time is normally wasted looking for traces and marks of the difficult discern aerial parts which indicate root-vegetable particularly during the late part of the dry season. In this connection, we should remember that most of the root-vegetable plants have a cycle of storing the nutrients in the roots during the dry season to feed on them. This ecological adaptation of plants, thus, serves in providing food-plants (e.g. root-vegetable origin) for people during a period when other sources of food-plants are scarce.

The practice of digging-out root-vegetables has usually been performed using some sort of tools. Such tools range from simply a branch or a wooden rod which can be found right at the site to very specialized ones. The digging stick is the most traditional tool and is distributed world-wide . It has to be long and thick enough to enable a good grasp by both hands (e.g. most often 3 metres long and 3 cm in diameter) and has a sharp slanting plane working end, which is sharpened whenever it becomes dull.

Most often picking and digging-out are simultaneously performed during any trip in quest of food-plants. The amount of food-plants gathered differs from one season to another. For instance, large quantities of food-plants are gathered without difficulty in a fairly short time after the rainy season as compared with the small quantities gathered over a fairly long time and with harder work required towards the end of the dry season.

6.2 Gathering Food-Plants in West Sudan (Darfur area).

The following ethnographic and ethnobotanic account about gathering food-plants will focus mainly on the Darfur region[2] (both northern and southern provinces). This region is situated in the far western part of the Sudan (cf. fig. no.20). The reasons for selecting Darfur in particular as an area for collecting my ethnographic and ethnobotanical data can be summarized in the following points:

1) Gathering of food-plants has a long tradition as a supplementary source of food and has remained unchanged for a long period of time.
2) This area can be considered as one of the extreme habitats in terms of the marked fluctuations of food resources from one season to another and shows the adaptability of the inhabitants to these fluctuations. Therefore, I believe this area can be used as a base-line for the comparison of adaptation strategies to contrasting environmental conditions and fluctuating resources.
3) North of this region is the desert zone where there are hardly any floral or faunal resources, south of it is the savanna woodland and grass zone where hunting is more

[2] Region here denotes the present political map of the area within the boundaries of the Sudan.

dominant as compared with the infrequent gathering of food-plants. Further east of this region is the riverain environment where gathering is only a leisure-time activity practised mainly by children. It remains, however, curious that practices of gathering in most parts of the Sudan are at present performed by children. A thorough study of this phenomenon might show this to be a remnant from the past of an old practice of gathering which was, then, mainly performed by adults. Subsequently, it gains a wider significance and implications when looking at it from an archaeoethnobotanical point of view.

4) This region is one of the areas which were badly affected by drought during the period between 1973-1985, and gathering was, then, almost the only means of subsistence and survival, particularly in the northern part of the region.

On the basis of the annual rainfall and the vegetation cover, the area of my study can be divided into two main zones: the semi-desert zone with an average annual rainfall of 150 mm situated at the northern part of the area and as far south as an arbitrary line east west crossing El-Fashir city (cf. fig. no.20), and the Acacia scrubs and short grass zone with an average of annual rainfall of 350 mm in the south. The desert type of sand-soil with mobile and immobile sand dunes (Ar. *qoz*) characterizes the composition of the soil in the northern part of the area (from El-Fashir and northwards). The alluvial cracking clay type of soil predominates in the southern part of this area (south of Kabkabyia and from Zalinji and southwards).

There are a very few *Acacia* sp. and other tree and scrub species that grow very sparsely on the extended sandy plateau in the north, however, the main vegetation cover grows in certain productive areas. These areas are confined to nearly all the seasonal streams and the low lying sandy depressions where different plant species grow in patches, clumps and groves. Naturally, these are the main sites where the gathering of food-plants is practised.

Edible seeds, grains, fruits, leaves, branches and root-vegetables of various grasses, perennials, herbs, shrubs and trees are gathered according to their availability throughout the year as a supplementary source of food. They also constitute the only source of food during difficult periods towards the end of the dry season and before the main cereal crops (sorghum and millet) are ripe or when there is a failure or insufficient harvest of the cereals cultivated. The recent drought in the region (specially in the northern part) has proven this for it was mainly the gathering of wild food-plants that made up for the food shortage.

Knowledge of the landscape and the localities within which these edible wild plants grow (specially those with edible root-vegetables), and of the nutritive value and distinctive quality(ies) of each of these is accumulated through the experience that each individual accumulates from childhood. More details on this will be presented in the following pages. Tables no.33 to 37 show an example of some of the plant species which are exploited as food plants in Darfur (Sudan) as well as some others which were also used for other purposes. The local names are included in these tables to illustrate that the inhabitants not only give names to different species but also that their knowledge and use is revealed by giving different names to different edible parts of the same plant species.

As regards the means used for gathering, they range from gathering by hand to the use of simple tools, e.g. sticks, branches and stones found at the site of activity, and of more specialized tools, e.g. digging sticks, knives, reaping hooked staffs, etc. Baskets made mainly from the roots of *Acacia senegal* (L.) Willd. (Ar. *hashab*) and the leaves of *Hyphaene thebacia* (L.) Mart. (Ar. *dom*) and pottery vessels of different sizes are the basic containers which are used while gathering.

No time is spent making, repairing or maintaining the simple gathering tools used; for they are always replaced by new ones picked up right on the site where the gathering is performed. Thus, this type of tool is always abandoned on the site after finishing the job.

On the other hand, the specialized tools are those which people deliberately spend time on by choosing the suitable raw materials and on making them. Therefore, they are always kept and resharpened (e.g. knives, digging-sticks) or repaired (e.g. hooked staffs) and continue in use until too worn out or impractical (e.g. too short) to use; they are then discarded and new ones made.

Digging-sticks similar to those used in West Sudan are reported to be in use among many other communities, for instance, the Tuareq (Nicolaisen, J., 1963: 174), the !Kung of Nyae Nyae in the Kalahari desert (Marshall, L., 1976: 98-100), the Hadza in Tanzania (Woodburn, J., 1970: 41-42), and the Aborigines of the West Australian Desert (Gould, R.A., 1980: x-xii). Moreover, the same type of reaping knife which has been used in Darfur until recently , is also reported from other areas such as the Nuba mountains in southern Kordofan (Sudan) and among the Kofyar of the Jos plateau in northern Nigeria (Netting, R., McC., 1968: pls.no. IVB and VA). Each of the above mentioned tools is used for specific function(s) in the different gathering activities as shown on table no.38.

Gathering grains of wild grasses, particularly those of *Dactyloctenium aegytium* (L.) Beauv. (Ar. *koreib*), and, infrequently, those of *Echinochloa colonum* (L.) Link (Ar. *difra*) and *Panicum turgidum* Forsk. (Ar.*tumam*) is a task which is very similar in procedure to harvesting domesticated cereal crops. Since the stands of these wild species are sometimes several kilometres away from the settlement areas, the women set up

Tables no. 33-37. Darfur (West Sudan): Some of the wild plant species which are common both as food-plants and for other purposes[1].

Table no. 33. Food-plants[2] frequently gathered.

Scientific name	Family	Local name	Period of availability	Gathering technique & preparation	Other uses
Dactyloctenium aegyptiacum (L.) Beauv.	*Gramineae*	umm asabe muohoyia	The grains are ripe a few weeks after the first heavy rains[3]	Grains are gathered or picked into baskets and/or swept from the ground. Preparation includes grinding and cooking the flour to produce porridge served with milk, stew or water and salt (and spices if desired).	The stalks and leaves are good animal fodder.
Echinochloa colonum (L.) Link.	*Gramineae*	difera	" "	Grains are gathered in a similar manner to those of *Dactyloctenium* sp. Preparation: includes threshing, winnowing, grinding and cooking the flour to make sweet porridge (similar to rice in taste) eaten with milk.	"
Panicum turgidum Forsk.	*Gramineae*	tumam	" "	Grains are gathered and prepared in a similar manner to those of *Echinochloa* sp.	"

1 This list is not a complete ethnobotany. It is based upon plants I collected. Identification of vernacular names was made by the local people. Dr. Iklas A-Bari assisted identifying the collection in the Herbarium, Dept. of Botany, U. of K.

2 Majority are wild grasses (of the *Gramineae* family). *Dactyloctenium* sp. and *Echinochloa* sp. are considered the best substitutes for millet and sorghum.

3 Two well spaced heavy rains (e.g. a week or two in between) are sufficient to bring these to maturity, two harvests can be gathered.

Table no.33 (cont.):

Scientific name	Family	Local name	Period of availability	Gathering technique & preparation	Other uses
Hibiscus esculentus L.	*Malvaceae*	bamia (fresh fruits) weika (dry ones)	July-November/ December The mature dry fruits can be stored a few years.	Both the green immature and the mature dry fruits of this perennial are gathered by picking them from the plant by hand. Preparation: The immature green fruits are normally cut into small pieces and cooked in boiling water, with salt (and meat if available) to make thick greyish stew eaten with thin baked bread[4] (Ar. *kisra*) or porridge[5]. The mature dry fruits are pounded and sieved. The fine flour extracted is cooked in boiling water, with salt (and meat if available) to make a greyish-brown, thick stew eaten with thin, baked bread or porridge. The green leaves of *Hibiscus* are sometimes mixed with those of *Corchorus olitorius* L., and cooked in water to make a stew eaten with thin bread or porridge.	
Thespesia garcheana (F.) Hofm.	*Malavaceae*	jakhjakh	Fruits are available most of the dry season and can be stored for a few years	The fruits of this scrub are picked by hand or using tools; the green mature ones are kept for some time to dry. Preparation: The fruits must be dry before any preparation starts. The petals are peeled from the fruits and pounded (often sieved). The fine flour extracted is cooked in boiling water	Frequently used as fire-wood

4 The baked bread is normally made from sorghum.

5 Porridge is made from both millet and sorghum, but millet porridge is the most common. At times of shortage of millet and sorghum, the baked bread and porridge are made from wild cereals.

Table no.33 (cont.):

Scientific name	Family	Local name	Period of availability	Gathering technique & preparation	Other uses
Thespesia sp.				with salt (and meat if available) to make a greyish, thick stew, eaten with thin baked bread[6] or porridge[7]. The taste of this stew is very much like that of the stew from the dry mature fruits of *Hibiscus* sp. (Ar. *weika*); therefore, it is mostly eaten as a substitute of *Hibiscus*.	
Corchorus olitorius L.	*Tiliaceae*	mulukhyia/ khudra	June-October	The green leaves and, infrequently, the dry ones of this annual are picked by hand or the whole stalk (stem) is uprooted and the leaves picked later. Preparation: The leaves (green or dry) are washed, then they are put in boiling water with salt, (spices if desired) and meat, if available, and is left to cook until it becomes a thick stew. It is usual to eat this stew with thin baked bread[8] or porridge[9].	All animals in the area graze on this plant.

6 See comments on preparation of *Hibiscus* sp.
7 idem
8 idem
9 idem

Table no. 34. Fruits[1] frequently gathered.

Scientific name	Family	Local name	Period of availability	Gathering technique & preparation	Other uses
Tamarindus indica L.	*Caealpiniaceae*	aradeib	Fruits are ripe in November/December. Leaves are ground round the year.	The ripe fruits of this tree are picked (from the tree or from the ground) and eaten. They can also be soaked in water for several hours and the juice extracted mixed with sugar and taken as a refresher. They can also be boiled in water and the cyrup extracted mixed with some millet-flour and cooked to make a sweet porridge (Ar. *madeida*).	Medicinal: The fruits are boiled in water and the extracted product is drunk to cure malaryia. Camels browse on the leaves. Handles of sickles, hoes, etc. are sometimes made from this tree sp.
Capparis decidua (Forsk.) Edgew.	*Capparidaceae*	humbog (fruits) tundub (scrub)	November-March/April	A scrub that bears pinkish-red berries which are picked from it and/or from the ground and eaten. Its fruits are never stored for lack of preservation conditions and equipment.	Its thorny branches are often used for fencing purposes
Capparis tomentosa Lam.	" "	murdo	November-April/May	A scrub whose orange-red berries are picked and eaten. Like the fruits of *C. decidua*, its fruits are unpreservable	Goats and camels browse on its leaves.

[1] All the fruits mentioned here are eaten either fresh or, infrequently, prepared as a proper dish, except for the berries of the *Capparidaceae* which have to be eaten (fresh) directly after gathering.

Table no.34: (cont.):

Scientific name	Family	Local name	Period of availability	Gathering technique & preparation	Other uses
Ziziphus spina-christi (L.) Willd.	*Rhamnaceae*	nabag (fruits) sidir (plant)	Fruit from November-April/May	A bush whose dark-yellow and pale-red drupaceous fruits are gathered by picking (sometimes from the plant but mostly from the ground). Hands, throwing stones and hooked staff are the means of picking. The fleshy kernels of the fruit (exocarp & mesocarp) are readily eaten fresh and/or dry, but they can also be prepared. Preparation: The fresh and dry fruits are cooked to produce dry, bread-like loaves (ar. *gurrasat-nabag*). The preparation consists of removing the fleshy parts of both the dry and fresh juicy fruits, mixing them and then shaping them (normally round flat pieces). Then they are placed in a pit dug in the ground. The floor and walls of the pit are matted with leaves of *Calotropis procera* Ait. (Ar. *ushar*). More leaves are used to cover the contents of the pit, which is then sealed with soil and topped with fire-wood and kept burning for a day. Then the pit is opened to collect the well cooked *Ziziphus* loaves. These loaves keep well for 10 years. Used at times of food scarcity and frequently eaten by herders during their long distance grazing trips. A piece is broken from the loaf when needed and eaten or ground, mixed with milk and served	Goats and camels frequently browse on the leaves and pods.

Table no.34: (cont.):

Scientific name	Family	Local name	Period of availability	Gathering technique & preparation	Other uses
Ziziphus sp.				as a mush drink or cooked with water or milk to make a porridge-like dish. The stones of the fruit (endocarp) are well boiled in water; and after their juice is extracted, they are removed. The extract is kept boiling until it is concentrated into a thick, honey-tasting, sweet syrup (Ar. *asal nabag*). Finally, the seeds are extracted from the stones, mixed with the syrup (honey) and served as a favoured dish.	
Grewia tenax (Forsk.) Fiori.	*Tiliaceae*	guddeim	October-January/February	A scrub whose yellow fruits (? berries) are picked from the plant, mostly by hand, and eaten; but they can also be prepared in the form of sweet mush. Preparation: This is done by soaking the fruits in water for a day. Then the mixture is boiled. After a moderately thick liquid is formed, the mixture is sieved; and the extract is mixed with a very small amount of millet-flour, some water is added, and the mixture is well stirred. When this is boiled again for a short period of time, a sweet mush will be produced.	Goats and camels browse on the leaves.

Table no. 35. Food-plants gathered due to food-shortage.

Scientific name	Family	Local name	Period of availability	Gathering technique & preparation	Other uses
Boscia senegalensis (Perr.) Lam. ex Poir.	*Capparidaceae*	mokheit	Fruiting period: November. Seeds may last until April/ May; but if they are gathered, they can be stored for quite a long period.	A scrub whose bitter-tasting seeds are gathered by picking (either from the plant or from the ground). Preparation: Before starting any preparations the seeds must be dry, so they will be left to become so. Then they are soaked in water in a pottery container for three days. The water is regularly changed whenever its colour becomes yellow until it is finally clear and the bitter taste of the seeds is completely removed. Then they are either boiled in water with some salt until they cook (Ar. *baleilla*) and can be eaten; or they will be left to dry, and then pounded or ground into flour, and cooked with water to make porridge. A local beer (Ar. *mareisa*) is also made by fermenting the seeds.	The spiny fruits are unpleasant for animals to eat but at times of fodder shortage, this plant species is eaten by goats and, infrequently, by camels.
Cenchrus biflorus Roxb.	*Gramineae*	haskaneit	Fruits ripe few weeks after the first heavy showers and can last for several years.	Fruits are gathered (by beating them into baskets or sweeping them up with, e.g., a branch from the ground). Preparation: The fruits with their thin needle-like hairs are ground and threshed, then sieved or soaked in water to remove the spines and the	

Table no.35: (cont.):

Scientific name	Family	Local name	Period of availability	Gathering technique & preparation	Other uses
Cenchrus sp.				woody husks. The seeds extracted are ground, and their flour is mixed with water and cooked to make porridge.	
Balanites aegyptiaca (L.)	*Simaroubaceae*	laloub (fruit)	Fruits are ripe in November/December. Dry fruits last for years. The leaves are green round the year.	The fruits of this tree are picked from it (by a hooked staff, by throwing stones and/or climbing and hand-picking) or from the ground. Preparation: The exocarp (skin) is removed and the flesh (mesocarp) is readily eaten. The endocarp (stones) is pounded into flour and cooked or mixed with millet or sorghum flour to make porridge. Cooking oil is also extracted from the endocarps by pounding them and heating them in a cooking-pot, the oil produced is separated from the residue by sieving. The leaves are gathered by hand-picking or with hooked staff. The food preparation from these include rinsing the leaves and cooking them in water after adding salt and sodium-bicarbonate (Ar. *atroon*) until they are properly cooked into stew. The stew can be eaten together with porridge or thin baked bread.	Medicinal: The extracted oil is also used to cure "cool" fever and head-aches by rubbing on the body. The meso-carp (fleshy part of the fruit) is a good medicine for stomach-aches.

Table no.35: (cont.):

Scientific name	Family	Local name	Period of availability	Gathering technique & preparation	Other uses
Tribulus terrestris L.	*Zygophyllaceae*	dereisa	Fruits ripen a few weeks after the first two heavy rains and can last for a few years.	The spiny fruits of this herb are picked either by beating them with a staff or a branch into a basket or by sweeping them from the ground using a broom, branches, etc. Preparation: The seeds are extracted by grinding the spiny husks and removing the residue by sieving and/or soaking in water. The seeds are then ground into flour and mixed with water and cooked into porridge which is eaten with milk or water.	During very severe periods of fodder-shortage, goats browse on the leaves.

Table no. 36 Root-vegetables gathered at times of food-shortage.

Scientific name	Family	Local name	Period of availability	Gathering technique & preparation	Other uses
Stylochiton kerensis N.E.Br.	*Araceae*	mogrod	The aerial part dries and weathers after the rainy season, but the tubers are available during most of the dry season.	The tubers of this herb are bitter to the taste and poisonous, yet they are dug out and eaten at times of food-shortage and famines. Any sizable branch or stick can be used to dig them out from sandy soils, but a more specialized tool (digging stick) is necessary when they are dug out from compact soils. Preparation: To get rid of the bitter poisonous contents the tubers are boiled in water (the local name for this process is *tamseikh*). Then they are pounded and the paste is spread out to dry. After that, it is stored; and when needed, some of it is mixed with water, baked as bread and eaten with water, stew or milk.	
Maerua sp.	*Capparidaceae*	kurdaleh	Available most of the year.	Sparsely growing in the area. The elongated corms of this species are gathered by digging them out in the same manner as *Stylochiton* sp. Preparation: The corms are left to dry, pounded and baked into a thin bread (Ar. *kisra*) and eaten with stew or milk.	

Table no. 37. Food flavours and fermentation-catalysts.

Scientific name	Family	Local name	Period of availability	Gathering technique & preparation	Other uses
Cassia tora L.	*Caesalpiniaceae*	kawal	From October to February/March	An annual underscrub, the leaves of which are picked by hand, prepared and used as flavours for different kinds of stews and porridges. Preparation: The preparation of the leaves consists of ferminting, drying and pounding or grinding. They are fermented as follows: Rinse the leaves, put them in a pot whose walls and floor are matted with leaves of *Calotropis procera* Ait. Add water after putting leaves of *Cassia* into the pot. Place the pot in a pit dug into the ground (the diametre and depth should match those of the pot). Seal the pot and bury it for several days. Then unearth it, sieve out the mash, and remove the leaves of *Cassia* and spread to dry. Then pound or grind into a coarse black powder. Add some of this powder to the stew and/or porridge to give the food special flavour when desired.	
Cymbopogon nervatus (Hochst.) Chiov.	*Gramineae*	mahareib	The dry culms are ready for use a month or two after the rainy season and last until the following rainy season. If stored, they can be kept for a few years.	The dry culms of this annual herb are uprooted. The green leaf-sheaths are spread to dry. Preparation: The leaf-sheaths are always rinsed before use. They are then added to tea to give it special flavour.	Medicinal: Cure for stomach-aches; chew the leaves and swallow the the juice or soak them in hot water (for short time) and drink

Table no.37: (cont.):

Scientific name	Family	Local name	Period of availability	Gathering technique & preparation	Other uses
Cymbopogon sp.					the liquid extracted. The juice extracted from the leaves (whether mixed with tea or as pure liquid) is also used to help digesting heavy food, particularly meat.
Solanum dubium Fresen.	*Solanaceae*	gubbein	The fruits are ready for use from October and remain available until the middle of the dry season.	The yellow berries of this woody herb are picked by hand and used as fermenting catalyst to snake sour milk. Preparation: The fruits are added to the milk while it is cooking. When the milk is finally cooked and removed from the fire, it gradually turns into sour milk. The exocarp (skin) and endocarp (seeds) of the fruits are separated by sieving before the milk gets sour.	

Table no. 38. An example of tools used in gathering food-plants.

Type of tool	Use of tools	Labour versus time required
Simple staff	1. Digging out shallowly buried vegetable roots in loose/sandy soil.	1. Always light labour since the soil is loose, but when root-vegetables are sparsely distributed or rarely available a long time is required to find them.
	2. Breaking off dry stalks and ears of grass-species.	2. Since stalks and ears are dry, little effort is put in. Time required depends on the density of the stands.
Branches	1. Beating off into baskets the ripe seeds and grains specially those small tiny ones, i.e. *Panicum*	1. Effort and time devoted depends on how dense and/or dry the harvest is.
	2. Serve as a broom to gather harvest of seeds fallen off.	2. Generally it is a light job except when dealing with small seeds and grains.
Digging sticks	1. Digging out root-vegetables.	1. Both the labour and time required depend on the compactness or looseness of the soil correlated with how deep in the ground the root-vegetable are and their density and distribution in the countryside and their distance from the settlement area.
Knives	1. Cutting off ears of seeds and grains.	1. Light job both in labour and time provided abundance of moderately dry ears.
Reaping hooked-staffs	1. Picking out of reach fruits, leaves, etc. either individually or collectively by hooking and shaking the fruits/leaves bearing branches.	1. Light effort but a long time is needed if resources are meagre.
Baskets & pottery	1. Collecting the harvest either by beating it directly into them or sweeping it from the ground.	1. Light job but less effort and energy are needed to carry around the empty and/or filled baskets compared with the relatively heavy pottery containers.

.temporary camps near the fields and stay for several weeks or a month (Tubiana, M.-J. and Tubiana, J., 1977: 14). It should be mentioned in this context that it is not only the harvesting that women perform, but also the threshing, winnowing and packing of the harvest. In this way they avoid the bother of daily travel between their villages and the fields and at the same time devote all their energy and time to their demanding task of reapening the grain ears within a specific period[3]. The harvest is brought to the village and stored in huge clay vessels (Fig. no.26) known locally as *dabanga* [4]. As among the Tuareq (Nicolaisen, J., 1963: 171), the gathering of wild cereals in Darfur in particular and the western Sudan in general is most important only in situations of rain or crop failure, but if their cultivated cereals,

> 'sufficed or there were good possibilities of obtaining enough, women would spare themselves the trouble of going off to harvest wild cereals' (Tubiana, M.-J., and Tubiana, J., 1977: 15).

Among some of the contemporary hunter/gatherer communities, the women are also saved the trouble of their gathering activities when hunting (which is exclusively a man's task) provides most of the food (Gould, R.A., 1980: 64; Marshall, L., 1976: 133, 140).

The stage of food-plant processing and preparation includes all the steps after the plants are gathered until the time they are served. Processing and preparing these food-plants depend to a large extent on which part of the plant is gathered, e.g. seeds, fruits, or root-vegetables, and also on the state of these edible parts, e.g. whether they are hard or soft, fresh or dry, bitter or sweet. Tables nos.33 to 37 show some of the food-plants which are eaten soon, after gathering them (at the site of activity or at the settlement area) without any processing or preparation. But some of these require a chain of processes before they are finally served. Some of these processes involve long, tedious and strenuous physical work and mental attention, some, necessary equipment (containers, grinders, etc.) and perhaps catalysts (for fermentation, taste, etc.) in some preparatory treatment.

The following example how the inhabitants of Darfur (northern part) prepare the seeds of *Boscia senegalensis* (Pers.) Lam. (Ar.*mokheit*) is perhaps a good demonstration of the above mentioned processing of food-plants which requires tedious and strenuous physical work and a long time for preparation. The preparation of the ripe seeds of *B.senegalensis* (Pers.) Lam. takes several days. After gathering the seeds from the trees, they are threshed and winnowed to get rid of the husks. Then the seeds are soaked in

---

[3] If the grain-ears are not harvested before becoming very dry, they shatter and fall to the ground, and so the task of picking them becomes more difficult and time consuming.

[4] Similar clay vessels (Herzog, R., 1957: Pls. no. 5 and 26) were used in the late 1960s for storing wheat and sorghum in e.g. Argo, Aniba, etc. north Sudan.

water to get rid of their very bitter taste. Soaking is repetitively done for three days in a burnt clay pot (Ar. *zeer*) filled with water. The size of the pot depends on the amount of seeds to be soaked, but it is normally a big one so as to soak as many seeds as possible and thus reduce the number of times of going through this long preparation process. Whenever the colour of the soil becomes yellowish, it has to be replaced by fresh soil until the water is finally clear indicating that the substance that gives the bitter taste is completely washed out and so the process of soaking comes to an end. The water has to be changed at least five times the first day and an average of three times the two following days. My informants from Kabkabyia and Aqud (cf. fig. no.20) emphasized that it is not only the unpleasant and bitter taste that they want to get rid of, but by proper soaking they avoid risks of getting stomach irritations, diarrhoea and being poisoned. Thus, it is always the case that experienced women will attend to (or supervise) this process. After the soaking stage the seeds are further prepared in one of the following three ways:

1) The seeds are cooked in boiling water then mixed with milk or some salt is added and then served.
2) The seeds are laid on a mat (made of *dom* tree-leaves) to dry. Stone grinders are used to turn them into flour. Finally, the flour is mixed with water and some salt and cooked to make a greenish porridge (Ar. *asidat mokheit*) which is eaten with either milk or stew.
3) The seeds are fermented to make the local beer called *mariessa* .

Another instance is the preparation of the spiny fruits of *Tribulus terrestris* (Ar. *dereisa*) and *Cenchrus biflorus* Roxb. (Ar. *haskanit*). The preparation of these starts with carefully threshing them and then soaking them in water so the spines now detached from the fruits float and can easily be removed. Both will then be laid on a mat to dry. As regards the *Tribulus* sp., when it is properly dry, it is subjected to another threshing and then sieved or winnowed to get rid of the husks. The next to last step is grinding the seeds into flour. Finally, the flour is mixed with water and cooked into porridge (Ar. *asida*).

If we extend our discussion beyond the descriptive observations, we shall acknowledge the following:

1) The inhabitants of Darfur (West Sudan) have a thorough knowledge of their local environments, the food-plant resources, their reproductive and life cycles, their properties and qualities and their nutritive values. The accumulation of such knowledge is gained by observation and participation in gathering activities that starts in childhood. When the adults (mainly women) repeatedly go on trips to gather food-plants they take their young children (both boys and girls) with them. These can observe their parents finding the localities where the plants grow. They can also observe what kind of plant species their parents look for, how to gather them, which

parts of these plants are gathered and the means and ways of gathering them. This has always been the background of how children (mainly girls) eventually start to participate in gathering with the adults until they are experienced enough to do it alone (Sharif, A. Harir: pers.comm.). Moreover, this is also the background with which the boys join (from the age of seven and onward) the adult male herders. It is an important background that help them to forage on the right food-plants when they are herding (ibid.).

2) Although the present day inhabitants of Darfur know the techniques of gathering, what is most important is their knowledge about what to gather and how to prepare it. This points to a long history of experimentation that may go back thousands of years before such knowledge was well established. "Hit and miss" ways of learning were the most likely manner to acquire knowledge about the properties of all the food-plants available in their environment and particularly about those which would otherwise be considered unpalatable such as the spiny or the poisonous species. My informants explained to me how irritating to the stomach and discomforting in digestion it is to eat *Tribulus* or *Cenchrus* porridge if the spines of the fruits are not completely removed and what the consequences are of eating improperly soaked seeds of *Boscia* sp. as I previously mentioned.

These are only two examples among many (cf. tables no.33 to 37) which, to a large extent, confirm my argument that the present knowledge of people about their food-plants must have been acquired through long, and probably hard, experience. For instance, it is likely that several cases of stomach infections, poisons and probably deaths took place before people realized the cause(s) and thus achieved a better understanding of the properties of their food-plants and learned the proper way to prepare them. Such cases of infections and poisonings were also the most likely reasons behind people's knowledge about medicinal plants (cf. tables no.33 to 37). This wide knowledge must have passed through many generations of different experiences before it became as we see it today.

As regards the division of labour and the social organization related to gathering activities, I observed that gathering wild food-plants for family consumption is the responsibility of women who are always helped by girls of an age from six years and upwards but infrequently helped by boys between five and eight years of age. Herders (always males) practise foraging and consume what they find while tending their herds in the vicinity of their base areas. Since they forage, they are normally selective in the sense that they only pick or dig out food-plants which are readily eaten without any preparation, e.g. fruit. But I was also told that when they go on longer herding trips (for weeks or months), they make use of their childhood knowledge when they collect food-plants that they have to prepare and cook before eating (Mahamoud H. Sharif, 1985: pers.comm.). The interpretation of this type of labour division is largely associated with the nature of adaptation of these people to their environment; the men are herders who are sometimes absent from the settlements, at times, for months before they are back. Women stay behind to rear the children and to tend to other household activities. Moreover, the cultural attitudes (e.g. superstition) and biological differences between males and females[5] seem to be some of the basic factors in determining what type of socio-economic activities an individual should do in the community.

A similar type of social organization and division of labour has been reported from many gatherer/hunter communities in different parts of the world, e.g., the Sandawe of Tanzania (Newman, J.L., 1970: 35), the !Kung in Kalahari desert (Marshall, L., 1976: 96-97), and the Aborigines of the West Australian Desert (Gould, R.A., 1980: 62, 67).

The adaptive strategy and the socio-economic organization of the inhabitants of north and south Darfur are also largely structured according to how resources fluctuate from one season to another. As regards the theme of this discussion, gathering of food-plants, its adaptive and socio-economic aspects can be summed up in the following:

1) Although cultivated millet provides most of their need for food-cereals, gathering of food-plants - particularly grains of wild grasses - still constitutes an important source of food, particularly in the northern part of Darfur. The practice of gathering, food-plants has - to a large extent - declined both economically and techno-culturally, but it is still a safe-guard[6] which can be resumed (as it was in the past decade) whenever food shortage occurs due to inadequate yields of cultivated millet, drought, etc. Except for the gathering activity itself (picking, digging-out wild food-plants), the subsequent stages (storing and food preparation) are similar to those pertaining to cultivated plants.
2) The knowledge that the Darfurians (at least the present older generation and some of the younger generation) have maintained about their landscape, local environment and wild food-plants is evident in the following:
   a) Their knowledge and memorization of the optimum localities and the types of soils where food-plants grow and the correlation of this information with occurrences of rain, water and the ripening time of the different species of food-plants in these localities.
   b) Their flexible adaptive strategy made it possible to cope with the extreme fluctuations in food resources from one season and/or period to another (rainy season versus dry season) and from one year and/or couple of years to others (if we take into consideration, e.g., the fluctuation from wet to dry in the last two decades). This flexibility and adaptability is reflected, for instance, in the special

[5] Asking about the division of labour and why only women and not men gather food, an old informant simply said to me "a woman is a female". Another more explicit one told me "as a woman that is part of her duty".

[6] Wild food-plants are fast maturing and tolerant to drought.

attention they give to storing equipment and facilities (large clay vessels, Ar. *dabanga*) and to granaries dug out of the ground (Ar. *matmourah*). Together with these, harvesting as much grain from wild grasses and root-vegetables as possible is for them a safe-guard when the dry season advances, shortly before harvesting the cultivated millet and sorghum in case these do not yield enough or when there are successive rainless years.

c) Their knowledge about the different edible plant-species and their reproduction cycle is well manipulated to exploit them throughout the year. Thus, there is always a wild food-plant of some kind to exploit during the different seasons, particularly during years of sufficient rains.

3) Women have the full responsibility for gathering of wild food-plants to be consumed by members of their households. This can be both difficult and time consuming when the dry season advances, and during successive dry years, when the plant resources are shrinking to a minimum of availability. Women gatherers then have to walk longer distances picking fruits, grains, etc. and digging out root-vegetables. It is not quite uncommon (during the last few dry years) to see a woman carrying a baby on her back (maybe older children following her), a basket, a digging stick and reaping knife in her hand, and walking 2 to 4 kilometres (at times up to 6 km) to gathering localities, in not omitting the time she spends and the range she roves.

4) It was underlined above that gathering food-plants is practised intensively at times of food shortage due to inadequate yields of cultivated cereals or during periods of successive rainless years. It is very important to emphasize that during such times gathering activities turn into a well organized and regular exploitation of wild food-plants. Both the division of labour and the economic structure are oriented towards achieving this goal. The women camp for weeks or months in the localities of wild food-plants to gather as much harvest as possible, while the young men and boys will frequent them with supplies of water and food from the settlement area. When they gather a good stock, it is the job of males to load the harvest on camels and transport it to their settlement area. A re-orientation of economic structure resumes during years of adequate yields from the cultivated cereals. Thus the fluctuating food-resources and supplies are met with flexible adaptive strategies which can always be adjusted or readjusted to cope with them. I shall in the following chapters elaborate on the value of this type of ethnographic and ethnobotanic information for interpretation of our archaeological data.

CHAPTER VII

Traditional Land-Use and Cultivation in the Central Sudan

7.1 Introduction.

In the following discussion the main emphasis will be on the traditional land-use and the production systems of the rainfed and irrigated cultivation in the Central Sudan. It is not my intention, however, to discuss the concepts and criteria used to define the traditional land-use and traditional cultivation terminologies, yet it is important to point out that these terms are intended to describe specific practices of land-use and cultivation. These practises include the following:

1) Application of labour in an area for productive purposes (O'Keefe, P., and B. Winser, 1972: 2) and that

> 'the balance people strike, between the productive and non-productive activities, is thus related to their consumption needs and to the return on their labour' (Haaland, G., 1982: 9).

2) Exploitation of land and land resources and performance of cultivation practices by simple but efficient means and technologies but excluding all the mechanized ones. That is to say, the existing traditional practices for which one will have material and/or documentary evidence dating back from a few hundreds to thousands of years before present.

The results of our observations on these and their analysis are intended to provide the basis for a discussion later on. This discussion will be mainly on the possible prehistoric and past forms of land-use and cultivation practices.

The area under survey is situated within the arid and semi-arid zones between latitudes 12°N and 18°N and longitudes 32°E and 34°E. The mean annual rainfall gradually increases from 70 mm in the north to 350 mm in the south. The main physical features are the Nile and the Atbara rivers in the northern and central parts and the White Nile in the southern part of the area (cf. fig. no.1).

The productive land is small and the exploited part (mainly in cultivation and/or grazing) is even smaller compared with the vast size of the area. This is basically due to the localized availability of water both in space and time. Accordingly, the seasonal behaviour of the White Nile and the main Nile rivers and the Atbara river influences the land-use practices along their narrow strips of land. Rainfall (as a critical climatic element in arid and semi-arid zones) explains the presence of settlements and land-use activities in areas situated away from the rivers, such as Shaqadud area.

The economy is predominantly agricultural. Due to the diversities of annual rainfall and the differences in the macro and microenvironmental conditions, there is a variety of agricultural practices which differ moderately or drastically from one area to another. On this background, we shall now discuss the land-use and cultivation practices in our study area.

7.2. Traditional Land-Use.

There are generally two types of lands: rainfed lands and riverain lands. The nature of the source and provision for distinguishing these two types of lands. Figure number 48 generally illustrates the traditional land-use in the Central Sudan.

7.2.1 Rainfed land.

This land constitutes the larger part of the Central Sudan. The soils range from very poor immature soils in the north to relatively rich soils with organic matter and nutrient element in the south. The main sources of water are:

1) seasonal rains which are characterized by their irregularity both in time and space particularly in the northern part of our area of study.
2) man-made reservoirs for catchment of rain-water (Ar. *hafir*) provide another temporary source of water which may last for a few months after the end of the rainy season.
3) wells dug to obtain ground-water (Ar. *beer*) constituting the main source of water for people and animals during the dry season.

Until recently the fallow period ranged between six to ten years. This rotational system relates to Boserup's 'Bushfallow' cultivation (Boserup, E., 1974: 15). However, the recent pressure on resources and intensification of cultivation led to a noticeable reduction of this period of fallowing even to the extent that it is completely abandoned in some parts of the Central Sudan.

The lands allocated for rainfed cultivation both in the northern and central parts are confined to the following areas:

a) low-lying sandy and sandy-silt plains,
b) depressions and patches along the beds and banks of seasonal water courses (Ar. *khor*), and
c) head-waters and fans of streams.

Fig.no.48 (A) Simplified illustration of the traditional land-use in Central Sudan.

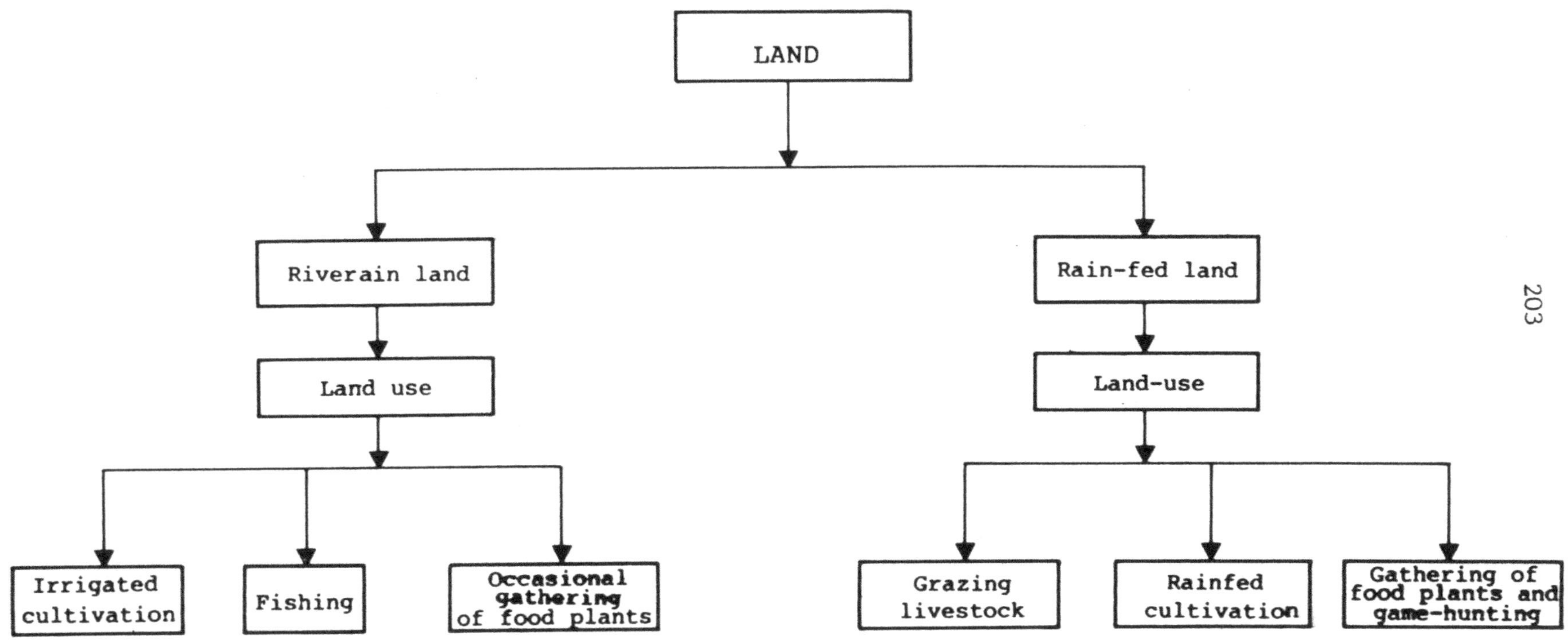

Determined by the topography, these cultivable plots are irregularly scattered all over the area. The southern part of the area - along the White Nile - is characterized, however, by having clayey plains and better water retention and hence it allows for larger cultivation-land than the northern and central parts. All rainfed cultivators in the Central Sudan share in common the tradition of not exploiting all the cultivable land. My informant in the Shaqadud and the White Nile areas explained this in terms of saving pasture for grazing during the dry season particularly during years of low rainfall. Since the main emphasis of the economy in the prehistoric Central Sudan since c. 6000 B.P. was on animal husbandry, it would also be expected that most of the land would be saved for grazing.

The present economy in the rainfed lands in the Central Sudan is based on pastoral-nomadism, transhumance as prevalent in the central part and semi-pastoralism as prevalent in the northern and southern parts. Rainfed cultivation is, thus, only a seasonal activity. Given the unreliability and irregularity of the rainy season, the yields from such cultivation can only contribute a supplementary source of food in pastoral economies.

In areas where there are semi-sedentary pastoralists e.g. Shaqadud area or where pastoral-nomads have access to rainfed cultivation, herds are allowed to graze on the crop-residue after its harvest. In such cases, there is a traditional linkage of exchange between agriculturists and pastoralists of the White Nile area. The latter camp for a few months in the vicinity of the cultivation areas specially during the dry season. The exchange includes items such as milk; its by-products and meat for sorghum and millet (McCown,R.L.,et al,1979: 299-300). But most important is the traditional ecological exchange related to land-use both by agriculturists and pastoralists. The linkage of exchange is reflected in,

> 'a kind of symbiotic relationship .... associated with crop residue grazing' (ibid: 300).

In exchange, the trampling and grazing of the pastoralists' herds partially help the agriculturists clearing the land at the same time the manure of the animals maintains the fertility of the soil. With regard to the food producers of the prehistoric Central Sudan, the main emphasis of their economy was on animal husbandry but they also practised rainfed cultivation of sorghum. Thus, it is not unlikely that they manured their cultivated plots in a similar manner to the present day semi-sedentary pastoral-nomads of the Shaqadud area.

7.2.2 <u>Riverain lands</u>.

These are alluvial cultivable lands situated parallel to or within the courses of the White Nile and the main river Nile and the Atbara river. As shown on fig. no.49 the

Fig.no.49 Subdivisions of riverain land:

1."gerif seluka"land 2.Saqia-land; A.high water level.B.low water level. C.river course

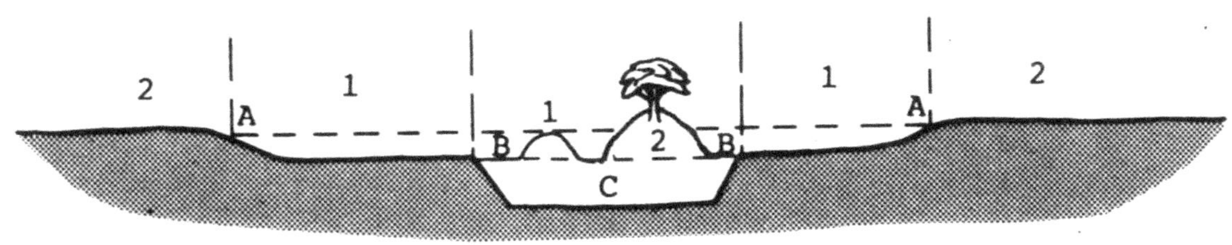

riverain lands are further subdivided to *gerif/seluka*[1] land and the *saqia* land. These different sub-types of riverain lands are distinguished primarily on the basis of the means of irrigation and secondarily on the basis of their location with reference to the courses of the major rivers; the White and River Nile and the Atbara river. Accordingly, the *gerif* land is irrigated by the annual flooding of these rivers. The *gerif* land(cf.fig.no.49) also includes those islands which are annually flooded and exploited in the same manner . on the other hand the *saqia* land is the land irrigated with water from the rivers or dug up well (Ar. *matara*) by means of *saqia* and/or *shadouf*[2]. This type of land is higher than the reach of the flood waters. It also includes those parts of the islands which usually remain dry round the year and exploited in the same manner as *saqia* land; only during exceptionally high flood waters, such as in 1946, may these islands be completely covered with water (cf. fig.49).

According to Boserup's classification of land-use (Boserup, E., 1974: 15-16) both the *gerif/seluka* and the *saqia* lands can be related to what she calls the "Annual Cropping" type. This is a rotational system of cultivation where the land is left fallow for several months between the harvest of one crop and the sowing of another (ibid). The peak of cultivation activities and thus the most intensive exploitation of both sub-types of the riverain land starts when the flood waters recede to the dry-season-course of the rivers (cf. fig. 49). Since the flood waters recede gradually from the *gerif/seluka* land, its exploitation goes in stages; when the water recedes the exposed land is sown. Its sandy-silt soil keeps sufficiently moist to be utilized throughout the period from September/ October to February, and sometimes later. The *saqia* land, being higher than the high water level, is characterized by predominantly cracking clayey soils. It is cultivated round the year with very short fallow periods between two to three months.

7.3 Traditional Cultivation.

There are two major types of traditional cultivation of food-plants (mainly cereals) in the Sudan at large and the Central Sudan in particular. These are rainfed cultivation and irrigated cultivation. The main criteria for classifying these types of cultivation are the sources and means by which the cultivated food-plants are watered. There are some general similarities but also regional differences between these types and within each of them with regard to the cultivation cycle, fallow systems, types of tools used and the manner in which these cultivation activities are performed. The similarities and

---

[1] The *seluka* is the Arabic name for a wooden tool similar to the digging stick. The flood-irrigated land *gerif* is frequently called the *seluka* land because it is the characteristic tool of cultivating this type of the riverain land.

[2] Detailed information on the *saqia* and the *shadouf* will be given in the following pages of this chapter.

differences between these activities would certainly widen our scope in the interpretation of the archaeological data.

7.3.1 Rainfed cultivation.

"Rainfed cultivation" refers to those activities or practices of crop cultivation which are essentially dependent on direct precipitation; the amount and distribution of rainfall are, thus, vital elements in deciding the success or failure of the cultivated crops. About 80% of the cultivated crops in the Sudan are produced in the rainbelt (Burnett, J.R., 1948: 289). It is extremely difficult to determine geographically the northernmost dry limits of rain cultivation due to the variability of rainfall and the broad range of fluctuation between the dry and wet phases. As for instance during good wet phases, the northernmost cultivation limit shifts about 100 kilometres towards the Sahara while this limit withdraws 200 kilometres southwards during the dry phases (Ibrahim, F.N., 1984: 115). Therefore the generally acknowledged dry limit of rainfed cultivation of eight months is not applicable in the Sahelian Sudan, including our study area. As a matter of fact the front-line of the dry limits goes beyond that to ten arid months and two semi-arid ones (ibid).

Due to rainfall fluctuating both in intensity and in spatial distribution combined with other factors such as overcultivation; it has been suggested that the actual limits of rainfed cultivation is 250 mm (in some areas it goes to 200-150 mm) and the ideal climatic dry limit at 1000 mm. Accordingly a "desertification hazard-limit" which empirically lies at 550 mm isohyets, was set as a boundary south of which the ecological system would not be disturbed if a fallow-cultivation cycle is observed (ibid: 116-117).

On the basis of the present annual rainfall isohyets in the Sudan (cf. fig. no.10), the northernmost dry limits of rainfed cultivation can be located approximately as follows (cf. fig. no.50):

1) The actual limit of rainfed cultivation (at 250 mm) is the line running southwesterly from Kassala to Wad-Medani; it continues through a point shortly south of El-Dueim extending straight north of El-Obeid to Um Keddada, then goes northwest to pass south of El-Fashir and finally falls in a southwesterly route to Tendelti.
2) The desertification hazard limit (at 550 mm) runs from a point between Gallabat and Gedarif southwesterly to pass south of El-Roseires and across the Renk to Dilling. Then it takes a northwesterly route with a curve-peak north of Jebel Marra and then falls in a southwesterly route that passes a short distance south of Zalinji.

Shifting cultivation, *hariq* cultivation and intensive cultivation are the three main traditional systems of rainfed cultivation in the Central Sudan. It is believed that the first two systems (shifting and *hariq*) have been practised in the Sudan for several millennia (Mubarak, M.O., et al, 1982: 143) probably as far back as the time when cultivation

Fig.no.50 The actual limits and the desertification hazard-limits of Rainfed Cultivation in the Sudan

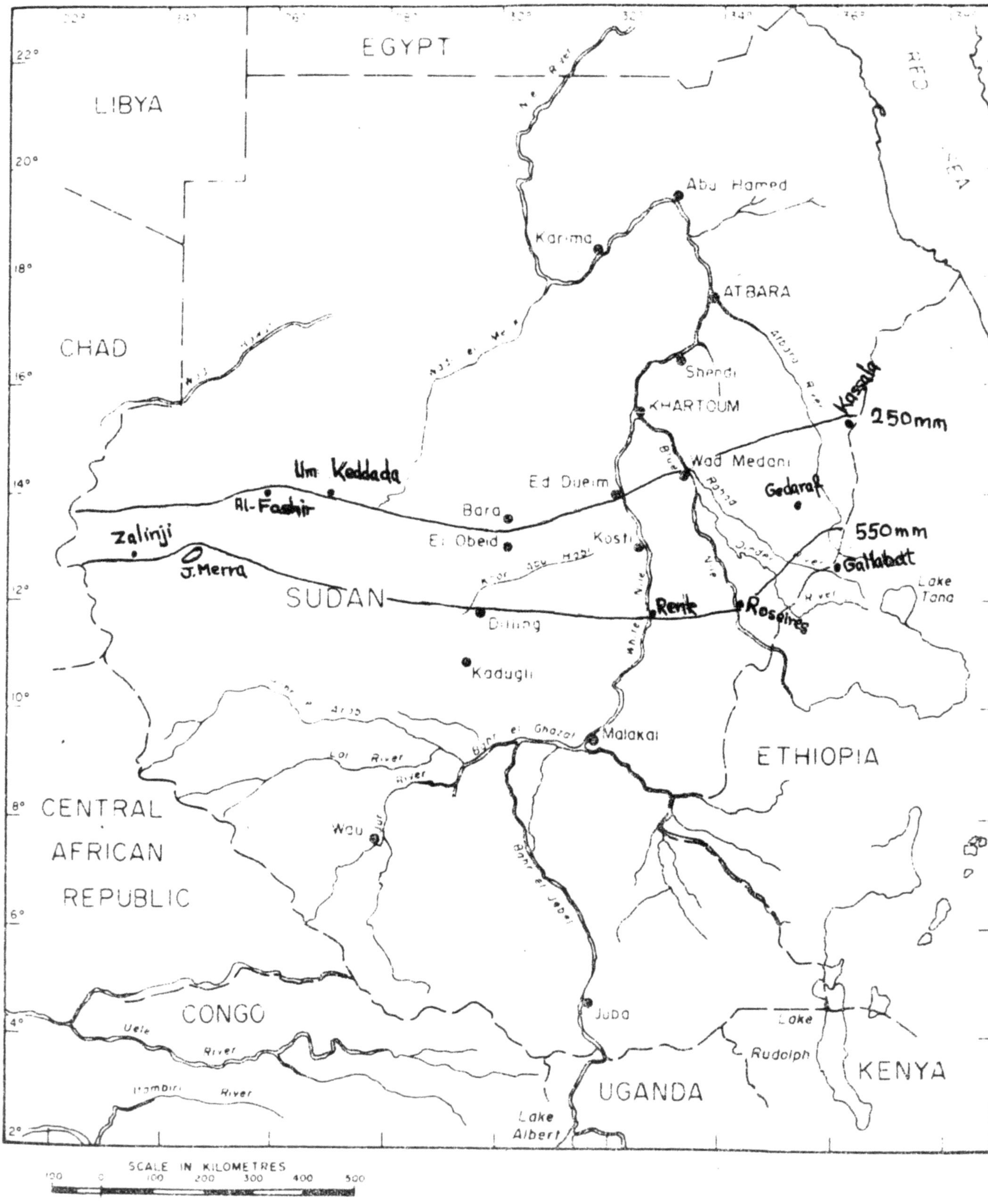

was first practised some 6000 years ago (cf. chapters VIII and IX). Since both systems are based on cultivation of a specific plot and then to abandon it temporarily or permanently and to move to another plot, these systems should essentially be practised in areas where land is far in access of any demand. The latter system (intensive cultivation) is, however, different and it does not seem to have been practised until perhaps the last few centuries.

7.3.1.1 Shifting cultivation.

In this system trees and bushes are cleared from the land which is brought under cultivation annually until the soil looses its fertility. Then the farmer moves to another area, mostly adjoining the old ones where the land is preferably untouched before (Burnett, J.R., 1948: 291; Mubarak, M.O., et al, 1982: 143). Meanwhile the abandoned fields are recolonized by grasses, herbs, scrubs and trees until fully regenerated by vegetation and thus becoming a potential for new fields (De Schlippe, P., 1956: 37). The average size of the cultivated land is 2.04 acres (Burnett, J.R., 1948: 291) and several plots may be cultivated during the same year(s) but under different crops.

Some studies have concluded that the farmer returns to the first plot he cleared after a period that varies from six to twelve years (ibid). But in practice, the shifting cultivators do not judge the value of the fallow period in terms of years or the duration of resting that the plot has undergone. They recognize the length of the fallow period and the state of the soil from the type of vegetation which covers the land. The shifting cultivator, thus, knows the stage of fertility and hence its usefulness for his crops (De Schlippe, P., 1956: 39). In this manner, there is a simple rotation system of crops with sufficiently long fallow periods.

Shifting cultivation has some advantages as well as some disadvantages. Its main advantages are as follows:

1) Despite cultivating the same plot successively to an exhausting point, it will be left fallow for a period long enough to regain its fertility and regenerate its vegetative cover (Burnett, J.R., 1948: 291).
2) The constant shifting of farms will indirectly impose control on diseases and pests (ibid) that can affect the crops.

The main disadvantage of shifting cultivation is that vast areas of cultivable lands are wasted; in most situations only one fifth of the cleared land is cultivated (ibid: 291-292).

7.3.1.2 The *hariq* cultivation.

The *hariq* cultivation[3] is another system of rainfed cultivation. It may well be called cultivation with the help of fire because fire is the basic tool in clearing a newly opened cultivable land. The *hariq* is a very suitable system in areas with rich clay plains, e.g. White Nile area where dense growth of strong and long grass exists. There may also be infrequent stands of woody bushes and trees which would be cut down during the dry season. When the first showers of the rainy season fall and a new mat of grass starts to grow under the old dry one, all are set on fire (ibid; Mubarak, M.O., et al, 1982: 143). Normally, the land is left fallow in some areas (e.g. Gedarif) for 2 to 4 years (Burnett, R.J., 1949: 292) and in other areas (e.g. Gezira and White Nile) it is left fallow for 6 to 12 years (Mubarak, M.O., et al, 1982: 143) as an ideal period to allow regeneration and accumulation of a mass of thick grass and bushes before these were burnt. But practically, when the *hariq* system of cultivation is applied on a plot which has been previously cropped, the time allowed for the land to regain its mat of grass is approximately proportional with the length of time it has been previously cropped (Burnett, J.R., 1948: 292). The *hariq* cultivation is at present practised in the southern parts of the White Nile and Upper Nile areas, southern parts of Kassala area (e.g. Gedarif) and some parts of Kordofan. The main crops of the *hariq* cultivation are sorghum which is the main crop in clayey soils and millet in the sandy soils.

There are three major merits of the *hariq* cultivation:

1) Setting fire to clear the land is a very light task if compared with the great man-power in-put required and the long time wasted in order to clear the cut down trees, bushes and remove these outside the cleared plot. Furthermore, this process of burning is equally advantageous in reducing the subsequent weeding.
2) The flames and heat produced from burning the dry bushes and grasses serve killing the new growth (Mubarak, M.O., et al, 1982: 143) and 'if the burning has been successful no subsequent hoeings should be necessary' (Burnett, J.R., 1948: 292).
3) The ash produced from the burnt bushes and grass serves as organic fertilizer that maintains and adds to the fertility of land and thus produces high yields.

In situations where woody bushes and trees have to be cut down, there are some subsidiary benefits such as the thorny bushes are used for fencing purposes, e.g. animal enclosures, farm fences, etc., and the branches and twigs are made use of as fire-wood and the big branches and timber are used for construction purposes, e.g. houses, huts, etc.

On the other hand, accidental, badly timed or unsupervised burning, particularly cumulative one, may lead to disastrous loss of land, destruction of the soil humus content, vegetation and hence result in food shortage (Burnett, J.R., 1948: 292; De

[3] It is different from slash-and-burn (Sweden) cultivation in the sense that *hariq* is often superfluous.

Schlippe, P., 1956: 199). There are, however, many traditional methods which are practised in the Central Sudan to control fire. The most efficient of these is to cultivate a considerably wide area, e.g. 100 metres directly round the *hariq* preserve (Burnett, J.R., 1948: 293). There are two other methods both based on the principle of fire-lines. One of these is the clearing of an area round the *hariq* preserve. The other is the burning of the grass surrounding the non-*hariq* forming by that fire-lines. The latter should be done before the grass of the *hariq* preserve has dried (ibid).

7.3.1.3 Intensive cultivation.

As I mentioned in the foregoing, intensive cultivation is in some aspect different from both the shifting and the *hariq* cultivation. The fundamental difference is that intensive cultivation is practised in areas where rainfed cultivable land is restricted and the same plot is successively cultivated year after year. Being so, it is not feasible to practise both shifting and *hariq* cultivation (ibid). Thus it is unlikely that intensive cultivation was practised in the Central Sudan during the period being studied in this work, c. 9000-1700 B.C. It was observed that the restricted areas of intensive rainfed cultivation are situated in the vicinity of densely populated areas and/or the areas which have been subject to an influx of population (ibid).

On the basis of this information and observations, it is feasible to suggest that the shifting and the *hariq* are the oldest systems of rainfed cultivation when the cultivable land was practically big and far in excess of any demand. The intensive rainfed cultivation emerged at a later stage when the increased number of population and their increased demand on the same exploited land imposed a new situation in which the use of land became more restricted. Eventually, the shifting and the *hariq* systems were replaced by a new one, intensive cultivation, in which the same plot is continuously cultivated.

7.3.2 Irrigated cultivation.

The available evidence of irrigation devices from the Sudan dates to the post-"neolithic" period; it is, however, most likely that the introduction of irrigated cultivation and irrigation technology co-existed with climatic deterioration[4], decrease in rainfall and increased population pressure on the available resources. These might also have been closely associated with the beginnings of the intensive and diversified cultivation practices. In addition, despite the fact that there is no material evidence so far that points to any connection between irrigated cultivation and the earliest evidence of domesticated sorghum, we have evidence that the peak of irrigation technology (water-wheel) was introduced during, a period, round the second or third century A.D. when cultivation of domesticated sorghum was already practised. Thus, irrigated cultivation did not have any

[4] The climate had started to become progressively drier 3000 B.P. (Wickens, G.E., 1982: 46).

role in the process of domestication of sorghum in the Central Sudan but perhaps enabled the inhabitants to practise cultivation the year round. I shall make only a brief survey on the present traditional practices of irrigated cultivation.

Traditional irrigated cultivation is practised mainly on the cultivable lands parallel to or situated within the courses of the main river systems and their tributaries in the Central Sudan (White Nile, Blue Nile, Atbara and the Nile rivers). There are also other cultivable lands which are situated far from the river systems but near sources of permanent or semi-permanent water, e.g. wells (Ar. *matara*), large water reservoirs (Ar. h*afirs*). Thus, these are artificially irrigated from such sources of water.

There are different methods and techniques which are used in the traditional irrigated cultivation. The main emphasis, in this paper, will be on the most common, widespread and relatively old methods and techniques in the Sudan generally and in the Central Sudan in particular. These are as follows:

1) Natural flooding.
2) Basin irrigation.
3) The counter-weighted lever (Ar. *shadouf*).
4) The water-wheel (Ar. *saqia*).

As I mentioned in the foregoing, these are all means of irrigation of cultivable lands associated with river systems, their tributaries, permanent and semi-permanent sources of water away from river systems. But some of these, namely, the natural flooding and basin irrigation are only seasonal means of irrigation, while the other two, *shadouf* and *saqia*, can be used throughout the year. There are also some merits as well as short-comings in each of these (or in comparison with each other) as regards their operation, reliability and efficiency as we shall discuss in the following pages.

7.3.2.1 <u>Natural flooding</u>.

Natural flooding is the simplest form of irrigation and most likely the oldest one. It is the common means to irrigate those parts of the riverain lands adjacent to the river courses in areas where the banks are not steep. During the high water season (July-September), the river courses expand and their water extend to flood these lands naturally (Ar. *gerif*). When the gradual recession of the high water level (flood) finally ceases and all the inundated land is exposed and sown, the well-soaked sub-soil keeps moist long enough to bring the sown plants to maturity. It is a method of cultivation which is very similar to the décure cultivation practised along the Niger river in Mali (Harlan, J.R., and J. Pasquereau, 1969: 70). Crops, vegetables and fruit-bearing creepers cultivated on the *gerif* land are normally of the fast maturing ones (cf. table no.39). The sowing of the seeds is the only activity before harvesting the yields. A wooden staff (Ar. *seluka*) with one pointed end (cf. fig. no.44) is used for sowing.

Table no. 39.

List of traditional means of irrigation and the main food-plants irrigated by each.

Means of irrigation	Archaeological evidence	Chronology	Cultural period[1]	Traditionally cultivated food-plants	Speculated main food-plants cultivated in the past in the Central Sudan
Natural flooding	No material evidence is found	Believed to be the oldest	(?) Pre-dynastic Egypt In Sudan (?) Shaheinab	Mainly the fast maturing plants such as: From crops: *Dolichos lablab* L. (Ar. *lubia*; few stands of *Pennisetum typhoideum* (Brum.) Stapf. & Hubbard and *Zea mays* L. (Ar. *aish-reef*). From vegetables: *Lycopersicum esculentum* Mill. (Ar. *tamatim*); mellons such as *Citrullus vulgaris* Schrad. (Ar. *battikh*); *Cucumis mela* L. (Ar. *shammam*) and *Cucumis dipsaceus* Erenb. & Spach. (Ar. *agur*). Marrows such as *Colocynthis pepo* (Ar. *qara*).	Cultivation of wild species of *Sorghum* and *Pennisetum*.
Basin irrigation	No evidence, only the basins (e.g. Kerma) believed to have been used for quite a long time.	Probably between 1500-1000 B.C.	The New Egyptian Kingdom. In Sudan; Kerma Culture.	Mainly crops: such as *Sorghum vulgare* (Ar. *durra*); *Pennisetum typhoidum* (Brum.) Stapf. & Hubbard; and *Phaseolus vulgaris* L. (Ar. *fasulyia*) in Shendi Basin[2].	*Sorghum* sp. and *Pennisetum* sp.

1 Adopted the distribution of the cultural phases in Trigger, B.G., 1965: 46.
2 Sources are Barbour, K.M., 1964: 143; Tracey, C.B. and J.W. Hewison, 1948: 743-744. Refer to the same sources as regards the cultivated crops in the Dongola Basin.

Table no. 39 (cont.):

Means of irrigation	Archaeological evidence	Chronology	Cultural period[3]	Traditionally cultivated food-plants	Speculated main food-plants cultivated in the past in the Central Sudan
The *shadouf*	No material evidence from the Sudan.	Earliest evidence in the Nile Valley comes from Egypt dating to 1500-1000 B.C.	The New Egyptian Kingdom. In Sudan: Kerma Culture.	Cultivation activities in the *shadouf* land in the Central Sudan are confined to vegetables and small orchards. Vegetables: *Lycopersicum esculentum* Mill.; *Solanum melogena* L. (Ar. *bazan-jan*); *Allium cepa* L. (Ar. *basal*). Fruit trees: small orchards of different *Citrus* sp.	*Sorghum* sp. and *Pennisetum* sp.
The *saqia*	Knobbed pottery vessels (*qadus*)	second, third and fourth centuries A.D.	In Egypt: Ptolemaic-Roman period. In Sudan: Meroitic Kingdom.	See details on page	Domesticated sorghum: *Soghum bicolor* Gossypium (Ar. *cotton*), *Phoenix dactylifera* L. (Ar. *nakhal*).

3 See footnote no. 1.

7.3.2.2 Basin irrigation.

The basin irrigation is closely associated with season floods of river systems. The Nile valley in Egypt is considered the homeland of the basin cultivation (Allan, W.N., and R.J. Smith, 1948: 621). This method of irrigation was probably well established in the Sudan during the period of the Kerma culture (c. 2000-1500 B.C.) (Trigger, B., 1965: 19-21). The Kerma basin has been used since then until about three decades ago. In this method, the low-lying tracts and depressions adjacent to the river courses or between and among hills/mountains (away from rivers) are flooded during the flood season or filled up during the rainy season. As in the case of basins adjacent to rivers, they remain filled up until the high water level begins to recede and so the water starts to retreat from these basins (ibid; Burnett, J.R., 1948: 289) back to the rivers. As regards basins situated between or among hills, their rain-water will eventually disappear due to its being mainly absorbed into the soil but also evaporated. Thus, the simplest form of basin irrigation is the annual flooding or filling of depressions either from rivers (Tracey, C.B., and J.W. Hewison, 1948: 743) or from run-off rain water from neighbouring hills or mountains. The ideal period of good soaking is 30 to 40 days before the basins are drained (Allan, W.N., and R.J. Smith, 1948: 622) or dried and the land is sown. Unlike the conditions in Egypt, the conditions favourable for annual flooding of depressions and basin cultivation in the Sudan exist in very limited areas along the Nile valley and other river systems, e.g. Atbara river, the White and Blue Niles. One of these areas is the Dongola reach including the Kerma Basin where the annual rainfall is between zero and 50 mm. The other one is the Shendi reach where the annual rainfall is between 60 to 160 mm. This reach stretches from the 6th cataract to Atbara (cf. fig. no.1). As it may be clear now, success of cultivation by means of basin irrigation is entirely dependent on fairly good floods (ibid; Lang-Anderson, R., 1980: 151) particularly those areas with low annual rainfall. The hoe and the plough are the basic tools in cultivation activities. The success of obtaining good harvests in lands irrigated by means of natural flooding or basin irrigation depends on the soil permeability of water until the crops come to maturity. All of these also require thinned and well-spaced sowing of seeds to avoid competition on moisture.

7.3.2.3 The counter-weighted lever irrigation (Ar. *shadouf*).

The counter-weighted lever (Ar. *shadouf*) is one of the oldest devices for lifting water to a desired but limited height from a suitable water source (river, pool, reservoir, canals, etc.). It is operated by man-power. The earliest record of the *shadouf* from the Nile valley comes from Egypt (cf. table no.39) dated to the second mill. B.C. (Abu Saleem, M.A., 1980: 5,45; Trigger, B.G., 1965: 21) but it is yet not exactly known when it was introduced in the Sudan. It is, however, certain that there was no change in the principal design or the operation of the *shadouf* since it was first made (Allan, W.N., and R.J. Smith, 1948: 631). It is made of two wooden posts or dried mud-pillars, a cross bar (lever) and a dipper. When the *shadouf* was first used, the dipper was made from pottery (Abu Saleem, M.A., 1980: 46) or leather (Lang-Anderson, R., 1908: 171).

The *shadouf* is used to lift water to cultivable land adjacent to the river courses, yet beyond the reach of the high water level. It is also infrequently used as a basic device of irrigation in areas which are situated away from river systems as for instance the Kheiran area in north Kordofan (Manger, L.O., 1982: 137; 1983: 51-52), where wells, *wadis*, etc. are the main sources of water. Although 0.52 acres to 0.69 acres is the standard size of *shadouf* cultivable plot, the exact area irrigated by means of *shadouf* per day would be correctly estimated and/or calculated only after providing some basic information such as:

1) the nature of the soil, i.e. whether it is sandy, clayey, clayey-sand, sand-clay, etc.
2) the season of irrigation, summer or winter, and even the time of the day (morning or afternoon period) when the *shadouf* is operated is also important; evaporation rate is a key factor in the context,
3) the type of the cultivated species, some demand a lot of water, others need very little.

The *shadouf* is widely distributed in the Sudan, but it is very common in the Central Sudan particularly along the White Nile, the Blue Nile and the Nile respectively. It is not as common further north due to the impracticability of lifting water by *shadouf* since the water-level is low and the Nile banks are relatively high. Although it has a long history in the Nile valley, the original design and out-put capacity have remained the same without any subsequent improvement and/or promotion. Thus its basic disadvantages and merits have remained unchanged.

A variety of food-plants are cultivated in the *shadouf* land (cf. table no.39); the hoe is the basic cultivation tool.

7.3.2.4 The water-wheel (Ar. *saqia*).

The *saqia* is an oxen-driven water-wheel which is used as a water-lifting device mainly for irrigation purposes. Its main parts consist of two toothed-wooden wheels (horizontal one to gear the vertical one). A third larger wheel has two long loops of ropes with a series of knobbed pottery buckets (Ar. *qadus*) (cf. fig. no.51) (Adams, W.Y., 1977: 346).

The ox is the most common animal used for operating the *saqia* but camels, horses and donkeys are infrequently used as well (Abu Saleem, M.A., 1980: 67-68, 184). The manufacture, maintenance and major repairs of the *saqia* are performed only by professional *saqia* craftsmen).

Fig. no. 51 The saqia in operation

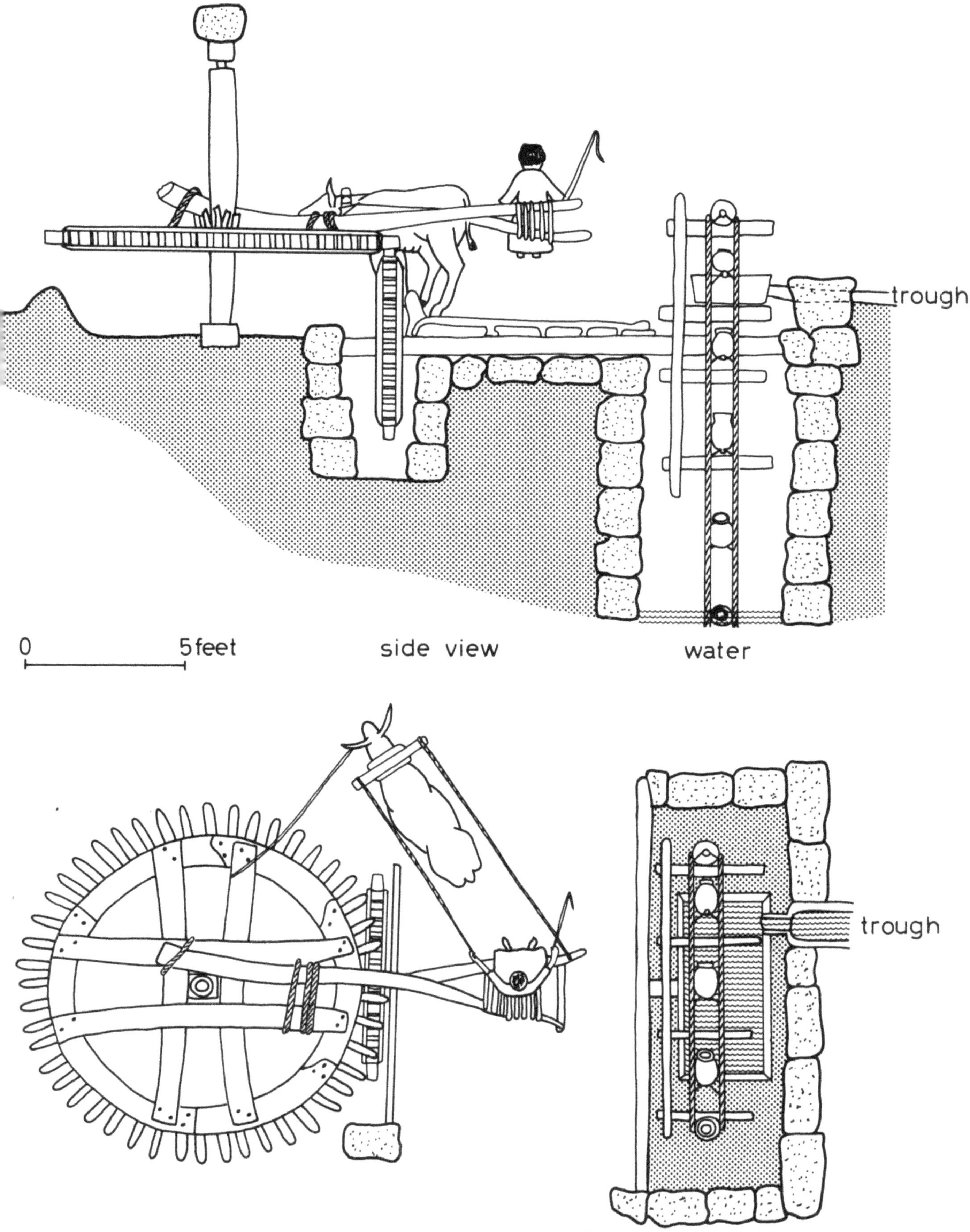

After Adams, W.Y., 1977: 347

Since the pharaonic Egypt had achieved very advanced technology in fields such as architecture, engineering, agriculture, etc., it has been assumed that the water-wheel (Ar. *saqia*) was also invented during the pharaohs' rule of Egypt. But, so far, there is no material and/or inscriptional evidence to confirm this assumption. As a matter of fact the origin of the *saqia* is still speculative (Adams, W.Y., 1977: 346; cf. table no.39). As regards its introduction in the Sudan, there is a strong belief that it came from Egypt (Trigger, B.G., 1965: 20) and that its spread must have been slow (Abu Saleem, M.A., 1980: 43; cf. table no.39). The archaeological evidence of the *saqia* in the Sudan consisted mainly of numerous remains of knobbed pottery vessels (Ar. *qadus*) which were found at different places in lower Nubia dated to the late Meroitic period; the A.D. second, third and fourth centuries (ibid: 198).

The largest number of water-wheels is reported from the northern part of the Sudan, e.g. Halfa and Dongola. As regards the number of *saqia* in the Central Sudan, the largest number of them is reported from Berber area, however, they are unevenly distributed; there are some parts in Berber where the *saqia* is hardly used due to either the cultivable land being very limited or the type of soil being too poor to sustain growth of crops such as the Rubatab area in Berber province. Although the *saqia* was reported from the White Nile and the Blue Nile provinces, in the Central Sudan, it was not popular. I was told by my informants in the area between Kawa and Kosti along the White Nile that the introduction of the *saqia* in the area was not met with success because the inhabitants are basically dependent on rainfed cultivation. Beside the technical difficulties of manufacturing, maintaining and operating the *saqia*, its introduction to the White Nile area was unsuccessful basically because the land irrigated by means of the *saqia* is too small if compared with the large rainfed-cultivated areas. Thus, the yields from the *saqia* land are very small if compared with those of the rainfed land. Furthermore, the former demands continuous work both to keep the *saqia* operating and to tend to the cultivated crops. In contrast to this, the annual rainfall (0-60 mm) in the northern provinces, Halfa and Dongola, is too low to sustain any rainfed cultivation. Thus the *saqia* is indispensable in the cultivation activities in these provinces.

As regards the areas situated away from the river courses, e.g. Kordofan, the *saqias* are very few due to the limited sources of superficial water and the unavailability of water in many of these during most of the year. Therefore, some of the *saqias* are only operated seasonally when the rainfall is fairly good so that ample supplies of water are trapped in special reservoirs. There is no standardized size for *saqia*-irrigated plots. But some results were obtained as regards the size of the plot in relation to range of the water lifting. However, it has been stated that the *saqia* can operate within 3 to 8 metres uplifting range but the higher the lifting within this range, the lesser the out-put capacity. The range of 3 to 8 metres will also make it manageable as to the deepening and readjusting the channels; as well as the resetting of the *saqia* platform. Accordingly, the plot that can be watered by the *saqia* has been estimated to be 5.2 acres (5 *feddans*) in average during the flood season when the lift is only 2 to 3 metres. During the dry season when the lift is 8 metres the estimated irrigated plot is only 2.08 acres (Adams, W.Y, 1977: 52; Allan, W.N., and R.J. Smith, 1948: 628; Barbour, K.M., 1964: 141; Trigger, B.G., 1965: 20). Regardless of the total size of the *saqia* cultivable land, the farmers have always kept a fallow cultivation system whereby 1/2 to 1/3 of the land (depending on the season) will be cultivated at a time. The *saqia* has, most importantly, enabled the farmers to perform different cultivation activities in all seasons round the year. For this reason, areas deserted for long periods were resettled and cultivated, e.g. the rehabitation of Lower Nubia during the second, third and fourth centuries A.D. Table number 40 is an illustrative example showing the different crops cultivated during different seasons of the year by means of the *saqia*.

Both the *saqia* and the *shadouf* have the advantage over the natural flooding and basin irrigation in that they can be used at different seasons while the latter two can be used only during the flood season. But there are some basic differences between them as shown below.

The *saqia*	The *shadouf*
1) High out-put of water, thus a big area can be cultivated (2.04 to 5.2 acres).	1) Low out-put of water, thus a very small area can be cultivated (0.52 to 0.69 acres).
2) The lifting range is wide, bet. 2 to 8 metres.	2) Lifting range is limited to maximum 3 metres.
3) Operated by animals (oxen) and does not demand human physical effort.	3) Operated by man-power and demands strenuous and continuous physical effort.
4) Although the *saqia* is made from locally available materials (wood, ropes, leather, pottery), its manufacture, assembling and major repairs require professional craftsmen.	4) Simple device (from wood, ropes and leather) that any farmer can make, construct and repair.

The plough and the hoe are the main tools used in the cultivation activities in the *saqia* land. The metal sickle is a tool commonly used in all of the natural flooding, basins, the *shadouf* and the *saqia* cultivation activities. The estimated date of the earliest archaeological evidence of metal sickles in the Sudan is c. A.D. 500-600, which coincides with the late period of the X-Group culture (Save-Söderbergh, T., 1982: 68). Weeding is practised mainly in the *saqia* and the basin irrigated lands. Table no. 39 shows when

Table no. 40. A classic example of cultivated plants and their cycle in the *saqid* land: from both North and Central Sudan.

Season	Duration	Cultivated Plants[1]			Comments
		Scientific name	Arabic	English	
Flood Ar. damira	June - November	1. *Sorghum vulgare* L. 2. *Dolichos lablab* Linn. 3. *Zea mays* L. 4. *Pennisetum americanum* 5. *Sorghum sp.* (cultivé) 6. *Hibiscus esculentus* L.	durra lubia durra shami dukhn own-koleib bamia	sorghum maize millet sorghum forage okra	damira is the most busy cultivation season in the year. The Sorghum sp. (no.5) is cultivated among the date palm trees as fodder, harvested before maturity. The palm trees' shade provide its being soft and has long stalks.
Winter Ar. shituri	November - March	1. *Triticum* sp. 2. *Hordeum* sp. 3. *Vicia faba* L. 4. *Phaseolus vulgaris* L. 5. *Allium cepa* L. 6. *Foeniculum vulgare* L. 7. *Coriandrum sativum* L.	gamih shair ful-masri fasulia basal shamar kasbara	wheat barley tick beans haircot onions fennel coriander	It is also a busy season, but less labour is required as regards irrigation; the evaporation is lesser and humidity is higher compared with the other two seasons.
Summer Ar. seifi	May - August	1. *Zea mays* L. 2. *Sorghum* sp. (cultivé) 3. *Hibiscus esculentus* L. 4. *Corchorus ditorius* Linn.	durra shami own-koleib bamia melukhyia	maize sorghum forage okra Jew's mellow	Most relaxed season as regards cultivation activities. The saqia is basically operated to irrigate the fruit trees[2] and and sorghum forage.

[1] The date palm, *Phoenix dactylifera* L. (Ar. *nakhal*), *Citrus aurantium* var. *sinensis* L. (Ar. *burtukan*), *C. paradisi* Macf. (Ar. *grainfriut*), *C. limonia* Oslo. (Ar. *limon*), *C. nobilis* Broun & Massey (Ar. *yousif-effendi*) and *Mangifera indica* L. (Ar. *mango*) are grown in orchards system and irrigated round the year.
[2] Idem.

each of these techniques of irrigation was introduced and the food-plants which are traditionally irrigated by them.

7.4 Summary.

The rainfed cultivation is most likely the oldest type of cultivation systems in the Central Sudan. The shifting and the *hariq* cultivation, which are the oldest of the rainfed systems, do not seem to have changed significantly since they were first practised. However, they have been subject to some modifications in some areas by force of circumstances, e.g. intensive rainfed cultivation. The earliest evidence of domesticated sorghum, *Sorghum bicolor* (L.) Moench from the Central Sudan was found in an area (Jebel Tomat) where people still practise rainfed cultivation of sorghum and millet.

With regard to the irrigated cultivation; the natural flooding is probably the oldest system of cultivation, it is however, uncertain whether it is older or younger than, or contemporary with, the rainfed cultivation. Basin cultivation is younger than natural flooding but both systems are only seasonally practised mainly after flood waters recede. The more advanced cultivation systems were established with the introduction of the *shadouf* and the *saqia* respectively. With the introduction of these systems the inhabitants of the Central Sudan were able for the first time to cultivate the year round. Since they were introduced relatively late in this area, the *shadouf* and the *saqia* do not seem to have played any role in the process of domestication of cereal crops, e.g. sorghum. But they have perhaps promoted the utilization of these from seasonal (rainfed) cultivation to cultivating them round the year.

The hoe is the main tool both in rainfed and irrigated cultivation. Metal hoes are today the most common types but wooden hoes (cf. fig. no.44) are still used in some parts of the Sudan, e.g. the Nuba mountains (cf. fig. no.1). Since no evidence of stone or metal hoes was found in the Central Sudan during the period being studied here, it is likely that wooden hoes (cf. fig. no.44) were used in the cultivation activities. Differences do occur from one area to another in the Central Sudan as regards other cultivation tools such as those used for sowing, weeding, harvesting ,etc. These range from simple wooden tools which are found in the vicinity of the fields, e.g. forked branches, to more specialized wooden tools (cf. figs.no.45 and 46A and B) to specialized metal tools. Although there is no evidence at present from the Central Sudan for wooden tools which were probably used in cultivation activities (between c. 9000-1700 B.P.), it is, however, most likely that the wooden tools used in the traditional cultivation are similar to or being modified types of, wooden tools which were used for the last few thousands of years. The earliest evidence of cultivation tools from the Sudan is a harvesting metal sickle (Ar. *moonjal*) dated to the late X-Group c. A.D. 500-600 (Save-

Söderbergh, T., 1982: 68). Relatively advanced cultivation tools, for instance the plough, are associated mainly with irrigated cultivation, e.g. the *saqia* and basin cultivation.

CHAPTER VIII

From Gathering to Cultivation of Wild Food-Plants to their Domestication

Section A

From Gathering to Cultivation of Wild Food-Plants

8.1 Introduction.

The last two decades have witnessed active archaeological research on the Early Khartoum and the Shaheinab traditions in the Central Sudan. The results of these studies have undoubtedly provided more detailed information and better understanding of the culture history of the Central Sudan during this period.

Current discussions of the gathering of food-plants mostly confine themselves to description and rarely deals with it as an economic activity essential for our understanding of the subsequent development in the utilization of food-plants that led to their cultivation and domestication. As regards the cultivation of wild food-plants, there are some few studies from the Central Sudan, but there is a tendency among some archaeologists to associate cultivation only with domesticated plant species, and accordingly to make misleading interpretations. Thus, I shall, in this study try to throw more light on these issuès and problems and suggest some guide lines as regards the definition and usages of terminologies such as cultivation of wild plants.

I will base my argumentation mainly on archaeoethnobotanical evidence which I collected jointly with Dr. Haaland and Dr. El-Mahi during our excavation of nine archaeological sites in the Central Sudan. In addition, I shall also incorporate relevant ethnographic evidence and also draw on material collected by others.

8.2 Gathering of Food-Plants during Early Khartoum Tradition.

The earliest direct evidence of remains of wild food-plants from the Central Sudan was recovered by A.J. Arkell from the classic site of Early Khartoum (Arkell, A.J., 1949: 80, 108) at present the Khartoum Chest Hospital.

Since then a number of sites of the Early Khartoum type have been excavated in the area (Caneva, E., 1983; Khabir, A.M., 1987; Mohammed-Ali, A., 1982: 41-45; etc.) including three sites we excavated during the seasons of 1984 to 1987 (cf. table no.41). Culturally, sites of this type came to be known as the Early Khartoum tradition[1] which lasted over 3000 years between c. 9000-6000 B.P. (cf. table no.41).

The archaeological remains recovered, both biological (flora and fauna) and material, from all of these sites indicate an adaptation in which hunting of wild game and gathering of food-plants were practised (Clark, J.D., 1984a: 115) and with an emphasis on the exploitation of aquatic resources (ibid; Haaland, R., 1987b: 48).

In a joint venture between Dr. Haaland of the University of Bergen and Dr. El-Mahi and myself from the University of Khartoum, we excavated two sites of the Early Khartoum type in the Atbara area (Abu-Darbein and Aneibis) and lately a third site in Ed-Damer (cf. fig. no.1); but analysis of the material from the latter is not yet finished.

The oldest and youngest radiocarbon dates from the Abu-Darbein site are 8500 ±100 B.P. and 7700±140 B.P. respectively. As regards the site of Aneibis, 7850±140 B.P. is the oldest date and 7290±150 B.P. the youngest (information on the physical environment is presented in chapter III).

The food-plant remains recovered from both the Abu-Darbein and Ed-Damer sites consisted mainly of immediately eaten food-plants that did not need to be processed or prepared. However, no evidence of any food-plants was recovered from the site of Aneibis. As is the case with the finds from other sites of the Early Khartoum type in the Khartoum area, the remains of food-plants from the Abu-Darbein and Ed-Damer sites consisted of large quantities of seeds of *Celtis integrifolia* Lam. and a lesser number of fruit-seeds of *Ziziphus spina-christi* (L.) Willd. (cf. table no.41).

Evidence of grains/seeds of grass-foods is very infrequent from sites of the Early Khartoum type and from the sites we excavated in particular. Clark mentioned that grain-foods were collected by the occupants of the site of Early Khartoum (Clark, J.D., 1984: 115) but no further specification is given of the nature of these grains. Furthermore, evidence of two mummified seeds which I identified as *Setaria* species were recovered from the site of Abu-Darbein. Since they were found at a depth of 30-50cm, they were probably contemporary with the archaeological remains, and that they are not

---

[1] Some archaeologists use the term the Khartoum Mesolithic.

Table no. 41. Culture history and economy as regard plants exploitation in the Central Sudan (9300 - 1700 B.P.)

Cultural period	Chronology years B.P.	Economic activities: Plants exploitation	Excavated sites	Biological (plant) evidence (impressions + mummified)			Material culture evidence		
				Gathering	Wild plants cultivation	Domesticated plants cultiv.	Gathering	Wild plants cultivation	Domesticated plants cultiv.
Early Khartoum Tradition	c. 9000 - 6000	"Broad spectrum economy" Gathering of immediately eaten food-plants	Abu-Darbein Aneibis Ed-Damer	*Celtis integrifolia* *Ziziphus spina-christi*			pottery grinders		
Shaheinab Tradition	6000 - 4000	Exploitation + cultivation of wild grasses + gathering of immediately eaten food-plants	Kadero[1] Um Direwia Zakiab Nofalab Shaheinab	*Celtis integrifolia* *Ziziphus spina-crhisti*	*Setaria* sp. *Pennisetum* sp. *Sorghum* sp.		pottery grinders	pottery grinders	
Post-Shaheinab Tradition + Meroitic	4000 - 1700	Cultivation of wild and domesticated grasses + gathering of immediately eaten food-plants	Rabak Shaqadud[2] Jebel Tomat	*Grewia tenax* *Ziziphus spina-christi*	*Pennisetum* sp. *Sorghum* sp.	*Sorghum* sp.	pottery grinders	pottery grinders	pottery grinders

[1] Excavated by a Polish expedition led by L. Krzyzaniak but one of our team, Dr. Haaland, joined in 2 seasons (1975-1976).
[2] Excavated by a joint U.of K. and S.M.U. expedition between 1981-1982. The present writer participated in the last season.

later intrusive trapped in a seed-burial. However, Clark's statement that collection of grains was practised during the occupation of Early Khartoum (ibid: 115) indicates one of the earliest (if not the earliest) dates for the utilization of grains in the Central Sudan. But no mention was made of the implications of this as regards, for instance the knowledge and technology required to process and prepare this type of food-plant (e.g. threshing, winnowing, grinding, etc.). Such technological knowledge must be acquired at the same time as the practice of consuming seed/grain-foods, otherwise stomach irritations and digestion problems are to be expected (cf. tables from 33 to 37). Furthermore, the beginning of the utilization of seed/grain-foods indicates that there was a developing interest in specific staple food-plants instead of just gathering a wide spectrum of plants. This may have been motivated by a number of factors, as I shall discuss later on. As regards the data we recovered, the only indirect evidence of the knowledge and technology required for the utilization of grains, consisted of remains of material-culture, such as grinding equipment and pottery (the latter would be essential for storage). However, the possibility that grinding equipment could have been used for other purposes and the fact that other evidence (direct or indirect) of the exploitation of food-plants is meagre open for a number of hypotheses. These are:

1) The occupants of the sites of the Early Khartoum type had the knowledge necessary for preparing grains/seeds as food but they began to utilize them at a later period after they had adopted a more extensive exploitation within which some food-plants were utilized more intensive than others. This extensive exploitation was most likely due to demographic pressure and environmental strains (especially during the last thousand years of the Early Khartoum tradition when climatic conditions witnessed a phasal deterioration).

or 2) They had a knowledge and technology which did not necessarily require the use of any specialized tools or equipment to, for instance, harvest, thresh, winnow, etc. the grains.

or 3) The occupants were utilizing seeds/grains as food all the time, but for one reason or another (cf. chapter IV) evidence of these in archaeological sites is either meagre or not found. The remains of material culture recovered, e.g. grinding equipment and pottery, does, however, indicate that the necessary knowledge required for processing, e.g. grinding, probably cooking or boiling etc., had already been obtained and hence, most likely utilized.

In this connection, it is not unlikely that grinding equipment was also used to process the fruits of *Ziziphus spina-christi* (L.) Willd. into thick dry cakes which can be stored (e.g. in pottery) and eaten when desired by the occupants of the sites of the Early Khartoum type. This assumption is based on evidence of split and fragmented seeds (stones) of *Ziziphus* sp. on almost all of the Early Khartoum sites. Such evidence is comparable to the ethnographic data from the eastern Sudan, Kassala-Khashm El Girba area (cf. fig. no.1). In this area, the present nomads crush the fresh and dry fruits of *Ziziphus spina-christi* (L.) Willd. with grinders. After removing the crushed seeds, they shape the mixture (of the dry and fresh fruits) into thick rounded cakes. When these cakes dry, they become hard and can be stored for years.

It may be clear from the foregoing discussion that the present state of our knowledge as regards the utilization of food-plants during the Early Khartoum tradition is still rather limited. I propose that more emphasis should be placed not only on what was gathered but how gathering and food preparation were done and what does this imply. In the chapter on my ethnographic study on gathering (cf. chapter VI) I attempt to raise and to discuss questions of this nature. The available evidence from the sites we excavated as well as from other contemporary sites indicates the following:

1) The Early Khartoum tradition had made its earliest appearance c.9000 years ago over a wide geographical area in the Central Nile Valley of the Sudan. This is evident in the radiocarbon dates (cf. appendix no.1), the distribution of the sites and the similarities of the diagnostic material culture evidence (i.e. wavy line pottery, bone harpoons). Differences are, however, observed between some of the sites in the Atbara area and those of the Khartoum area. For instance pottery is very abundant at the sites of Aneibis as compared with what was recovered from the site of Saggai (cf. fig. no.1). 'There is [also] a clear difference in the near absence of wavy line [pottery] at Aneibis' (Haaland, R., 1987b: 50).
2) The sites of the Atbara area and (the preliminary results from) Ed-Damer indicate (in agreement with other Early Khartoum sites) a sedentary mode of fishers-hunters and food-plant gatherers.

   The size of the sites, the proximity to permanent sources of water (mainly rivers, cf. fig. no.1) and the accessibility of a variety of food resources both aquatic and terrestrial were the most likely reasons both for the sedentary mode of life and the relatively large size of the sites [e.g. Early Khartoum site was 5600 m^2 (Clark, J.D., 1984a: 115), the Aneibis site is 7000 m^2].
3) The exploitation of food-plants was largely based on those which can be eaten without preparation. The recovered evidence of grain/seed-foods is very meagre. But such an evidence, together with the presence of pottery and grinding equipment and other circumstantial evidence provides the basis for suggesting that the occupants had the requisite knowledge and equipment about food-plants in general or at least about those which required processing and preparation such as (?) grain-foods.
4) The sedentary mode of the Early Khartoum fishers-hunters-gatherers and the "relatively" large size of the sites indicate that there was an increase in the population over time (Haaland, R., 1981: 209; Shaw, T., 1978: 69). On the assumption that

productivity in a previously unoccupied niche would be high, Haaland suggested that population growth was rapid rather than gradual (Haaland, R., 1987b: 51).

It is thus likely that this would have led to an increased pressure on land resources, e.g. on food-plants and wild game. As the archaeological remains indicate, these resources constituted an important component of the diet. As a result of such pressures, a trend towards more diversified and broadly oriented exploitation both of floral and faunal resources would have been developed; a trend known in archaeological terminology a "a broad spectrum economy".

In short, the data presented above generally indicate that through more than 3000 years (between c.9000 and 6000 B.P.), exploitation of aquatic resources was important in the subsistence economy of the Early Khartoum tradition (including the sites we excavated) and that there was also an increasing tendency towards a broadly oriented exploitation of land resources. Most likely, these tendencies culminated in a utilization of the potential staples (seed/grain-foods) that developed into a more intensified exploitation which in turn led to the beginnings of cultivation of wild food-plants (mainly wild grasses) in the Central Sudan.

8.3 Cultivation of Wild Food-Plants.
Cultivation: Definition and Criteria.

It is extremely difficult to put forward a standardized definition of the term cultivation that is universally applicable or to establish a general pattern for the development of cultivation and cultivation practices. However; it is likely that the cultivation of plants did not require new knowledge such as the discovery that plants grow from seeds. But when compared with gathering, the beginning of cultivation initiated a new type of relationship between man and the plant species with which he was already well acquainted.

This new relationship evolved when man began to depend upon a more steady supply of food of vegetative origin. Consequently he started to devote more effort, than was required in mere gathering, to secure his demands from this type of food.

Cultivation in its broadest meaning can thus be described as the 'promotion of favourable growth conditions [for food-plants]' (Helbaek, H., 1969: 403). It includes all "deliberate" activities of caring for plants (Harlan, J.R., et al, 1976: 9) in the expectation of a subsequent harvest or reward. These activities are distributed over various intermediate and transitional stages which range from indifference towards to tolerance of, to protection of, to encouragement of to full attendance to plant growth. It is important to note here that "cultivation does not intrinsically imply that the cherished species will die if the care of the cultivator comes to an end" (Helbaek, H., 1969: 403). It has even been suggested that attitudes towards plants generally amount to some form of cultivation regardless of whether these are intentional or unintentional (Yarnell, R.A., 1971: 523).

The literal meaning of the term cultivation is 'preparation [of land] for crops by ploughing, etc; 'help' [crops] to grow e.g. by breaking up the soil around them, killing the weeds, etc.' (Hornby, A.S., et al, 1971: 238).

The combination of the intermediate and transitional stages with all the efforts of initiation and furtherance of plant growth by means of land clearance and preparation, tilling the soil and concentration of growth by sowing/casting and the exclusion, and the removal and avoidance of unwanted plants is referred to as "full cultivation"[2] (Harlan, J.R., et al, 1976: 9; Helbaek, H., 1959: 365-366; Yarnell, R.A., 1971: 552-553). However, it is not always in practice the case that all the various activities are performed and/or needed as I shall give examples of later on.

8.4 The Transition from Gathering to Cultivation of Wild Food-Plants.

Unfortunately, the available evidence does not in any direct or indirect way show the actual transition or "shift" from the gathering of wild food-plants to their cultivation and domestication. But if the description of cultivation proposed is correct (cf. above), it does mean as well that cultivation is an activity that can be exercised on wild, incipiently or fully domesticated plants. In this connection, the palaeoethnobotanist Helbaek was among the first scholars to distinguish between cultivation of plants which are wild and those which are domesticated (Helbaek, H., 1969: 403-404). Furthermore, regular performance of cultivation activities on wild plants over a period of time can be considered among the crucial factors that lay the foundation for the genetic change (domestication) of these plants. But if genetic change does not occur, this does not mean the wild plant (in question) was not cultivated (Higgs, E., 1976: 33). Logically and biologically

> '.... a cultivated plant need not necessarily be domesticated - can not by any means so be from the outset - while on the other hand, a domesticated plant can exist only as a cultivated plant' (Helbaek, H., 1969: 404).

Accordingly, I suggest that partial, sporadic and eventually undivided cultivation activities are to be considered as the transitional stage between the gathering of wild food-plants and their domestication. If this suggestion is correct, then, I think the cultural, demographic,

[2] Full cultivation, fully accomplished, undivided cultivation activities are terms which meant technically to denote the combined performance of all cultivation activities.

environmental, etc. factors which are thought to have led to plant domestication (Alexander, J., 1969: 123-127; Bender, B., 1975: 19-20, 37-38; Binford, L.R., 1971: 48-49; Boserup, E., 1974: 16-17; Braidwood, R.J., 1982: 115-126; Bray, W., 1977: 234-243; Butzer, K.W., 1976: 106; Clark, J.D., 1976: 88, 90-91, 96-97; Flannery, K.V., 1969: 73-100; Hassan, F.A., 1981: 217; Helbaek, H., 1959: 365; Meyers, J.T., 1971: 101-121; Newman, J.L., 1970: 47-48; Pfeiffer, J.E., 1976: 23-28; Shaw, S., 1977: 93-111, Washburn, S.L., and C.S. Lancaster, 1968: 293-95; etc.) may, as well, be seen as regards the Central Sudan, as primary interrelated factors and preconditions for the beginning of the cultivation of wild plants. Among these, the most important ones are:

1) Culmination of man's knowledge (that he gained over thousands of years) in the last c. 9000-6000 years. This is reflected in the Central Sudan - as well as in other areas in the Sahel zone - in the gradual transition from a narrow to a broad spectrum type of economy. The more intensive utilization of food-plants, including grains and seeds of grasses, was an essential development within the frame of the new broad spectrum economy. Exploitation of wild grain seed-foods started with a wide range of grasses, however, some of them were exploited more intensely than others and that led to their cultivation and eventually to their domestication.
2) Demographic pressure on resources (given the technology at the time) coupled with environmental strains have also been considered as some of the pre-conditions and factors for the transition to cultivation. In an attempt to test the relevance of the above mentioned factors with regard to the Central Sudan, we shall see that there is evidence for an exploitation of a wide variety both of faunal and floral resources particularly during the late Early Khartoum tradition (c. 7000-6000 B.P.) and the early Shaheinab tradition (c. 6000-5000 B.P.). This coincided with a tendency toward phasal deterioration of climate (Wickens, G.E., 1975: 45-65) during the same period (7000 onwards) when compared with the earlier period of occupation of sites of the Early Khartoum type, some 2000 years earlier. Although there is no direct evidence for an increase in population density and consequent pressure on resources, this is indirectly indicated by the large size of the sites (cf. chapter III), the quantitative increase of remains both of material culture (Haaland, R., 1981: 60-194) and of the fauna (El-Mahi, A.T., 1982: 34-40; El-Mahi, A.T., and R. Haaland, 1984: 28-32).
3) Specialized tools associated with extensive exploitation and cultivation of plants were believed to have emerged during this period and hence are regarded as one of the pre-conditions.

As for the Central Sudan, there is no direct evidence from this period of any tool(s) the first invention or introduction of which can be associated with, e.g., cultivation (cf. tables no.41 and 42). But, on the one hand, there is evidence of a newly(?) introduced or invented tool, namely the gouge, but it is not yet certain if it is associated with cultivation (cf. chapter V: 5.3.6). On the other hand, there is quantitative evidence for a significant increase of tools, e.g., grinders; which have already existed and were used to process a variety of foods including cereal-grains (for more details, see chapter V).

8.5 The Central Sudan Case: The Shaheinab and the Post-Shaheinab Traditions

The cultivation of wild food-plants should not be interpreted as a mere transitional stage from gathering to domestication, which has undergone a process with various evolutionary stages, but should be seen as a new and important economic activity that initiated a totally different man-plant relationship, from that of gathering. Furthermore, cultivation began within a context of food-plants gathering economy; and its different intermediate stages, e.g., protection and most basic activities (cf. table no.42), have been individually, selectively or collectively conducted until all (or most) of them were regularly practised. This interpretation is derived mainly from archaeoethnobotanical evidence recovered from eight sites of the Shaheinab (c. 6000-5000 B.P.) and post Shaheinab (c.5000-1700 B.P.)[3] type in the Central Sudan (cf. table no.41) as well as from ethnografic evidence from the Sudan and elsewhere.

Archaeoethnobotanically, there is evidence which indicates that the occupants of the Central Sudan started to cultivate wild grasses, e.g. *Sorghum,Pennisetum* and*Setaria* sp. about 6000 years ago within a predominantly gathering type of economy (cf. table no.41). Utilization of immediately eaten food-plants such as *Celtis* sp. and *Ziziphus* sp. was practised at least a thousand years after the cultivation of wild grasses had started (cf. table no.41). Not only that but there is evidence of *Grewia tenax* (Forsk.) Fiori. from the site of Shaqadud (cf. fig. no.1) which is a post Shaheinab type of site. The fruits of this species and those of *Ziziphus spina-christi* (L.) Willd.[4] were gathered in addition to the cultivàtion of wild pearl millet (cf. table no.41). The large numbers of seeds of this fruit indicate that it was intensively exploited, particularly towards the later period (the upper third of the 3.5 m culture deposits) of occupation of the site. It also seems that the gathering of this fruit-food (for which no evidence was found from the older sites along the river systems of the Central Sudan) had gradually replaced the gathering of *Ziziphus* fruits (cf. tables no.7 and 8). A direct implication of this is that gathering of food-plants was pursued in different parts of the Central Sudan along and away from permanent sources of water (i.e. rivers) long after the cultivation of wild grasses was first started. An indirect implication is that gathering of wild food-plants has always been an important

[3] The post-Shaheinab period includes the Rabak/Shaqadud between c. 5000-2500 B.P. and the Meroitic period from c. 2500-1700 B.P.

[4] The fruits may also be processed to make a sweet dish; for more information see table no.37.

Table no. 42. Archaeological evidence from the Central Sudan and possible comparable ethnographic evidence of some basic cultivation activities.

Activities	Archaeological evidence for wild food-plants cultivation	Archaeological evidence for domesticated food-plants cultivation	Possible ethnographic analogies for	
			wild food-plants cultivation	domesticated food-plants cultivation
Land preparation	no evidence	no evidence		Branches found at the site of activity and their abandonment afterwards in W. Africa
Sowing	"	"	The Shoshoni "Red Indians" (no tools are mentioned)	Pointed wooden tools by Turkana nomads in N. Kenya
Weeding	"	"		Wooden tools: Nuba mountains in S. Kordofan, Sudan
Harvesting	"	"	Use of hands or branches in Zaghawa area in W. Sudan	Use of hands: El-Gedarif area in the Sudan
Storing	pottery	pottery	Large clay jars (Ar. *dabanga*)	Hanging heads (sorghum) in (kitchen) hut: Ingassana in E.Sudan. Storing in dug up pits and large clay jars: Darfur, W. Sudan.

component in the economic activities especially those related to the exploitation of plants. This is indirectly evident in the continuous exploration for alternative sources of gathered food-plants, e.g. fruits of *Grewia* replacing *Ziziphus* at Shaqadud.

In short, although it is hardly possible to find in an archaeological context any direct evidence of how the various cultivation activities evolved, our evidence from the Central Sudan does indicate that the gathering of food-plants evolved into cultivation of wild food-plants. These two types of activities were carried on for thousands of years even after the occupants of the area were already cultivating domesticated plants, e.g. sorghum (cf. table no.41).

Ethnographic observations and studies from the Sudan and elsewhere provide some very interesting demonstrations of different activities which I classified earlier as cultivation activities. I shall cite here those which most likely bear some resemblance to those practised in the prehistoric and early historic Central Sudan. An example which demonstrates that tolerance can be one form or intermediate stage in cultivation activities , is Ethiopian oats which are related to the weedy *Avena barbata* L. The oats grow as an acceptable weed in the fields of emmer wheat and barley. As a result of this tolerance, these weeds started to produce semi-shattering and non-shattering races. The non-shattering ones are harvested along with the emmer wheat and barley, and the mixture is sown and harvested the next season (Harlan, J.R., et al, 1976: 10). Another example that demonstrates how the protection of wild plants can also be one form of cultivation is given from the forest zone in West Africa where wild yam grows (Harris, D.R., 1976: 335) and is believed to have been, as it still is, protected, harvested and replanted. In this connection, it has been suggested that sowing and replanting wild food-plants is one of the cultivation practices which was developed due to necessity (Stemler, A.B.L., 1980: 515).

A third example which does not demonstrate deliberate sowing but rather self-sown seeds and grains and how these might have been harvested in prehistoric and early historic times is derived from ethnobotanical observations among Tuareq camps in Air. It was observed that millet and sorghum were self-sown round deserted camps as a result of loosing some of the harvested grains while preparing them for food (Nicolaisen, J., 1963: 184). Since this observation only reported that the stands of self-sown grains of millet and sorghum occur round (ibid) and not within the camps, it is most likely that these grains were lost while threshing and winnowing them. This interpretation is based on ethnobotanic observations from the Sudan where threshing and winnowing are always done in an open area in the vicinity of but never within settlement area. If so, we can elaborate on this interpretation by adding that protecting and harvesting this type of self-sown grain round camps probably goes back to the prehistoric times. This perhaps marks the earliest occurrence of our present day traditional rainfed cultivated fields in the vicinity of the settlements but there is virtually no chance of finding sites or recovering archaeological evidence of such a nature. Similar small cultivated fields in the vicinity of settlements are now very common in the western Sudan (Nuba mountains), the southeastern Sudan (Ingassana area) and the southern Sudan (e.g. the Nuer area).

An example of deliberate sowing of wild plants is derived from the Shoshoni Red-Indians of east central Nevada. They sow certain wild seeds, especially those of *Chenopodium* and *Mentzelia* (Steward, H.J., 1938: 33). Moreover, the Shoshoni of Owens Valley often irrigated certain wild seeds (ibid: 32).

As regards the development of cultivation practices, both botanical and ethnographic inferences have also been given to demonstrate that active harvesting and subsequent sowing of wild food-plants were the principle stages that constituted basis of incipient domestication. An example of this comes from the protoagricultural stage of food procurement when wild yam was regularly harvested in the forest zone in west Africa (Harris, D.R., 1976: 335). Other examples of sporadic cultivation of wild and semi-wild oil-palm can be seen in west Africa where harvesting and replanting were sporadically practised in the area from Senegal to Cameroon (ibid: 325, 335). It has even been stated that by the time emmer wheat and barley were domesticated in Egypt 7000 years ago, the Saharan population was "raising" food-plants (Stemler, A.B.L., 1980: 509). This statement and the assumption of the same author that 'early methods of cultivation of wild grasses may have involved .... sowing of seeds' clearly presuppose that these "raised" food-plants were wild and not domesticated.

The archaeological and historical records are, unfortunately, incomplete with regard to the development of the cultivation of wild plants in the Central Sudan. This is partly due to the scarcity of reliable evidence and partly to the lack of proper studies on palaeo land-use. However, the available fragmentary evidence from the Sudan suggests the following:

1) A little evidence of the initial steps which may have led to extensive exploitation of cultivation from the Sudanese frontier to Middle Egypt was dated back to 11000 to 14000 B.P. (Wendorf, F., and R. Schild, 1976: 270-271), but evidence indicative of such activities began to exist abundantly by c. 9000 B.P. This evidence consists of large numbers of grinders and (for the first time) of pottery recovered from sites of the Early Khartoum type in the Central Sudan (Clark, J.D., 1984: 116). The economy was, then, based on the utilization of terrestrial and aquatic food resources and the gathering and harvesting of wild food-plants. Although evidence of grain-foods was mostly inferred from the Early Khartoum site, the possibility that seeds of wild grasses (*Gramineae* family) were harvested and perhaps protected during their period of growth, should not be excluded until future research should prove otherwise.

2) The earliest archaeoethnobotanical evidence of cultivation of wild cereals, e.g. *Sorghum* sp., *Pennisetum* sp., and *Setaria* sp. from the Central Sudan probably dates to c. 6000 years B.P. (Haaland, R., 1981:195-197;Krzyzaniak, L., 1978: 159-172; Magid, A., 1982: 105-107; Shaw, T., 1976: 125). This evidence consists of impressions of cereal grains in pottery (cf. table no.41; Haaland, R., 1981: 195-197; Magid,A., 1982: 90-94). It also consists of abundant grinding equipment and pottery as compared with what was recorded from the same area in the period between c. 9000-6000 B.P. But no evidence is found of any tool that can be directly associated with such cultivation activities as harvesting or sowing; however, the use of cultivation tools of perishable nature, e.g. wood, is quite possible (cf. figs. no.45 and 46).

It also seems likely from the available indirect cultural and botanical evidence that cultivation of wild cereals did not start all of a sudden, but it was rather a process that passed through many developmental stages. This process was set in motion by a prior knowledge of wild grasses and their behaviour and the ways and means of processing and preparing them.

8.6 Summary.

The main points put forward in this paper about cultivation can be summarized in the following:

1) Cultivation is a process that demarcates a new type of utilization of plants and hence a more intimate relationship between man and the plants that he was already exploiting and/or was familiar with. This relationship started with the gathering of food-plants. It is, however, very difficult and would be artificial to set dividing lines for the transition between the gathering type of economy and the beginning of the cultivation of wild plants. Gatherers have also manipulated wild food-plants in various ways in response to many factors such as environment, demography, etc. As a matter of fact, gathering was probably practised in conjunction with cultivation for several millennia in the Central Sudan (cf. table no.41) and elsewhere (Bender, B., 1975: 1-5). It has been postulated, on the basis of ethnographic and ethnobotanical data, that gatherers may also have made little clearings or even planted wild food-plants (ibid).
2) As far as the prehistoric and early historic cultivation of food-plants in the Central Sudan is concerned, I find it inappropriate and highly biased to associate cultivation activities only with domesticated species. Generally speaking it is often the case that active cultivation of wild plants results in their genetic change (domestication). But if this process of cultivation stopped for a considerable period of time, then the process of domestication would also come to a halt. This would not mean the plant had never been cultivated. Furthermore, it is not always the case that cultivation activities are

regularly practised and/or undividedly performed. As I previously mentioned, archaeoethnobotanical evidence from the Central Sudan indicates that the occupants of the sites were cultivating wild cereals (Haaland, R., 1981: 213-215; Krzyzaniak, L., 1978: 159-172). This supports my argument that lack of evidence of morphologically changed species and perhaps lack of evidence of cultivation tools do not necessarily mean that the species in question was not cultivated (cf. tables no.43 and 44).

3) Cultivation is a practice that consists of a number of interrelated activities which can be practised altogether or individually, yet the fact remains that a form of cultivation is performed. It is very difficult to list all these forms particularly with regard to the present traditional cultivation. I assume, however, that there were three different yet interrelated forms of incipient cultivation. Regardless of their chronological order, these forms can be listed as follows:
   (a) a state of merely protective attitudes towards, e.g., sorghum in which several activities such as fencing, weeding, isolation from the free breeding gene-pool (Magid, A., 1982: 105) were partially or fully practised.
   (b) a state of protoculture, when the harvesting of seeds and the infrequent sowing of what was left after immediate needs were satisfied, e.g., round settlement area, were practised (ibid).
   (c) a state of protection but most important, reproduction of the species by simple sowing (ibid) to provide denser stands of the wild species and hence to yield a greater harvest.
4) There is no direct evidence - so far - of the cultivation of wild species in the prehistoric Central Sudan, however, the correlation of the remains of the material culture and the botanical data we obtained from several archaeological sites in this region (cf. table no.41), indicates that such activities may have been practised since 6000 years B.P. (ibid: pp.103-105; Haaland, R., 1981: 195-197, 213-215). Hence, it is not unlikely that some form of cultivation of wild plants started there at an even earlier period. This interpretation is supported by the present traditional cultivation in West Africa which show that most of the cultivation tools are of wood and branches. These tools are abandoned at the site of the cultivation activities after the job was done (Alexander, J., 1969: 123-125). Thus tools of similar nature may have been used and disposed of in the same way in the past and then simply decomposed and perished.
5) Cultivation activities are, indeed, labour demanding and time consuming when compared with gathering. There were a number of factors that led the Central Sudan and sub-Saharan populations in general to undertake such strenuous cultivation tasks. Among these the most basic are believed to be an increase in the population and consequent pressure on resources, climatic deterioration and environmental strains.

This could have resulted in the hunters-fishers-gatherers being pre-adapted in that they already developed protective attitudes towards plants.

6) Finally, there is no definitive evidence or technological criteria - so far - to set a traditional boundary between gathering to cultivation of wild plants (mainly cereals) and between the domestication of these in the Central Sudan and in the Sahel zone at large. As a matter of fact the available evidence strongly indicate that the three activities involved in plant exploitation (gathering, cultivation of either wild or domesticated plants) had come into existence, they continued to co-exist until the present day, and that each one was the pre-condition for the emergence of the next in sequence.

Section B

The Domestication of Cereals in the Central Sudan: With emphasis on sorghum[5]

8.7 Introduction.

In the light of my discussion of the cultivation of wild plants (see section A), I shall now focus on a more advanced stage of this practice and try to explain why and how this has led to genetic changes in plants (domestication) with the main emphasis on sorghum. I shall also discuss the essence of harvesting and sowing and the assumed crucial role of harvesting tools in the process of domestication. My interpretations on these issues are derived mainly from the archaeoethnobotanical material we recovered from the Central Sudan but also from other relevant archaeological and ethnographic evidence from this area and elsewhere.

8.8 The Process of Domestication: Definition and Criteria.

The domestication of plants has been defined as an evolutionary process - not a revolutionary event - in which plants are genetically changed through human activity (Bender, B., 1975: 1; Harlan, J.R., 1977: 358; Harlan, J.R., et al, 1976: 6; Higgs, E., 1976: 34; Stemler, A.B.L., 1980: 521; etc.). It is basically a biological process but one that would not have been initiated without the interference of man. This does not, however, mean plant domestication started deliberately; it was most likely an unintended consequence of certain intended human activities (Haaland, R., 1981: 212) such as sowing or harvesting to obtain sufficient supplies of food of vegetal origin.

Selection pressures are exerted by man when he plants what is harvested. If the same gene pool is actively cultivated with sufficient intensity for a sufficient period of time, changes will occur in the morphology and ecological adaptation of the wild species (Bender, B., 1975: 1; Harlan, J.R., et al, 1976: 6). In this connection, a distinction has been drawn between the terms "cultivation" and "domestication". Cultivation is thought

---

[5] Since there is no biological evidence so far of domesticated pennisetum, this robust cereal shall be cited when appropriate.

to be 'a matter of directing ecology, whereas domestication depends mainly upon some physiological inefficacy in the plant, which is advantageous to man' (Helbaek, H., 1969: 404).

The domestication of plants is an ongoing process (Harlan, J.R., et al, 1976: 6). It should be noted that this process rarely leads to 'the evolution of new species in the biological sense' (Harlan, J.R., 1977: 358) and that although the domesticated plants are dependent for their growth and survival on man the accumulated genetic differences tend to be relatively modest (ibid). Appeal to external morphological changes as one criterion for distinguishing domesticated plants has been criticized by those who have suggested that changes in, for instance, the shape and the size of grains may result from natural selective processes (Higgs, E., 1976: 32). With regard to this issue, the results of the statistical study I conducted on several recent wild and domesticated species (cf. chapter IV) show that the size and sometimes the shape of the grains depends on many factors such as their position in the inflorescence, the amount of water the plant receives during its growth and reproduction period, the type of soil and nutrients, etc. (for more information see chapter IV). It is also essential to correlate the biological evidence (morphologically changed species) with evidence from material culture and the context in which these are found. In addition, it is also important to retrieve and study any evidence indicative of the climate. Then one may be able to evaluate the ecological advantages and disadvantages dependent on natural or man made conditions determining the growth and reproduction of the species in question (Fægri, K., 1987: pers. comm.). For instance, there are many hybrids which do not survive under natural conditions but may stay alive due to man's creation of a suitable ecology (Helbaek, H., 1969: 404). This is particularly true as regards widely used cereals such as free threshing wheat and maize (ibid: 405); but its validity has not been established in the case of *Sorghum* and *Pennisetum*. In short, changes in morphology - as a diagnostic character - of the domesticates is essential; however, its reliability depends on how cautious and thorough the examination of this character is and also on our ability to perceive morphological changes attributable to domestication.

8.9 Domestication of Cereal Crops.[6]

In the process of cereal domestication, much of the genetic change is automatic and results from regular planting and sowing of harvested grains (Harlan, J.R., et al, 1976: 6). Unless the grain-stock for the next year's sowing is taken from the harvest of the previous year the process of domestication will always be set back to the beginning

---

[6] Domestication of tree and tuber crops and less robust cereals than *Sorghum* and *Pennisetum* is beyond the scope of the present work.

(Stemler, A.B.L., 1984: 129). In this manner of regular harvesting and sowing, strong selection[7] pressures are at work; and two populations start to emerge. One population is tended by man, and the other maintained by nature, 'with selection going in opposite directions' (Harlan, J.R., et al, 1976: 6).

The natural selection would proceed as it always did favouring dispersal of propagules (ibid: 6-7). That is to say, the ecological adaptation of wild cereals, e.g. *Sorghum* and *Pennisetum*, enables them to release the seed bearing structure and the propagules to be disseminated (Stemlers, A.B.L., 1980: 515) away from the parent plant to a relatively open type of habitat. Thus, they can germinate more freely without competition for sunlight, water and soil nutrients from the parent plant of any other members of the species (Stemler, A.B.L., 1984: 129). If the cereals do not disseminate - under natural conditions - all the grains will fall and germinate in the same place but none of the progeny will reproduce due to lack of light and competition for water and soil nutrients. There has always been in the wild population of cereals - as in all wild plants - strong natural selection for efficient mechanisms of seed dispersal (Stemler, A.B.L., 1980: 515; 1984: 129). Such dispersal mechanisms are facilitated by the formation of an abscission zone which consists of several layers of specialized cells at the base of the seed-bearing structure. With the maturity of the wild grains/seeds, the cells in the abscission zone become weak as a result of the dissolution of their wall. Hence, the stalk of the seed bearing structure detaches from the parent plant (cf. diagram no.8; Stemler, A.B.L., 1980: 515). One to several genes[8] control the formation of the abscission zone and the dissolution of the wall of the cell (ibid)As regards the population which is tended by man, it has been observed that the needs of the farmer are the contrary to those of the wild plant population (ibid). Man would consciously start to select non-brittling or semi-brittling heads with less efficient dispersal mechanisms and save the harvest for the next sowing. The remaining part of the field is indiscriminately harvested for consumption. In other words, the components that contribute to the next generation are those which are chosen by the cultivator; hence the rest will be eliminated from the gene pool (Harlan, J.R., et al, 1976: 8). This process is still going on in traditional cultivation in the Sudan. For instance, in the Ingassana area in the east, the Nuba mountains in the west and among the Nuer in the south, compact inflorescences with relatively large grains are selected and stored[9] as grain-stock for the next year's sowing. This manner of harvesting and sowing

---

[7] Man selects for colours, different flavours and tastes, storage or ease in processing, etc. (Harlan, J.R., et al, 1976: 6).

[8] According to Doggett, the number of the basic chromosome in the genus *Sorghum* is 5 but this number increases at least four times in what he defined as cultivated species such as the diploid sorghum and 2n=40 in tetraploid etc. (Doggett, H., 1965: 51, 54). It was also stated that the tropical ecotype weed of sorghum is 2n=20 chromosome as evident in southeast Asia and south India (deWet, J.M.J., and J.P. Huckabay: 1967: 79).

[9] The heads are normally stored in the kitchen-hut where the smoke from cooking scares away insects and flies which could otherwise destroy them.

Diagram no.8 **DIFFERENCES BETWEEN WILD AND DOMESTICATED CEREALS: SPIKES**

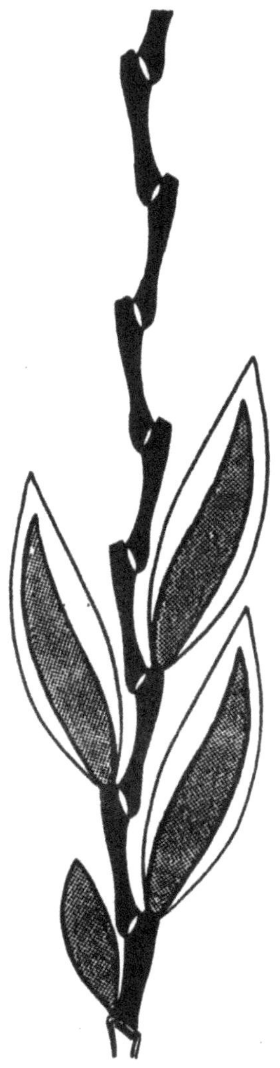

Analytic diagram of cereal spike. The black, composite column is the axis or rachis. Its component parts, the internodes, are separated by white interstices denoting the points at which, in the wild cereal, the axis comes apart in sections or units consisting of the internode with attached spikelet(s) (brittle spike). These disintegration points may become solid by mutation; then the axis does not fall apart, but remains one tough column. The toughness of the spike is the most significant property in domestication.

The spikelet consists of two glumes (the hatched area) enclosing one or more florets (white area). A vestigial spikelet is shown at the base of the axis.

The floret consists of two inner bracts (paleas) within which the ovary, styles and anthers are placed, eventually developing one fruit (caryopsis or grain). Anatomically the grain is a seed vessel containing one seed, but the two organs, the seed case and the seed coat, are fused into one (grain shell).

After Helback,H.,1969: 390

will help promoting anautomatic and rather quick appearance - in the actively cultivated population - of mutants in which

> 'cells in the region where an abscission zone normally forms, do not dissolve their cell walls and the seed bearing structure does not detach from the parent plant' (Stemler, A.B.L., 1980: 515)

It is these mutants (which are disadvantageous to the wild population because they hold on to their grains[10]) that are most often the ones to be chosen for harvesting and to be sown the year after (Harlan, J.R., et al, 1976: 7)

The rates of morphological and ecological changes from wild to domesticated species depend largely on the regularity, the intensity and the length of time since the grains are first harvested and sown and that the most modified part(s) of the domesticated plant is usually being the part(s) most used by man (Stemler, A.B.L., 1980: 515). In the case of sorghum and pearl millet, the most modified parts are the grains (cf. fig. no.52). The more the above mentioned activities are performed, the more the plants become dependent on man for their survival; and the morphological and ecological differences between these and their wild progenitors gradually accumulate and become more and more prominent (Helbaek, H., 1969: 402-405). Some of the results of experimentation with cultivating wild plant-species show that ecological and morphological changes can be established after only ten years (Fægri, K., 1987: pers. comm.). It has also been stated that with 50% gene flow, a polymorphic population of *Drosophila* was obtained after thirty-six generations, and with 25% gene flow, a polymorphic population with a greater divergence was obtained in twenty-five years (Dogget, H., 1965: 67). Despite lack of evidence it was suggested that domesticated sorghum possibly arose from the wild one in this manner (ibid). An example of the most basic differences between wild and domesticated cereals as seen in *Sorghum* and *Pennisetum* is given on table no.45 and diagram no.8. The grains are separately described (cf. table no.43). This is because most of the archaeoethnobotanical material from the Sudan consists of fragmentary but quite often carbonized grains and glumes or their impressions (on pottery) (cf. chapter IV). Grains are separately described because they are the most modified part of the plant (cf. fig. no.52). The most significant external morphological differences between wild and fully domesticated cereal become most apparent when wild and domesticated plants with mature grains are compared (Stemler, A.B.L., 1980: 514).

---

10 In genetics non-shattering of inflorescences of domesticated cereals is considered a recessive character; therefore, in such genotypes with non-dispersal traits, the genetic mechanism to dissolve the walls of the cells in the obscission zone is defective and hence is not carried out (Stemler, A.B.L., 1980: 515).

Table no.43. Some basic diagnostic differences between wild and domesticated cereal. Example: *Sorghum* & *Pennisetum*.

Wild race	Domesticated race
Grains	
(1) Shattering* of heads and ears and dispersal of grains.	(1) Non-shattering of heads and non-dispersal of grains.
(2) Usually enclosed in glumes at maturity (grains have tightly fitting husks).	(2) Usually exposed by the gaping and half-opened glumes.
(3) Generally, grains are small in size.	(3) A trend towards large grains.
(4) Maturity of heads of grains is not uniform due to growth of many branches from the stem producing many inflorescences over a period of time (normally several weeks).	(4) Uniform maturity of heads and grains from one head due to sowing seeds that were harvested at a single time.
(5) Grain-dormancy is common.	(5) Grain-dormancy is much more reduced or eliminated.
(6) Relatively elongated grains.	(6) More rounded-oval grains.
Other diagnostic characters: Whole plant	
(1) Plant perennial with well developed rhizomes.	(1) Plant annual without rhizomes or at most short ones.
(2) A few number of panicle branches or rows of spikes and relatively long panicle branches and internodes.	(2) Multiplied (increase number of panicle branches or row number of spikes) and condensed (shortening) of panicle branches and internodes) inflorescence.
(3) Intermediate grow with maturation over a long period of time (grain production continues as long as conditions permit several branches with several inflorescences).	(3) Single growth with one maturation period (one stem with one inflorescence).
(4) Loose inflorescence.	(4) Compact inflorescence.
(5) Brittle rachis.	(5) Toughening of rachis.
(6) Tough glumes.	(6) Loss of tough glumes.
(7) Racemes are tough en persistent sessile spikelets.	(7) Racemes are articulated at maturity
(8) Leaves are narrow and short.	(8) Leaves are broad and long.

* Shattering (spontaneous release of grains) is one of the most critical and diagnostic characters in distinguishing between wild and domesticated races.

Fig.no.52 **Sketch of five basic spikelet types of domesticated sorghum.**

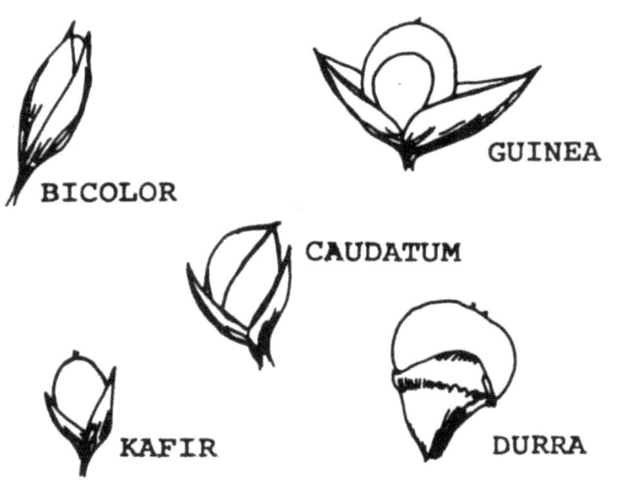

After Harlan,J.R.-and J.M.J.de Wet 1972: 173

It is clear from the shape of the spikelets that the S.bicolor race is the most primitive type of domesticated sorghums;having long clasping glumes, elongated seeds and open panciles.

There are several intermediate morphological forms between wild and fully domesticated cereals, but it is quite difficult and sometimes an impossible task to identify or to determine accurately the extent of the morphological change, for instance to recognize an early or middle stage of change in individual or fragmented parts of plants. A whole plant should, thus, be an ideal specimen for identification because different characters (e.g. leaves, internodes, grains, etc.) can be compared and integrated in the final result of the identification. Since the archaeoethnobotanical evidence is mostly found as individual or fragmentary parts of plants, the procedure for identifying these emphasizes the indispensable role of reference collections of recent plants for comparative purposes. It is also ideally and practically necessary to have a reference collection of different species or subspecies of the same genus both in their wild and domesticated states. I have found from my experience that it is immensely useful to prepare carbonized and impression-reference collection from fresh parts of different species and subspecies of the same genus both in their wild and domesticated states. It is equally useful to determine statistically the percentage of change in the size and external shape of the collection after carbonization of the specimens (cf. chapter IV). In this manner, it will be possible to account for the expected change in the size and external morphology of similar evidence recovered from archaeological sites.

Not all actively cultivated plants are morphologically changed. An example of this is the large number of wild species of grasses such as *Setaria*, *Panicum* sp., etc. which have been - and some still are - cultivated or harvested in sub-Saharan Africa and the Central Sudan (cf. table no.41 and chapter VI). But none of these has become a domesticate of significant economic or dietary value. Harlan and others have referred to Jardin study of 1967 which shows that more than 60 species of grasses in Africa were regularly harvested but that only five of them were domesticated (Harlan, J.R., 1977: 373; Harlan, J.R., et al, 1976: 9; Stemler, A.B.L., 1980: 516; 1984: 129). Another classic example is sesame, *Sesamum indicum* L. Although there is evidence that this crop species has been subject to regular and continuous cultivation from the 5th century B.C. (Bedigian, D., 1984: 73-74) until the present this plant species is still naturally releasing its seeds from their capsules and spilling them on the ground if not harvested at the right time (ibid: 74). This is obviously contradictory to the fact that the loss of ability to disperse the seeds and fruits is the most common diagnostic character indicative of genetic change in plants (Helbaek, H., 1969: 404). One may, then, ask why only a few species of this large number of grasses were domesticated. Why do some still retain some of their wild form characteristics although they are regularly cultivated? The answers to these questions are probably many and complex but Stemler attributed part of the answer to the method of harvesting and the introduction of what she described as "good, effective, more advanced cutting tools" (Stemler, A.B.L., 1980: 521-522; 1984:

129-131). Although I agree that the manner in which harvesting is done is essential in the process of domestication of cereals I have my reservations as regards the role of harvesting tools. I shall discuss this in detail later in this chapter.

Another part of the answers to these questions is based on ecology. That is to say, although some plant species are regularly harvested and sown, they still retain all or some of their wild-form characters, because they did not succeed in habitats created by man.

> 'Fitness to a primary habitat is one thing; fitness to a cultivated field is another' (Harlan, J.R., et al, 1973: 323).

The ability to thrive in a man-made habitat is considered as the second basic characteristic of domesticated cereals (Hawkes, J.G., 1969: 19-21; Stemler, A.B.L., 1980, 517). Furthermore, reasons still unknown some plants do not genetically respond to man's activity and their genotypes remain constant without change for quite a long period of time even though the plant has been subjected to continuous regular cultivation (Fægri, K., 1987: pers. comm.). An example of such is sesame, *Sesamum indicum* L. A third possible answer to the first question may be related to cultural preferences. Both ethnographic studies and archaeological evidence indicate that people often begin by experimenting[11] with cultivating several cereal species. Eventually, they would continue to cultivate only one or two species while the others would be dropped or returned to only at times of need. My ethnographic observations among the inhabitants of Darfur in western Sudan indicate their strong cultural preferences for pearl millet even in habitats best suited for cultivation of sorghum. Only at times of food shortage[12] will the inhabitants eat sorghum.

Archaeoethnobotanical evidence recovered from the Central Sudan shows that wild species of *Sorghum*, *Pennisetum*, *Setaria*, *Panicum* were probably cultivated between 6000-4000 B.P. (Klichowska, M., 1978: 42-43; cf. table no.41). But about 4000 years later, it was only *Sorghum* and *Pennisetum* sp. that were still cultivated and by then domesticated (cf. table no.41). Thus, it is also possible that during prehistoric and early historic periods in the Central Sudan cultural preferences were one of the factors that led to the domestication of some species through their continuous cultivation and the abandonment of such regular interference as regards other species and so they remained in their wild state.

---

[11] Experimentation is a term which is used in this context to denote an unconscious choice of trying many "species" at the same time.

[12] I obtained this information during field work in 1985 when there was a severe drought and serious food shortages that lasted over five years between 1980-1985.

8.10 The Role of Harvesting and Sowing in the Domestication of Cereals.

8.10.1 Importance of harvesting.

The centre of attention in the following discussion is on the manner in which harvesting and sowing must be performed in order to initiate the process of cereal domestication. I shall outline the generally held interpretation as regards this issue and my own interpretation in the light of the results I obtained from my research in the Central Sudan.

The process of cereal-domestication did not start before there was a regular and continuous pattern of selecting heads of grains from the field for the next year's sowing from the present year's harvest (Harlan, J.R., et al, 1976: 8, Hawkes, J.G., 1969: 22; Stemler, A.B.L., 1980: 515; 1984: 129). The ideal manner in which this activity is performed is described as follows:

> 'a common procedure at harvest time is for the cultivator to go through his field and carefully select heads to be saved as stock seed for the next year planting' (Harlan, J.R., et al, 1976: 8).

Since the heads of wild grasses shatter at maturity, the harvesting of these must take place during a carefully decided time during the brief period between the ripening of the seeds and the shattering of the heads (Cohen, M.N., 1977: 145).

Although the cultivation of wild cereals is an essential step in the process of domestication, this by itself will not lead to domestication if the sown grains are harvested by stripping or tapping them off the plant or sweeping up fallen grains from the ground (Stemler, A.B.L., 1980: 515, 516-517, 521; 1984: 129) because in this manner the harvester will not affect the population of cereals differently from the forces of natural selection. Thus, there will be no change in the selection for mechanisms of natural dispersal (Stemler, A.B.L., 1980: 516). Therefore it is essential,

> 'in order for cereal population to be changed genetically, the human harvester must set up a situation in which natural dispersing plants are not selectively favoured. This is exactly what happens when whole inflorescences are harvested and transported [away from the field] to storage areas' (ibid).

This method of harvesting and handling the yield would allow ingathering of cereals that do not disperse their grains (ibid). This means (conscious or unconscious) selection would favour the mutants that are constantly produced in any natural population which do not disperse its seeds (Stemler, A.B.L., 1980: 516; cf table 44). Less efforts is

Table no.44. Adaptation syndromes resulting from automatic selection due to planting harvested seed* with emphasis on *Sorghum* and *Pennisetum*.

SELECTION PRESSURES ASSOCIATED WITH HARVESTING RESULT IN:

A. Increase in % seed recovered
   Adaptations: (1) Nonshattering
   (2) More determinate growth
   (a) Growth Habit I. - Cereals whose wild races have lateral seed-bearing branches; sorghum, pearl millet. There is a trend toward apical dominance resulting in: fewer inflorescences, larger inflorescences, larger seed, greater daylength sensitivity and more uniform ripening.

B. Increase in seed production
   Adaptations: (1) Increase in % seed set.
   (2) Reduced or sterile flowers become fertile.
   (3) Increase in inflorescence size, especially in maize, sorghum, pearl millet.
   (4) Increase in number of inflorescences especially in wheat, barley, rice, etc.

SELECTION PRESSURES ASSOCIATED WITH SEEDLING COMPETITION RESULT IN:

A. Increase in seedling vigor
   Adaptations: (1) Greater seed size.
   (2) Lower protein; higher carbohydrate.

* After Harlan, J.R., et al, 1973: 314 with some modifications.

devoted and fewer grains are lost in this method of harvesting and transportation. As I mentioned in the foregoing, the traditional cultivation practices in the Sudan include regular selection of heads of sorghum and their storage for the next year's planting. Evidence of whole carbonized inflorescences is one of the earliest indications of domesticated sorghum (*S. bicolor* (L.) Moench) from the Central Sudan. They were recovered from the site of Jebel Tomat (A-Bari, E., 1987: pers. comm.); but no evidence of whole inflorescence of wild sorghum from this area is available at present. However, since evidence of domesticated sorghum-inflorescences was found, there is reason to believe that inflorescences of wild sorghum were also harvested and possibly kept and perhaps taken along when the people moved from one area to another. More detailed information is presented in the following chapter.

8.10.2 Importance of harvesting tools: Re-evaluation.

The foregoing stresses the essence and consequence of harvesting selected whole inflorescences of cereals and the storage of these for the next year's sowing. These conditions were probably all the requirements needed for the domestication of festucoid grasses such as wheat and barley. But it has been suggested that the requirements as regards the domestication of panicoid grasses are somehow different,

> 'harvesting panicoid grasses .... such as [wild] sorghum and and pennisetum requires a good cutting tool ....' (Stemler, A.B.L., 1980: 520).

This statement clearly emphasizes the necessity of a cutting tool in harvesting *Sorghum* and *Pennisetum*. The same author added that,

> 'harvesting with a cutting tool exerted a new set of selective pressures on populations of grasses that were sown and harvested that resulted in plants that did not naturally disperse their grains' (ibid: 523).

Although the last statement chiefly emphasizes the necessity of harvesting tools, its focus is undoubtedly broader than the previous one because it repeatedly stresses that domesticated cereals in general are developed due to the introduction of cutting tools in the harvesting of wild cereals.

Many scholars working on the transition of food production in sub-Saharan Africa support the foregoing statements (Clark, J.D., 1976: 97; Bender, B., 1975: 52-53; Harlan, J.R., et al, 1976: 17-18; Higgs, E., 1976: 33; etc.). They conclude that although the exploitation of wild African cereals probably started about 10000 to 6000 years ago, domestication did not take place before 3000 to 2000 years ago. They attributed this delay in the domestication of African cereals mainly to the late introduction of a new harvesting method which employed a very good and effective cutting tool (Stemler,

A.B.L., 1984: 129-131), which was much better than any tool 'produced until relatively late in the history of man in Africa' (Stemler, A.B.L., 1980: 520). The introduction and the use of such a "good and effective" cutting tool in harvesting is seen as 'a crucial step in the process of domestication of African cereals' (ibid: 523). While, however, at one point the necessity of cutting tools in the process of domesticating all wild grasses was clearly stated (without exceptions), at another this was indicated as being insignificant in the process of domesticating small festucoid grasses (such as wheat and barley). But it was stressed that regular sowing and harvesting are essential in the domestication of these festucoids. To quote our source,

> '.... domestication did not begin until people began harvesting whole inflorescence, ... Harvesting inflorescences of small festucoid grasses such as wheat and barley is far easier than harvesting panicoid grasses such as sorghum and pennisetum .... Harvesting whole inflorescence of wild sorghum and pennisetum requires a very good cutting tool ......' (ibid: 520).

It appears to me that these statements (by the same author) are selfcontradictory since, on the one hand a general rule is established while on the other some exceptions to this rule are made without explaining why. They also contradict what stated elsewhere, namely, cutting tools were essential in the process of domestication of festucoid grasses, e.g. wheat and barley (Flannery, K., 1965: 1247-1256). Thus it is unclear whether the role of the cutting tool is crucial in the domestication of all grasses or only of some of them. Furthermore, does such a contradiction of statements mean a cutting tool was not at all important if the harvester managed to detach the inflorescence without using any tools? Finally, nothing has been mentioned as regards the type of tool used. It would have been useful to have a description of such a good, effective cutting tool e.g. the raw-material from which it was made, specifications of the working edge(s), e.g. whether these were serriated, dull or had a razor/knife-like sharpness, etc.

I have more than one reason to argue that cutting tools were not essential in the harvesting of cereal populations in order to exert pressures favouring non-shattering heads. This argument is based on the available archaeoethnobotanical evidence from the Central Sudan (cf. tables no. 41.and 42), ethnographic studies conducted in sub-Saharan Africa and my own studies in the Sudan. As regards the archaeoethnobotanical evidence, no evidence of a cutting tool that can be associated with harvesting has been found in those contents where evidence of domesticated cereals, e.g. sorghum, has been recovered. *Sorghum bicolor* (L.) Moench was found at the site of Jebel Tomat in the Central Sudan (cf. fig. no.1) and dated to A.D. 245±60 (Clark, J.D., and A.B.L. Stemler, 1975: 588-591) but no evidence of harvesting tools was reported from this site. The remains of domesticated sorghum from the Jebel Tomat are one of the earliest pieces of evidence found - so far - in the Sudan. On the basis of this and of faunal and material culture evidence, Clark stated that the occupants of the site combined herding with hunting, fishing and the cultivation of sorghum during the rainy season (Clark, J.D., 1984: 123). A small scale excavation was carried out (cf. chapter III) at the same site (El-Mahi, A.T., and R. Haaland, 1984: 28-32) but no evidence of a cutting and/or harvesting tool of any kind was found (cf. tables no.41 and 42). Another type of evidence is derived from a rock engraving at Jebel Qeili (cf. fig. no.1) dated to the first century A.D. (round A.D. 41-45) in which domesticated sorghum[13] (cf. pl. no.1) is depicted (Dunham, D., 1957: 7; Shaw, T., 1977: 98-99; Shinnie, P., 1967: 51,96,159).

Chronologically, despite the fact that this rock engraving at Jebel Qeili is almost two hundred years older than the site of Jebel Tomat they both fall within the late Meroitic period. Culturally, both sites are also associated with the late Meroitic period. That is to say, the theme of the scene of the rock engraving of Jebel Qeili is about king Sherkarer triumphing over enemies (cf. pl. no.1). According to Dunham's chronological chart of Meroitic kings, Sherkarer most likely ruled between years A.D. 41-45 (Dunham, D., 1957: 7; Shinnie, P., 1967: 50, 61). In addition Shinnie maintained that the site of Jebel Qeili is one of the most southerly Meroitic monuments (Shinnie, P., 1967: 50).

As regards Jebel Tomat, the remains of material culture recovered from the late occupational period of the site are similar to those from the late Meroitic period. This is evident from similarities in pottery (Clark, J.D., 1984: 123). Furthermore, the luxury items such as imported stonebeads, faience, copper and iron certainly confirm its Meroitic affinities (ibid). Thus, both chronological evidence and remains of material culture show that these two sites seem to be contemporary with the Meroitic civilization at a period when iron technology was well established (Shinnie, P., 1971: 80-107; 1985: 28-36), but no evidence is found of iron cutting tools or any other type of cutting tool that might be associated with harvesting. Earlier evidence of domesticated sorghum has been reported from the area of Meroe (cf. fig. no.1) and dated as far back as 200 B.C. (Clark, J.D., 1984: 123).

According to Shinnie, iron hoe was the main cultivation tool (Shinnie, P., 1967: 162). But despite the presence of iron hoes and other types of large and small iron implements such as tweezers, pins, arrow-heads, shears, daggers, knives, adzes, axes, chisels, etc., in the area of Meroe and further north at Barkal, Kurru, Nuri and Kawa (cf. fig. no.1; Addison, F., 1949: 75, 107, 152, 153, 163, 172, 173, 178; Crawford, O.G.S., C.B.E., and F. Addison, 1951: 32, 33, 38, 74, 86, 98, 99, 100, 128, 129; Dunham, D., 1955: 12, 23, 46, 166, 211; 1957: 26, 29, 44, 170, 173, 177, 192; 1963: 35, 118, 155, 206, 275, 180, 334, 380, 383, 442; 1970: 70; Save-Söderbergh, T.,

[13] On the basis of the distribution of the wild and early domesticated races of sorghum (Harlan, J.R., and A.B.L. Stemler, 1976: 465-488), I think the evidence from Jebel Qeili is most likely *S.bicolor* (L.) Moench and not guinea corn as Shaw suggested (Shaw, T., 1977: 98).

1982: 70; Shinnie, P., 1967: 162-163; Shinnie, P., and R. Bradley, 1980: 222-294), no evidence at all of a metal or (even) a stone tool that might have been used for harvesting is reported in any of these publications. The earliest evidence of iron sickles (cf. pl. no.12) - to my knowledge - was found in site 19 at Faras station in the northern Sudan (cf. fig. 1) and dates to the late X-Group period ca. A.D. 500-600 (Save-Söderbergh, T., 1981: 68). This evidence is a few hundreds of years later than the earliest evidence for domesticated sorghum from the Central Sudan.

Stemler's suggestion that shears [diagram no.9(a)] might have been used for harvesting sorghum (Stemler, A.B.L., 1984: 128) is not supported by any archaeological or ethnographic evidence. In addition, being familiar with shears, I can say that they are impractical [cf. diagram no.9(a) and (b)] and very disadvantageous to use as tools for harvesting of sorghum. That is because the cutting blades would be diagonally superimposed on one another when the tool is closed on an object [cf. diagram no.9(b)] with a gap of a few millimetres between them [cf. diagram no.9(c)]. Thus if this tool is used for harvesting, its blades will make only two alternating grooves or unfinished cuts in the sorghum stalks [cf. diagram 9(d)], otherwise several attempts at shearing across the same section would be needed before the stalk was finally cut. Such a repetitive process (along the same section) necessarily requires concentration while cutting, and is both time and labour consuming. In addition, the shape of the shears as such and of their handle in particular would not provide either an appropriate grip [cf. diagram 9(a)] or an efficient performance while harvesting the tough stalks of sorghum. According to my knowledge shears are used at present only for shearing sheep, goats and donkeys (in that order of preference) in most parts of the Sudan. Thus, they could have also been used for the same purpose in the past. In addition, since there is evidence for a textile industry, cotton and cloth from e.g. Meroe (Dunham, D., 1963: 76, 143, 146, 150, 206, 283), it is not unlikely that small shears were also used as scissors[14] for cutting cloth.

Ethnographic observations from El-Gedarif in the Central Sudan (cf. fig. no.1) show that the rainfed traditional cultivation practices in this area include harvesting of fruiting cobs of domesticated sorghum by hand. We do not know how old this method of harvesting is or if there is a continuity of tradition from (?) the Meroitic period as exemplified by the site of Jebel Tomat (about 1700 years ago) to the present day. Harvesting of sorghum by hand was one of the methods probably used from prehistoric times up to the present where it survives in the traditional methods of cultivation in the Sudan.

[14] The available published materials from the Meroitic civilization do not indicate any evidence of scissor or scissors-like implements other than these shears.

Plate no.12 Iron-sickle: late X-Group period

After Save-Soderbergh,T.,1982:pl.90, no. 1

Diagram no.9 Sketches of iron shears from different views.

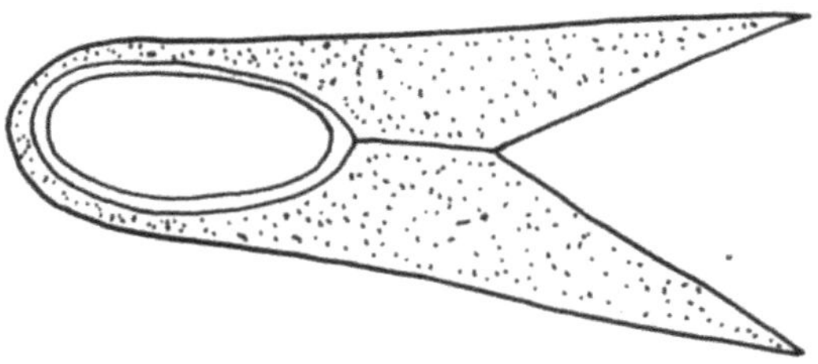

a) top view (after Shinnie, P. 1967: 162)

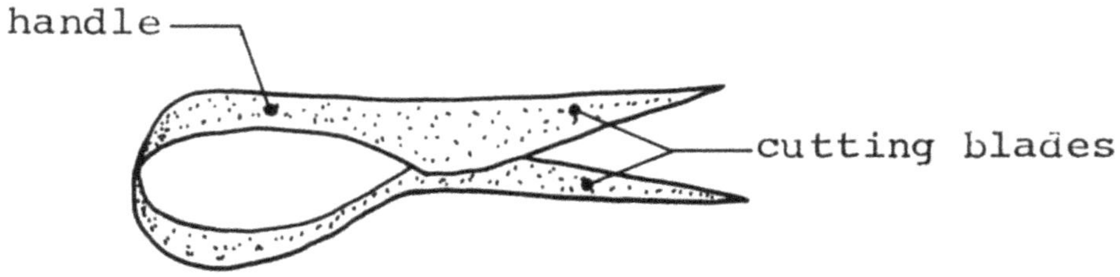

b) side view: diagonally superimposing blades

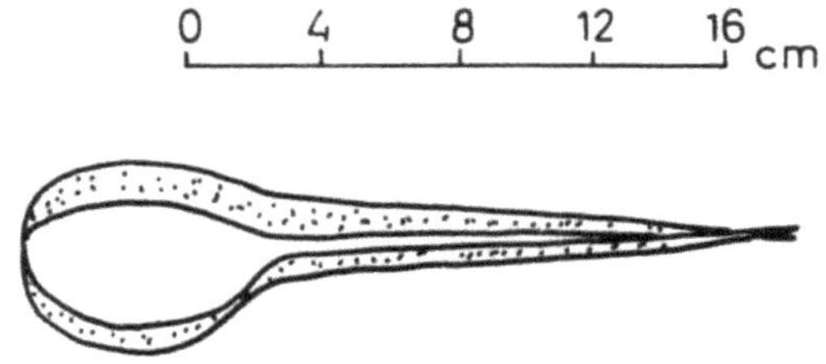

c) side view: Blades closed on each other (note the gap between them)

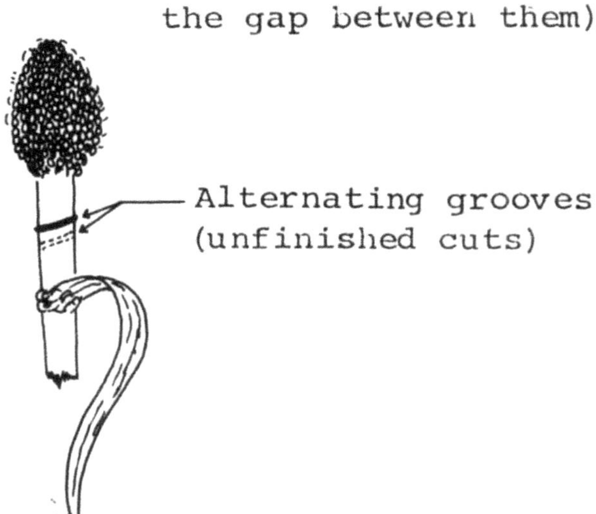

d) Effect of cutting (shearing) blades on stalk of sorghum

Another method of harvesting (pearl millet and sorghum) in traditional cultivation uses a short iron blade hafted in a wood handle[15]. This type of tool was commonly used in the western Sudan and is locally called *'Tigda'* (Arkell, A.J., 1937: 306-307). The earliest available evidence of the *tigda* in the Sudan comes from the ruins of a town at Uri in northern Darfur which was dated to between A.D. 1225 and 1475 (ibid: 300).

This brief outline of the presently available evidence supports the interpretation that cutting tools were not essential in the process of domestication of *Sorghum* and *Pennisetum* but offers no explanation for the late dates for the beginning of cereals domestication in the Central Sudan. The lack of harvesting tools on archaeological sites - in the Central Sudan - where there was evidence both of wild and domesticated sorghum (cf. tables no.41 and 42) indirectly supports my argument. Cutting the heads of sorghum by hand was probably much easier than using tools (lithic or metal). A practical reason for not using cutting tools in the harvest of sorghum is the fact that some (and sometimes most) of the grains will drop off the head while the stalk is being cut as a result of the friction between the cutting tool and the stalk regardless of whether the tool has a serrated or a knife-like working edge. This is particularly true with wild cereals and incipient domesticates which were still shattering or had not yet completely lost their natural shattering and dispersal mechanism as in the incipient domesticates. Another observation which gives further support to my argument is from such villages round the city of Atbara as El-Bauga (cf. fig.no.1). The farmers in these villages use iron sickles with serriated working edges to cut the stalks of sorghum just above the ground before piling them in the field for drying. But when they separate the "now" dry heads of sorghum, they prefer to use their hands because - I was informed - it is much easier and faster. In short, the archaeological evidence presented here supports my view that harvesting tools did not play an essential role in the domestication of *Sorghum* (and possibly *Pennisetum*) in the Central Sudan and can not explain the late date of their domestication.

In the light of the foregoing discussion, I would like to stress the need for conducting further research with emphasis on finding archaeoethnobotanical material which would support or falsify this interpretation.

8.10.3 Importance of sowing.

The focus of this discussion will be on the manner in which sowing should be done and its significance in the process of cereals domestication.

According to Bender "accidental" sowing of regularly harvested seeds would also lead to genetic change (domestication) of wild cereals just as purposefully done sowing would do (Bender, B., 1975: 53). But in making this suggestion he did not justify or

---

[15] This tool is more efficient and mostly used for cutting the heads of pearl millet rather than of sorghum ones because the latter are too hard to cut with it (Arkell, A.J., 1937: 306).

explain how regular harvesting could be effective in the process of domestication of the yield (fruiting cobs) was not derived from regularly sown seeds. It is even more important to mention that this suggestion contradicts the generally accepted view that regular sowing of grains/seeds from the previous year's harvest is as essential in cereal domestication as regular harvesting of whole selected inflorescences (Harlan, J.R., et al, 1976: 8; Stemler, A.B.L., 1980: 516; 1984: 129). That is to say '.... if domestication of cereals is to be achieved there must be regular sowing and harvesting' (Stemler, A.B.L., 1980: 515). Furthermore, it was strongly emphasized that the specific manner in which the grains/seeds are sown is just as important as regular sowing. This manner of sowing consists of,

> 'an active process which involves the careful retention of seeds and the concept of placing them in the more or less prepared soil' (Hawkes, J.G., 1969: 22).

Sowing[16] in this manner automatically results in larger seeds (deWet, J.M.J., et al, 1976: 461). The seedlings produced will also have competitive advantages of soil nutrients (cf. table no.44) and sunlight if the field is infested with weeds[17]. Therefore deliberate sowing of harvested seeds, season after season, creates a completely new set of adaptation; and the selection pressures change in favour of non-brittle spikes (Hawkes, J.G., 1969: 23; Stemler, A.B.L., 1980: 515-516; 1984: 129-130).

As a result of practising a regular and continuous harvesting and sowing the same gene pool of wild cereals in the manner mentioned in the foregoing (beside other practices to be discussed in the forthcoming section), the domesticated African cereals were finally developed. There are at present several races of domesticated sorghum in Africa, a simple classification of these, their intermediate races and their probable wild progenitors and the distribution of all these are illustrated in figures number 2, 53, 54A and B; plates number 13, 14A and B; and table number 45. It should, however, be noted that there are neither conclusive genetic, morphological (deWet, J.M.J., and J.P. Huckabay, 1967: 799-800) or concrete archaeological data (Doggett, H., 1965: 67) that shed light on the issue of the probable wild progenitors. As regards *Pennisetum*, current usage includes all domesticated pearl millets and their spontaneous relatives within *Pennisetum*

---

16 In the décure cultivation (commonly practised in the Delta of the Niger river), the crop must on the moisture stored in the soil, so the seedlings are first established in the sandy bed and after the flood water recedes and the land becomes available, the seedlings are uprooted and individually transported and replanted in the newly exposed moist land (Harlan, J.R., and J. Pasquereau, 1969: 15). This method of sowing and resowing (transplantation) is quite a recent one in Africa and rarely practised in the Sudan (A-Bari, E., 1982: pers.comm.). Thus it is only a remote possibility that transplantation was practised or known in the prehistoric, early historic or even the recent past of the Sudan.

17 Weeds are not necessarily from the same genus of the plant: all undesired plants in the crop-field are called weed.

Figure no.53. Simplified classification of the domesticated and spontaneous taxa of *Sorghum bicolor* * (L.). Moench.

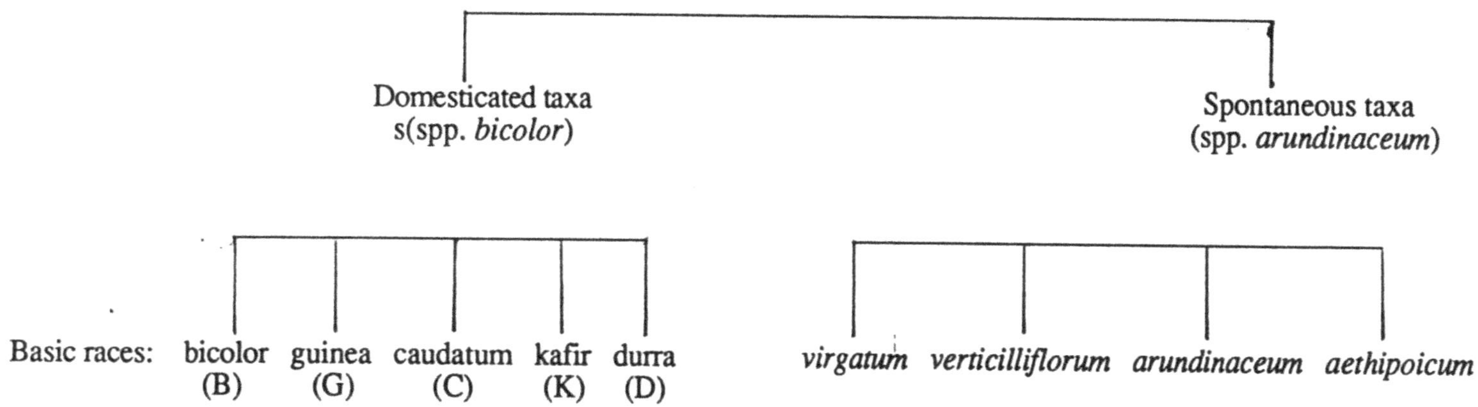

Intermediate domesticated races are products of the combination of the basic domesticated races; these are: (GB), (CB), (KB), (DB), (GC), (GK), (GD), (KG), (DC) and (KD) after Harlan, J.R., and J.M.J. deWet, 1972: 173.

* *S. bicolor* (L.) Moench include both domesticated and spontaneous taxa. Recent taxonomic studies found it best to treat wild and weedy sorghum complexes as one species and hence treated as subspecies *arundinaceum* and the domesticated ones as subspecies *bicolor* (Harlan, J.R., et al, 1976: 453).

Fig.no.54-A Distribution of the basic and intermediate taxa of <u>Sorghum bicolor</u> (L:)Moench (Guinea and caudatum)

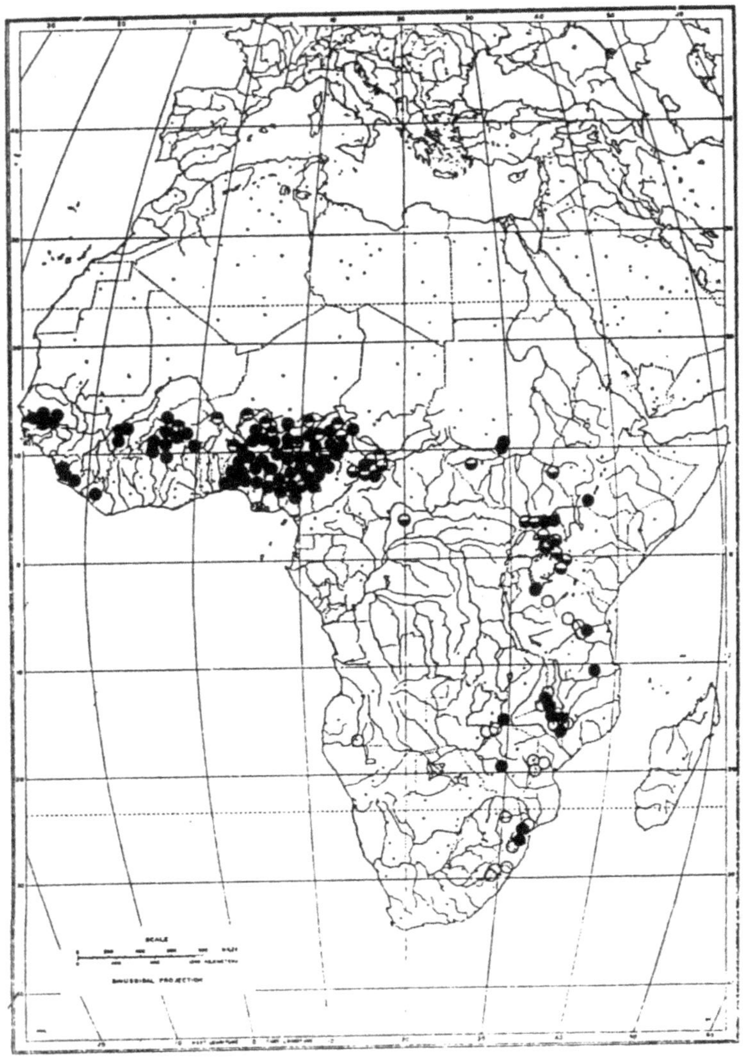

The guineas and half-guineas of Africa

Key: Solid circles: guinea
Open circles: guinea-kafir
Solid bottoms: guinea-caudatum
Solid tops: guinea-durra

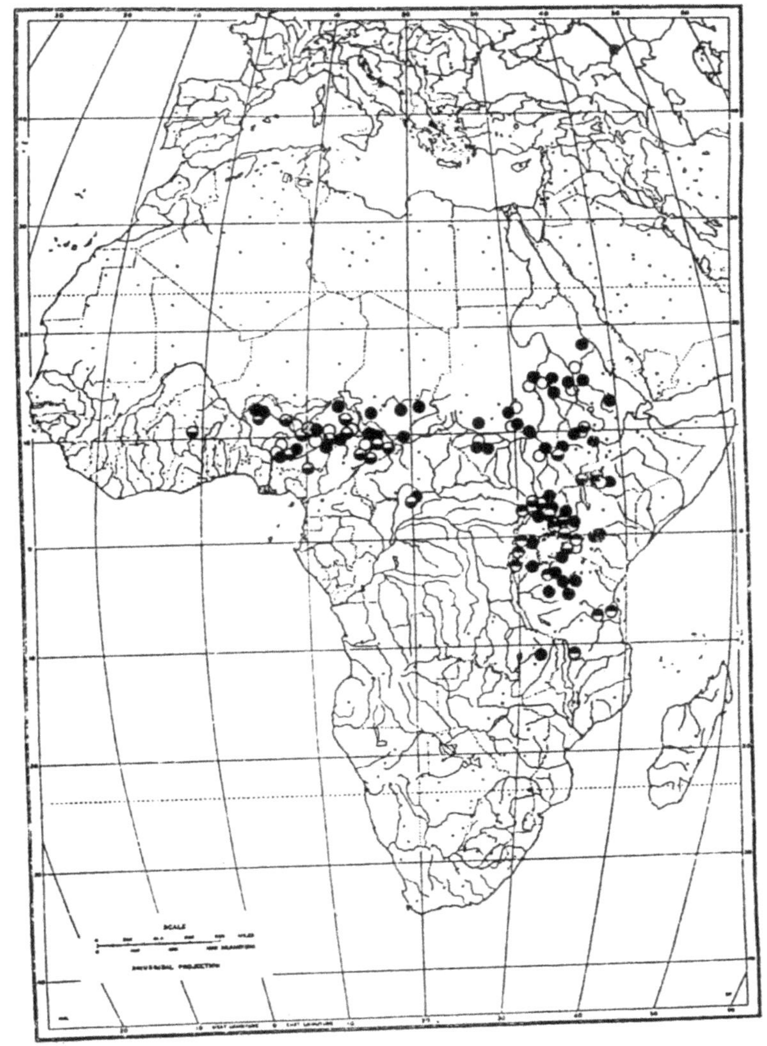

The caudatums and half-caudatums of Africa

Key: Solid circles: caudatum
Open circles: durra-caudatum
Solid bottoms: guinea-caudatum
Solid tops: kafir-caudatum

Fig.no.54-B

Distribution of the basic and intermediate taxa of <u>Sorghum bicolor</u>(L.)Moench (durra and kafir)

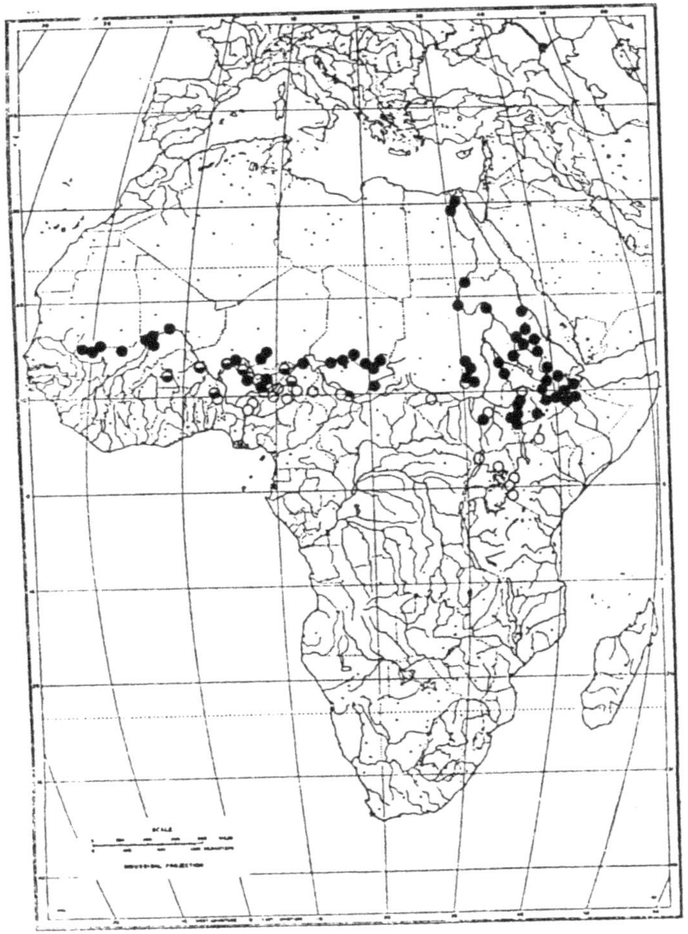

The durras and half-durras of Africa.

KEY: Solid circles: durra

Open circles: durra caudatum

Solid bottoms: durra guinea

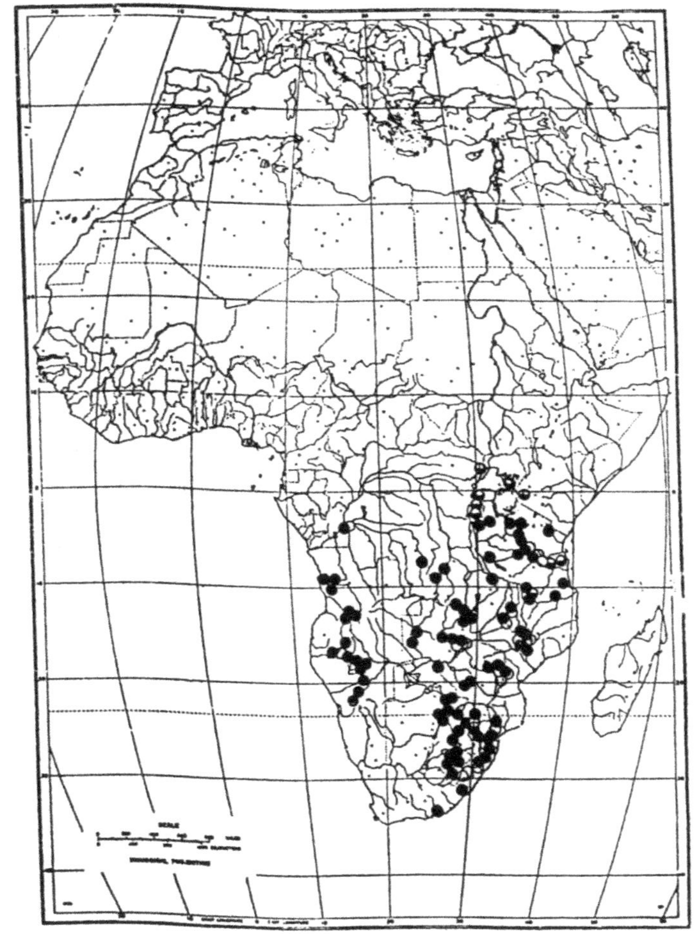

The kafirs and half-kafirs of Africa.

KEY: Solid circles:kafir

Open circles: guinea kafir

Solid bottoms: kafir-caudatum

After Harlan,J.R.&A.B.L.Stemler,1976: 470-471

Plate no. 13
Sorghum bicolor(L:)Moench:spontaneous taxa,
(illustration of inflorescences)

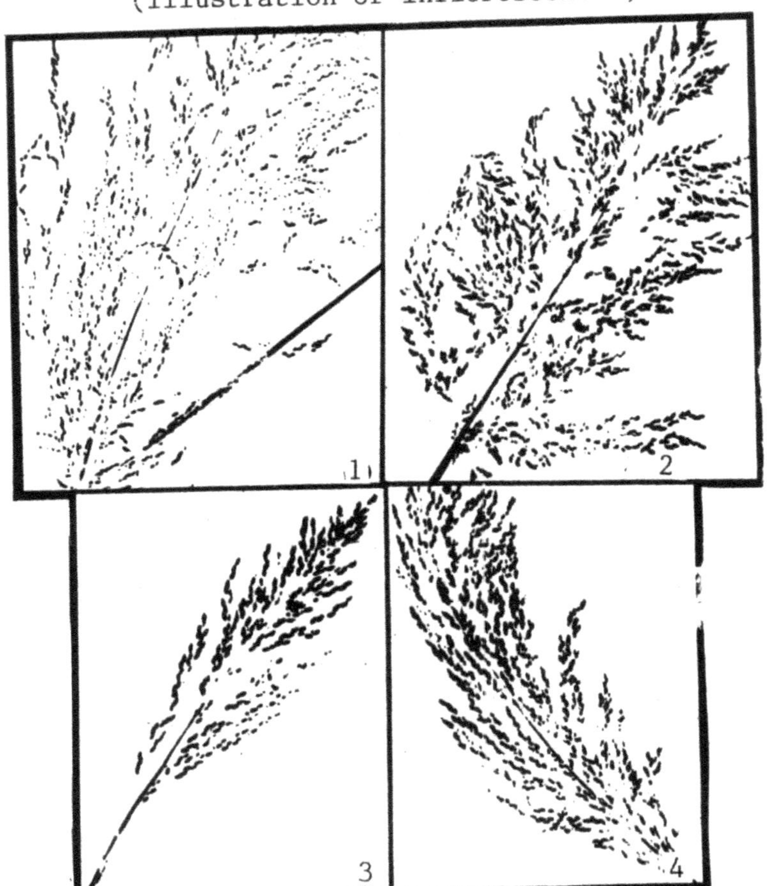

1-2:S.arundinaceum;3:S.aethiopicum;4:S.verticilliflorum.
(After deWet,J.M.J.,and J. P. Huckabay,1967: 793).

Plate no.14:A and B   Sorghum bicolor(L.)Moench:domesticated taxa(illustration of spikelets).

1 :race bicolor
2 :race caudatum
3 :race durra
4 :race guinea
5 :race kafir
6 :race drummondi(a highbrid between wild & domesticated races)

(After deWet,J.M.J.,et al,1976: plates no. 1 to 6).

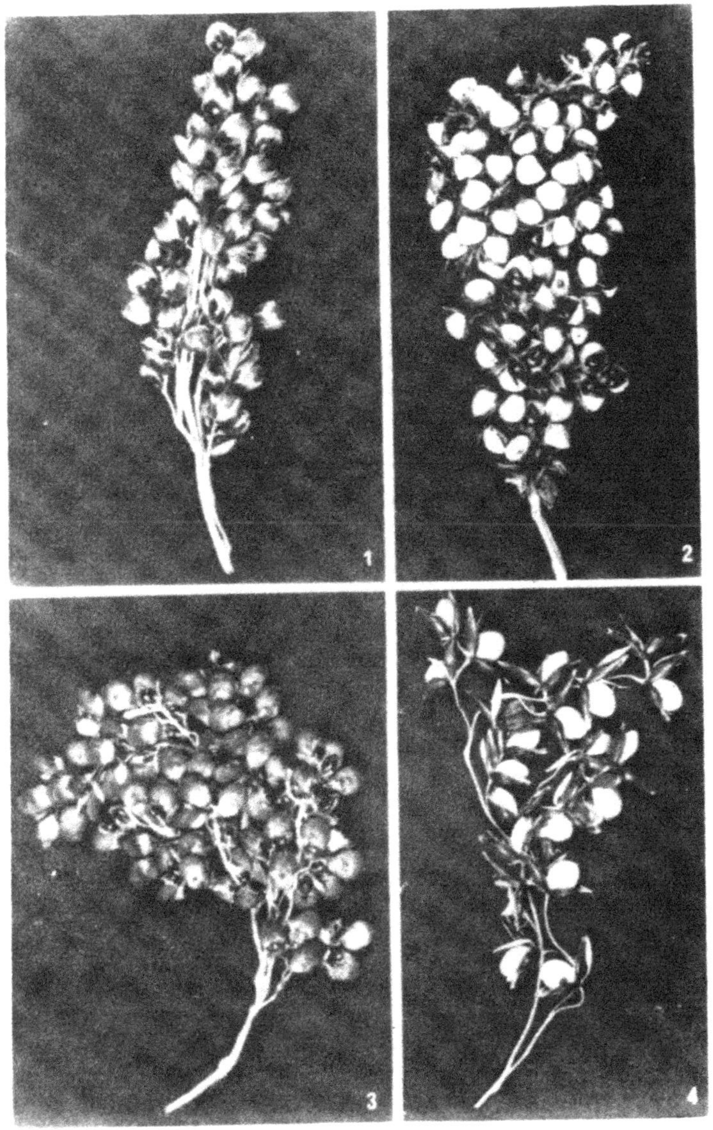

Plate no.14-A

Plate no.14-B

Table no.45. Some of the most commonly used domesticated sorghums in the Sudan at present.*

Basic race	Intermediate race	Local (Sudanese) name
caudatum - guinea		Feterita-Gedarif
-	" "	Dibakri
-	" "	Gesabi
-	" "	Wad-yabis
-	" "	Zirazira
-	" "	Matouk
guinea	-	Mayo
"	-	Wad-fahal
"	-	Mayo Jubal
dūrra	-	Milon
"	-	Zinari
"	-	Kutum-agbash
-	durra-caudatum	Daber
-	" "	Safra
-	kafir-caudatum	Umbenien
-	" "	Agono

* Dr. A-Bari, E. of the Botany department, U. of K. provided the names of most of the species of this list.

CHAPTER IX

Section A

Some Obstacles in the Process of the Domestication of Cereals in the Central Sudan: With emphasis on sorghum

The late date for domestication in the Central Sudan (and perhaps in Sub-Saharan Africa as a whole) may be explained in terms of some factors and obstacles which have either hindered or slowed the process of domestication of cereals. These factors and obstacles are - respectively - irregular cultivation practices, hybridization and back-crossing of the cultivated species with their close wild relatives.

9.1 Irregular Cultivation Practices.

Regular sowing of the previous year's harvest is one of the basic factors that triggers the process of genetic change in the cultivated gene pool. It is most likely that the acquisition of the habit of regular sowing and harvesting was both a gradual and a slow process. Although both harvesting and sowing were known, the prehistoric and early historic population of the Central Sudan first took the step of harvesting only; and this seems to have been practised for a long period of time, probably thousands of years. One of the basic reasons for keeping to it was the availability of wild stands [of sorghum] that yielded sufficient harvests without the need for sowing '.... why should anyone [sow] a cereal where natural stands are as dense as the cultivated field?' (Flannery, K., 1969: 80). But 'harvesting by itself, will not lead to domestication of cereals' (deWet, J.M.J., et al, 1976: 460). It would only be when the wild stands did not yield enough, due to, e.g., population increase or climatic strains such as successive years of insufficient rain, that the people would have started to sow part of the harvest to secure their next year's supplies of vegetal diet. Even after they started to harvest and resow part of the harvest, these activities would most likely have been performed on an irregular and discontinuous basis. For instance harvesters might have not kept to the same gene pool in the cycle of harvesting and sowing; each year a new wild population might have been harvested and part of the yield sown the next year. In this manner, the process of domestication, even if it was in fact initiated, would always have been set back due to sowing population which - entirely or greatly - disperse their grains (Stemler, A.B.L., 1984: 129-130). If we follow this line of reasoning we find that it would only have been very late in the history of the Central Sudan that man would have started to select whole inflorescence on a regular basis and to resow them the next year.

In sum, perhaps one of the basic factors for the late date for the domestication of cereals, namely sorghum and perhaps millet, in the Central Sudan and Sub-Saharan Africa may have been the manner in which cultivation was conducted. This includes the harvesting only and irregular[1] sowing and harvesting which were apparently practised until a very late date. In addition, if both irregular and regular cultivation activities were first practised within the natural habitat of the wild cereals, (which was most likely the case), there would have been one basic obstacle that would have either hindered or slowed the process of the domestication; namely the active process of interbreeding, hybridization and backcrossing between the incipiently cultivated cereals and their closely related wild and weedy species. How does this hybridization happen and how much does it affect the process of domestication of cereals-crops?

9.2 Hybridization, Backcrossing and Cereal Domestication.

Hybridization[2] is defined as 'the result of crossing of more or less dissimilar parents' (Swartz, D., 1971: 238). In other words, hybridization 'is crossing between individuals belonging to populations which have widely different adaptive requirements' (Stebbins, G.L., 1971: 131). The result of hybridization is the formation of hybrids (Blackmore, S., and E. Toothill, 1984: 182). Thus, a hybrid is 'an off-spring of two different varieties, species or genera' (Swartz, D., 1971: 239). The definitions clearly indicate that the parents of the "hybrid" may be conspecific but belong to different ecotypes, may belong to closely related species or to widely different species or even genera (Stebbins, G.L., 1971: 131). Backcrossing[3] is 'the cross of a hybrid with one of its parental races, a mating between a heterozgote and a homozgote' (Swartz, D., 1971: 56).

Grasses are both autogamous (self-pollinating) and allogamous cross-pollinating (Bender, B., 1975: 54). Thus in the absence of isolation (in which selection should ideally be complete), grasses can cross-pollinate and new hybrids be produced, not only

[1] Irregular is meant to denote here haphazard sowing and harvesting and sporadic discontinuous sowing of the previous year's harvest.
[2] The main emphasis in this discussion is on natural hybridization and backcross processes.
[3] Idem.

among wild species but also among wild-cultivated species and weedy-cultivated species. This is where the process of incipient domestication of cereals slows down or is set back to the beginning. This is particularly true in the case of sorghum (and pearl millet) as I shall briefly discuss below.

9.2.1 *Sorghum bicolor* (L.) Moench. A case study.

I have chosen *S. bicolor* (L.) Moench as the main case study because we have archaeological evidence from the Central Sudan (the area being studied here) of both wild and domesticated species. Furthermore, the domesticated *S. bicolor* (L.) Moench (of which we have evidence in our archaeological record) is the least specialized of the basic races of domesticated sorghum and has the widest distribution in the African savanna (deWet, J.M.J., et al, 1976: 455) and in the Central Sudan (cf. figs. no 54A and B). Finally, *S. bicolor* (L.) Moench is large and variable[4] including both domesticated sorghums and their closest spontaneous relatives (ibid: 455) and thus provides demonstrative examples.

It has been found most convenient and practical to treat the crop-weed/wild complexes as single species of *S. bicolor* (L.) Moench (Harlan, J.R., and J.M.J. deWet, 1971: 509-516) within which domesticated and spontaneous taxa are included in separate subspecies (ibid; cf. fig. no.53). As regards the wild sorghums of *S. bicolor* (L.) Moench, they are all included in its subspecies *S. arundinaceum* (Desv.) deWet and Harlan with its four basic taxa *S. aethiopicum* (Hack) Rupr., *S. arundinaceum* (Desv.) Stapf., *S. verticilliflorum* (Steud.) Stapf. and *S. virgatum* (Hack.) Stapf. (deWet, J.M.J. et al, 1976: 457; cf. pl. no.13). Since sorghum is both self and cross pollinated, these wild taxa 'hybridize freely in nature whenever they are sympatric' (ibid; Doggett, H., 1965: 55; Purseglove, J.W., 1985: 258) producing variable morphological varieties and some segregating hybrids (cf. diagram no.10) which in turn may start backcrossing.

Although it is very detailed and rather complicated Snowden's classification system of the domesticated sorghums, *S. bicolor* (L.) Moench (Snowden, J.D., 1936) is considered as the earliest complete work which all subsequent classifications have either adapted or modified (Harlan, J.R., and J.M.J. deWet, 1972: 172). One of the latest and most practical classifications of these is the one proposed by Harlan and deWet (ibid: 172-176). In this system of classification the domesticated primary gene pool *S. bicolor* (L.) Moench is concluded in five basic[5] races and their intermediate races as illustrated in the previous section on figure number 53 The latter are frequently produced

---

[4] The greatest variation in the genus *Sorghum* is found in the N.E. part of Africa north of latitude 10°S and east of longitude 25°E (Purseglove, J.W., 1985: p.260); this including the greatest part of the Sudan.
[5] Cross-pollination among domesticated sorghum races is usually around 5 per cent but this varies with weather conditions and type of panicle (Doggett, H., 1965: p.63).

Diagram no.10

Scatter diagram indicating the range of morphological variability of the wild varieties of S.bicolor, and some segregating populations of hybrids among them. The vertical axis represents peduncle-sheath length in cm.
Each dot represents the average value of 10 plants studied belonging to a natural collection. The morphological values of the hybrids are based on measurements of individual plants.

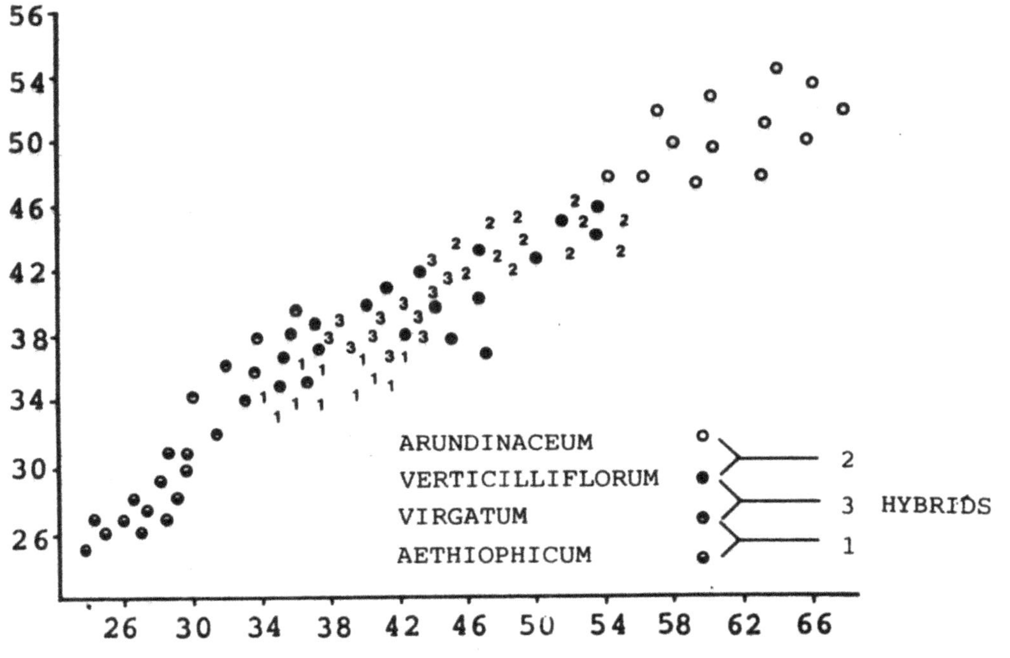

After de Wet,J.M.J.,et al.,1970: 706

by hybridization 'where two [basic] races become sympatric' (deWet, J.M.J., et al, 1976: 457).

It is also often the case in cultivated fields that members of the wild sorghum (spp. *arundinaceum*) readily cross with the domesticated races (spp. *bicolor*) (ibid: 459). 'The ease with which the varieties of cultivated [domesticated] sorghums cross .... with *S. arundinaceum*, is a headache for the taxonomists' (Purseglove, J.W., 1985: 268). On the basis of this fact, I assume that hybridization between wild and incipiently cultivated species and the backcrossing of hybrids [and/or the sterility of F3 generation, (ibid: 260)] were factors that might have hindered the process of domestication of sorghum in the Central Sudan. In addition new hybrids would be produced and compete (and probably over-run the incipient domesticates) but there is also backcrossing in which:

> 'wild-type gene, ... , can be transferred to the cultivated [domesticated] sorghums.... Transfer of cultivated-type genes to wild populations is less obvious' (deWet, J.M.J. et al, 1976: 459-460).

Thus the transfer of genes is "almost" always from the wild types to the domesticated ones because the former are more resistant and competent to survive. For instance, hybridization between wild/weed and domesticated races of sorghum can happen as follows:

(1) *S. verticilliflorum*	is sympatric and crosses with all bicolor domesticated races,				
(2) *S. arundinaceum*	"	"	"	"	with guinea and caudatum,
(3) *S. aethiopicum*	"	"	"	"	with caudatum and durra,
(4) *S. virgatum*	"	"	"	"	with durra (ibid: 459-461)

The race *drummondi* [cf. pl. no.14-B(6)] is proposed collectively to include all the hybrids and hybrid derivatives between wild and domesticated subspecies of *S. bicolor* (L.) Moench (ibid: 459). Its inflorescence and grains look more similar to the wild forms than the domesticated ones.

It may now be clear that the problem with such hybridization between wild and domesticated races of sorghum and the backcrossing processes is the production of populations which are closer to the wild race. Consequently, these populations, ecologically speaking, are competitors with and better survivors than the domesticated races. For instance,

> 'derivatives of hybridization between subspecies arundinaceum and subspecies bicolor are not always easy to distinguish on the basis of morphology from

> wild sorghums. Spikelet morphology is often of the wild type, and inflorescences of these hybrid derivatives frequently resemble those of their wild parents' [deWet, J.M.J., et al, 1976: 459-469; cf. pl. no.14-B(6)].

These hybrids do not always develop in the fields into hybrid-swarms due to the frequent and rapid-successive backcrosses that eventually lead to the recovery of the parental types. But - as we previously mentioned - it is always the case that the genes from the wild type are transferred to the domesticated type but this is rarely recognizable and less obvious in the opposite direction.

Since the hybrids are morphologically similar to the wild race and since they recover (by crossing back) their wild parental type much faster than the cultivated one (due to their very basic ecological adaptation and reproduction mechanism), it is often the case among extant races that only a small portion of the total yield will be harvested in the fields of sorghum. Bearing in mind what the harvester would look for (compact non-shattering or semi-shattering heads) and on the basis of observations made in the field, it was noticed that only 50% of the total yield is harvested from the cultivated sorghum fields in El-Gedarif area in the Central Sudan (A-Bari, E., 1986: pers.comm.). It is therefore not unusual for these fields to look, literally, as if they had not been harvested at all (ibid).

Hybridization and production of hybrids and backcrossing are still reported to be present in the sorghum fields, for instance in the El-Gedarif and the Nuba mountains areas. The tendency of weeds 'to mimic the cultivar with which they are associated' (Harlan, J.R., et al, 1973: 319) is still evident in these areas (cf. chapter VII). This is mainly due to lack or absence of techniques to isolate effectively the domesticated races from the wild/weedy ones and the difficulty of stabilizing the mutant strains - when both wild and extensive hybridization (deWet, J.M.J., and J.P. Huckabay, 1967: 793). Since a situation of this nature still exists 2000 years after the earliest evidence - so far - of sorghum domestication in the Sudan, there are good reasons to believe that the same processes and probably more active ones have always been going on in the prehistoric and early historic periods.

As regards pearl millet, *Pennisetum americanum* (L.) Schum; it has also been proven that cross-pollination normally occurs among different domesticated and wild species. Self-pollination can also occur if there is some overlapping of the female and male phases in the same inflorescence or in different inflorescences on the plant (Purseglove, J.M., 1985: 204, 206). It is now known that the styles come out (in millets) before the anther from the tip of the base of the false ear.

> 'When the styles dry the anthers come out, beginning around the middle of the false ear. The result of this phenomenon of protogyny is that it allows cross-fertilization' (Portéres, R., 1976: 435).

Consequently, hybrids, which are to a large extent similar in their reaction to those of sorghum are produced.

Just as all other African cereals, both sorghum and pearl millet have weedy races beside their wild and domesticated ones (Harlan, J.R., 1977: 358). In ecological terms a weed is a plant species which is 'adapted to open disturbed or unstable habitat....' (Hawkes, J.G., 1969: 18). In this connection, the wild races are different from their weed relatives in that they are adapted to primary (little or undestroyed) habitats. The difference between the weeds and the domesticated races is that while the weeds can grow without the aid of man in habitats such as ditches, paths, and along or in cultivated fields, the domesticated races 'are restricted to cultivated fields and cannot survive without the aid of man' (Harlan, J.R., 1977: 358).

In biological[6] terms, the weed races are produced as a by-product of domestication .... being formed through hybridization wherever wild and domesticated races are growing together' (Harlan, J.R., et al, 1976: 8-9). They can as well be formed due to man's activities which characteristically alter the balance of competition for survival in favour of weeds by suppressing those forces of natural selection that would have otherwise tend to eliminate them (Yarnell, R.A., 1971: 553). In terms of horticulture, a weed 'is an unwanted plant competing with those we are trying to cultivate' (Hawkes, J.G., 1969: 18). Competition of weeds with crops (for i.e. sunlight) is a serious threat that frequently causes substantial crop losses (Harlan, J.R., et al, 1976: 9). Thus, it is clear from the foregoing that weedy forms are the primary result of hybridization of hybrid derivatives between wild and domesticated species/subspecies as illustrated in the following example of some of the most common races in African cereals complex.

.The primary hybrids (weeds) can also interbreed (backcross) with both or either parental types and occasionally very complex second and third generations are produced (deWet, J.M.J., et al, 1976: 459; Harlan, J.R., et al, 1976: 9; Harlan, J.R., and A.B.L. Stemler, 1976: 466). Backcrossing of weeds with the domesticated races may result in effective introgression, but their hybridization with the wild races is also highly possible. Since it is often the case that wild and weed races grow in the same field together with the domesticated race (sorghum), both backcrossing and primary hybrid populations are produced. Thus the outcome can even be a more complex population. This is well demonstrated in the present pearl millet which is basically a domesticated species, though most of the west African population of the crop comprises a certain

---

[6] It is often possible to identify weeds by their intermediate morphologies (Harlan, J.R., et al, 1976: p.9).

amount of shattering types which continuously maintain themselves as weeds in the fields (Harlan, J.R., et al, 1976: 10).

Crop	Domesticated races	Wild races	Weed races
SORGHUM	*Sorghum bicolor*	spp. *arundinaceum*	*S. drummondi*
RICE (African Rice)	*Oryza glaberrima*	*O. barthii*	*O. stapfil*
TEFF	*Eragrostis tef*	*E. pilosa*	*E. pilosa*
FONIO			
Common Fonio	*Digitaria exilis*	*D. exilis*	*D. exilis*
Black Fonio	*D. iburua*	*D. iburua*	*D. iburua*
MILLETS			
Pearl Millet	*Pennisetum americanum*	*P. violaceum*	*P.americanum*
Finger Millet	*Eleusine coracana*	*E. africana*	*E. africana*
Guinea Millet	*Brachiaria deflexa*	*B. deflexa*	*B. deflexa*

Modified, after Harlan, J.R., 1977: 359.

> 'since the crop is cross fertilizing, the genes for shattering are carried along genetically in the non-shattering forms' (Harlan, J.R., et al, 1976: 10-11).

In addition a hybrid swarm is infrequently produced between the wild and the domesticated species.

If one bears in mind the fact that we are not dealing with 'direct ancestors' but rather with the wild and weed ancestral races of the present domesticated sorghum and pearl millet, we may understand how much more complex the primary hybrids and their second and perhaps third generation could have been when incipient cultivation of these cereals first started. This illustrates how difficult it was to isolate the early cultivated races from their wild/weed relatives. Bearing in mind these complexities, we can explain that hybridization was one of the major reasons for the late domestication of, e.g., sorghum despite of the fact that its cultivation had been practised for a long period of time. It was only after isolating the regularly cultivated wild sorghum from the free breeding gene pool that its domestication process was initiated (as I shall discuss in the following pages).

Section B

An Alternative Explanation for the Late Date of Cereal Domestication in the Central Sudan.

In the previous chapter I argued that the hypothesis which states that the late development of domestication of African cereals (in sub-Saharan Africa) was a consequence of the late date of introduction of good-cutting "harvesting" tools, e.g. iron tools, is incomplete and does not consider one of the essential conditions necessary for the emergence of changes characteristic of domestication. My challenge to this hypothesis is based on both the available archaeoethnobotanical and ethnographic evidence from the Sudan. In this chapter, I shall use this evidence in order to formulate an adequate explanation: if it is man's activity which exerted the selection pressure which over time favoured reproduction of genetically different species, it still remains to account for how the selection pressure could produce changes in the gene pool of a plant population as long as it operated only on a fragment of a total free-breeding population. A necessary condition for the domestication of cereals is the isolation of a regularly cultivated cereal population, e.g. sorghum, from the gene pool of the free-breeding one. I will argue that although it is difficult to see how this condition could be satisfied within the natural habitat of the free breeding cereal population, it was fulfilled after livestock, mainly cattle, became the basis of the economy. In the context of the savanna environmental conditions, this frequently implies migratory and/or seasonal movements (Haaland, G., 1968: 60-70). If we assume that early pastoralists also exploited plant resources, this would have created a problem because the areas which are best for, e.g. cattle, during the rainy season are the drier areas which probably were marginal for the growth of the free-breeding cereal exploited. A possible solution to this problem would have been for the pastoralists to bring with them a stock of heads of grains and cultivate them in the drier rainy season areas.

Before I present my model for this alternative explanation, I shall briefly discuss the general concept of isolation, its different types and which of these was the factor that was most likely to have initiated the process of domestication in the Central Sudan.

9.3 Isolation: A Key to the Domestication of Cereals.

I discussed in the foregoing part of this chapter how grasses can - in general - be self-pollinated and/or cross-pollinated (cf. section A). The main emphasis of the current discussion about isolation and its effects in the process of the domestication of cereals will mainly be on sorghum and pearl millet. Sorghums are also self-pollinated grasses (cf. appendix no.6). It has, however, been shown that cross-pollination among domesticated sorghums is generally about 5 per cent though this is subject to fluctuations due to weather conditions, (i.e. winds) and the type of panicle (Doggett, H., 1965: 64). More active cross-pollination is encountered among wild races of sorghum and to a large extent between these and the domesticated races (Purseglove, J.W., 1985: 264,268). Differences in the inflorescences (e.g. dense, expanded) and flowering between wild and domesticated races and ecological adaptation are counted among the basic reasons for the variability of pollination in different forms of sorghums (cf. appendix no.6). As regards pearl millets, I mentioned previously (cf. section A), that both cross-pollination and self-pollination normally occur (cf. appendix no.6).

Isolation of a cultivated plant species can be defined as the establishment of barriers to gene flow between the cultivated species and their spontaneous races (Harlan, J.R., et al, 1973: 323). In this connection, there are several isolation mechanisms which operate either collectively or individually and result in a restriction of crossing between related species. These can be summarized in the following simplified mechanism factors of isolation (Heslop-Harrison, J., 1973: 69-70):

1) External factors. These operate outside the plant and (tend to) prevent cross pollination between different related species. Emphasis in this paper will be on those factors which were imposed by man and which resulted in spatial, temporal and ecological isolation and hence were effective in the reproduction cycle and genetic change of the cultivated plant species.
2) Internal factors. These are the ones operating inside the plant tending to prevent fertilization and the development of the hybrids to maturity.
3) Factors which operate to prevent or (at least) reduce the viability of hybrids and their chances of reaching the stage of reproductive maturity and/or reduce or eliminate their fertility.

Fægri has divided the reproductive barriers for isolation, e.g. against hybridization, into two groups: - vegetative isolation and reproductive isolation (Fægri, K., and L.Van Der Pijl,1979: 25).

9.4 Forms of Isolation.

Since the main focus of this chapter is on how isolation might have been an effective means in the process of domestication of cereals, I have attempted - for mere conventional reasons of presentation and argumentation - to draw an abstract distinction between what are in reality inseparable and realistically indistinguishable forms of isolation. I combined these in two simplified forms: man-induced isolation and natural "normative" isolation. These forms often occur at the same time and operate simultaneously; yet I believe that man-induced isolation was most effective and has played (and still is) a greater role in the process of cereal domestication[7] (namely, of sorghum and millets) in the Central Sudan. Thus, the main emphasis here will be on man-induced isolation.

Among the most likely "incipient" forms of man-induced[8] isolation are those which take place within or outside the natural habitat of the cultivated cereal, e.g. *Sorghum* or *Pennisetum* (cf. figs. no.2 and 3).

9.4.1 Isolation within the natural habitat.

Man-induced isolation within the natural habitat may be in the form of separation in space and/or season of the cultivated cereal from the free breeding gene pool (cf. fig. no.55). Separation in space within the natural habitat is done unintentionally by setting (small) fields of cultivated species at a distance from the area of the wild growing relatives, thus partially isolating these fields in space from the wild growing sorghum fields.

Separation in space is best demonstrated in the traditional cultivation in some parts of the Sudan, such as among the Nuer in the southern Sudan, the Ingassana in south-eastern Sudan and the Nuba in the Nuba mountains (cf. fig. no.1). These communities cultivate sorghum and/or pearl millet in small plots beside or in the vicinity of their settlements. The prehistoric and early historic populations may also have cultivated small plots nearby their settlement areas due to their need for greater yields of e.g. grain within the same natural habitat as that of the wild species. As I observed from the location of the sites in the Central Sudan, it was most likely that the settlement area and hence these cultivated plots were at a distance from the wild growing stands, thus partially isolating the cultivated species from the free breeding gene pool.

Since grasses (e.g. sorghum, pearl millet) are wind pollinated (Heslop-Harrison, J., 1973: 71) and considerable horizontal transport can be achieved, separation in space

---

[7] It is not always the case that isolation (in general) will lead to domestication of all plant species; some have a negative response to man-made habitat (for more information see chapter on domestication).

[8] It should always be observed that man-induced isolation was unintentionally done when sorghum (and pearl millet) were first domesticated and still is unintentionally done in traditional cultivation practises, and that isolation without regular cultivation does not lead to domestication of cereals.

Fig. no.55. Some "incipient" forms of man-induced isolation of the cultivated population.

Area of cultivation / Period of cultivation	Within the natural habitat of the free breeding pool	Outside the natural habitat of the free breeding pool
Before or after the flowering period of the free breeding pool	Seasonal and spatial isolation	Seasonal and spatial isolation
Within the flowering period of the free breeding pool	Spatial isolation	Spatial isolation

Cultivation of wild population

(of the small plots) should ideally be beyond the range of wind transport of grass-pollen to avoid crossing and hybridization between the wild stands and the cultivated ones. But it is often the case in reality that isolation is not necessarily complete. For instance, if a partial isolation of 60% were achieved, that would have been sufficient to initiate the process of domestication (Fægri, K., 1987: pers.comm.). Perenniality of wild grasses, however, makes it rather difficult to avoid cross pollination between these and the cultivated population if the latter is grown at the same period and in the same area as the wild/weedy stands are growing. The earliest evidence for the exploitation of sorghum in the Central Sudan was found within its natural habitat, but no morphological features of domestication (cf. table no.43) were found. This external mechanism factor, separation in space, has apparently been of no significance, at least during the early stage, in the process of cereal domestication in the Central Sudan.

In this connection weeding practices might indirectly be related to the spatial separation in that the cultivated population is grown at the same period as its wild relatives are, but it differs in that the location of the cultivated field can be both within or at a distance from the wild growing stands. The removal of the wild relatives would secure sufficient sunlight and soil nutrients long enough for the cultivated population to reach it reproductive maturity. But it is as well essential to do weeding at a time well before the closely related wild and weed races (to the cultivated population) reach their flowering[9] stage and hence to avoid or reduce risks of cross-pollination and hybridization. I do, however, think that weeding was practically difficult and most likely was not performed during the prehistoric and early historic periods in the Central Sudan. This is especially true as regards the difficulties of weeding fields of cultivated cereals such as sorghum and millet due to the striking similarities between these and the closely related wild and weedy races at the stage when weeding is to be done. This interpretation is also supported by archaeological, ethnobotanic and ethnographic evidence. Taking into consideration the archaeological evidence, although bare-hands could have been used for uprooting and/or cutting the weeds, no artefacts which can be associated with weeding were found in the prehistoric Central Sudan. As regards the ethnobotanic and ethnographic evidence, although the traditional cultivators in west Sudan, both in the Kordofan and Darfur regions, weed their fields of sorghum and pearl millet, they also acknowledge[10] the enormous difficulties they encounter in distinguishing between the crop and the weeds especially when they conduct the first weeding (Ar. *al-Mur*[11]). Tracey and Hewison

---

[9] It is often the case (in present traditional cultivation) that weeding is done more than once, mainly to secure sunlight and nutrients; but this practise unintentionally serve the purpose of reducing cross-pollination.

[10] This piece of information was gathered when the present writer was doing a field work (in 1985) focused on the traditional cultivation practises in Darfur (W. Sudan).

[11] *Al-Mur* literally means bitter. The peasants use this term metaphorically to show how difficult the first weeding is.

cited some sources which referred to what has been described as 'the oldest form of agriculture' along the Nile in north Sudan and dated to 1900 years ago, that is simply casting the seeds and harvesting the yields without doing any weeding[12] (Tracey, C.B., and J.W. Hewison, 1948: 745). In short, neither separation of cereal population in space nor weeding crop fields seem to have been practised when cultivation first started in the Central Sudan.

As for isolation in time within the natural habitat (cf. fig. no.55), it is essential that the grains i.e. of sorghum should be sown and/or cast at a time when most, if not all, of its wild and weedy close relatives are not germinating. In this manner, differentiation in flowering period and pollination time will be set between the cultivated population and the wild ones. Thus crossing and interbreeding will either be prevented or at least reduced, and the process of domestication will proceed without serious obstacles, provided that cultivation is regularly practised. Since all of the wild grasses are perennials that grow during the rainy season, sowing or casting the desired species should be at a time long enough before or after the growth (from germination to reproductive maturity) of the wild relatives to avoid cross-pollination. For instance in the case of sorghum it should be either early summer or late winter, e.g. long before or after the rain-period. This requires some artificial irrigation of the crop. The earliest direct archaeological evidence from the Sudan for artificial irrigation is the water-wheel (cf. 7.2.3.4 in chapter VII) dated to the second century A.D. (Abu-Saleem, M.A., 1980: 43). Although there is not any evidence - so far - for a direct connection between artificial irrigation (water-wheel) and the earliest evidence of domesticated sorghum from the Central Sudan, its introduction[13] would definitely have made it possible to practise cultivation at any time of the year. It would also have made it possible for the first time in the history of the Central Sudan to cultivate cereals (such as barley and wheat) which in the absence of artificial irrigation require winter rains. There are also the water-reservoir (Ar. *hafir*), e.g., in the western Butana round Shaqadud and Naqa (cf. fig. no.1) which are now used for trapping the rain-water to be used towards the end of the rainy season or when the rain-water is insufficient to water the sorghum plots. These *hafirs* are believed to date to the Meroitic period. Their exact dates are, however, unknown; and whether they have been used for watering cultivated cereals during the Meroitic period is also uncertain. The results I obtained and the interpretations I arrived at in the present study, have indeed raised more questions than the ones I attempted to answer and have also stressed the necessity of testing them by conducting more excavations and perhaps also initiating some experimental studies.

---

[12] Evidence which shows when weeding first began was not encountered.

[13] As regards the impact of the water-wheel on the socio-economic structure, see 7.3.2.4 and Adams, W.Y., 1977: 291, 319.

9.4.2 Isolation outside the natural habitat.

Regular cultivation of wild cereals outside, or even on the periphery of, their natural habitats is the ideal situation for initiating the process of their domestication. It is only in this type of situation that almost all the mechanism factors of isolation (cf. 9.3) would be set at work simultaneously. That is to say, since the cereal(s) is (are) cultivated outside their natural habitats, cross-pollination with their spontaneous relatives is prevented or reduced. Furthermore, the reproduction system of the now cereal cultivated in isolation tends to change from allogamy (cross-pollination) to autogamy (self-pollination) or to the reinforcement of the autogamous pollination (Fægri, K., and L. Van Der Pijl, 1979: 137). In a new environmental setting the "biological clock" of the cultivated species tends to alter both in the timing and periods of germination, maturity and the reproductive maturity of the plant (Krzywinski, K., 1987: pers.comm.). Consequently cross-pollination, backcrossing; and the development of hybrids[14] is reduced or prevented, hence speeding up the process of domestication. The end result is often domesticated species which are ecologically dependent on man and are genetically stabilized beyond recall (e.g. to wild/weedy form) even if these are brought back to the natural range of the spontaneous relatives.

Man-induced isolation is most effective in the process of domestication when man moves[15] from one area to another (cf. fig. no.55). These movements can be due to one or several factors. Among these, the main factors are change in economy, e.g., to increased emphasis on animal husbandry, climatic deterioration and/or population pressure. An ideal situation for most effective isolation is to move the cultivated species from its natural habitat[16] to a peripheral "marginal" habitat (where conditions are unfavourable for the wild species to grow naturally) and hence its regular cultivation becomes a necessity - if supplies of vegetal diet are required. In this manner most of the pressures of natural cultivation are changed, with the activities of man playing an increasingly important role in order to maintain plant-reproduction outside its natural habitat. It should be stressed that while selection pressures induced by man are continued, most of the pressures of natural selection are reduced or removed from the cultivated gene pool (Bender, B., 1975: 5). It should, however, be stressed that selection pressures induced by man must be an inseparable part of the regular cultivation activities, otherwise there would always be the risk of adopting spontaneous relatives (wild/weedy species) into the cultivated gene pool. Perhaps weeding out spontaneous relatives in this type of situation is equally important in isolating the cultivated gene pool. This should not, however, be understood as being essential in exerting any selection pressure.

---

14 Some of the mechanism factors of isolation take effect in hybrid progeny to an extent which can reduce or destroy its fertility (Heslop-Harrison, J., 1973: 70).

15 Movements can be temporary (trans-humance) or on permanent basis, e.g. migration.

16 The natural habitat may also shift geographically due to, e.g. environmental and climatic changes.

Scholars such as Doggett (Doggett, H., 1965: 67), Purseglove (Purseglove, J.W., 1985: 264) and Flannery (Flannery, K., 1965: 1247-1256) have generally mentioned the importance of isolation in the process of domestication of such cereals as sorghum, pearl millet, wheat and barley. It has also been indicated that domestication of plants, in general, most likely occurred outside the predicted nuclear areas (Jarman, H.N., et al, 1972: 39). Harlan and Zohary have even given a cultural explanation that supports the idea of biological isolation as an important factor for explaining the beginning of cereal domestication. They stated that,

> 'domestication may not have taken place where the wild cereals were most abundant. Why should anyone cultivate a cereal where natural stands are as dense as cultivated fields?.... farming[17] itself may have originated in areas adjacent to, rather than in, the regions of greatest abundance of wild cereals.' (Harlan, J.R., and D. Zohary, 1966: 1075-1080).

As in other forms of man-induced isolation (cf. fig. no.55) this one was done unintentionally in prehistoric and early historic times. However, in this type, crops were moved as food outside their natural habitat and were probably cultivated during the rainy season along fans and head-waters of streams, along and within the beds of seasonal streams and low lying areas (where rain water is trapped) and rivers. There are good reasons to believe that man-induced isolation of the cultivated cereals (mainly sorghum) outside their natural habitat was a basic - if not the only - factor that provided the necessary condition for their domestication in the Central Sudan. Biological, climatic and material-culture evidence recovered from this area support this interpretation as I shall present in the following account.

9.5. An Alternative Explanation for the Late Date of Domestication of African Cereals: The Case of the Central Sudan.

The available archaeoethnobotanical evidence from the Central Sudan show that over 4000 years passed between the beginning of the cultivation of wild sorghum and the earliest evidence for its domestication. An important question arises here: How was the process of domestication initiated?

The most likely answer (as we previously summarized) is that it was only after the cultivated cereals, e.g. sorghum, were isolated from the free breeding gene pool that the process of their domestication was initiated.

[17] Harlan & Zohary (ibid), however, have not considered that factors such as population pressure and climatic deterioration which may lead to farming within the natural habitat of the wild stands.

Man-induced isolation outside the natural habitat of the wild-growing sorghum (cf. fig. no.55) was the most likely basic factor inducing its domestication[18] in the Central Sudan. The new area or region in which sorghum was domesticated was "marginal" as to its natural habitat in the sense that the wild species could not - ecologically - have been able to survive (successively) and/or compete with other species under the new natural conditions. In this manner possibilities of cross-pollination, interbreeding, hybridization, backcrossing, etc. between the cultivated species and their closely related wild relatives were either largely reduced or eliminated. Thus, the process of genetic change (domestication) evolved and continued. But it is worth mentioning in this connection that regular sowing of the previous year's harvest is equally important. Since the new habitat was outside the range of the naturally growing wild species, it was most likely that the cultivated population was sown in naturally flooded and rain-water trap areas to provide the necessary moisture until the crop reached maturity. In this type of situation, as I observed in some traditional practices in the Sudan, if weeding is properly and timely done, it could serve as an effective means of isolating the cultivated population by eliminating their competitors. However, for our purpose here, it has no role in exerting selection pressures in favour of genetic change (domestication) of the species. Obviously the regular cultivation of the wild sorghum was deliberately conducted to provide a sufficient food stock of vegetal origin in a habitat where the wild stands did not naturally grow. But at the same time selection pressures in favour of genetic change were unintentionally imposed and accelerated through some basic activities. These are: harvesting of heads with relatively intact grains to facilitate easier transport and storage for a long period without loosing many of the grains. In addition, storage of the heads for the next year's sowing. In this manner, both human beings and the new "marginal" habitat (in which the wild sorghum is regularly cultivated) had indirectly yet integrally operated towards the unintentional exertion of new selection pressures which favoured mutants with non-dispersing mechanisms in the cultivated population. This, I hypothesized, initiated the emergence of a domesticated population of sorghum. If we follow this line of reasoning, the picture will not be complete without answering the following questions:

1) Why did it take such a long time to isolate the cultivated population outside the natural habitat of its wild growing relatives?
2) What were the factors (direct or indirect) which caused the isolation to take place?
3) What were the factors and/or conditions that forced and/or motivated the people to move into these new, yet, peripheral areas?

---

[18] There is not, at present, confirmed evidence of domesticated pearl millet from the Central Sudan during the period between 6000-1700 B.P.

These questions were first answered hypothetically, and the hypothesis was then tested using archaeological and ethnographic evidence as I shall show in the following pages.

As regards the first question, the archaeoethnobotanical evidence from the Central Sudan since the 9th mill. B.P. strongly indicates that initially man did not exert a heavy pressure on food resources in areas where wild stands of sorghum were naturally growing. The diet of the prehistoric population of the Central Sudan was based on broad-spectrum resource-utilization. It consisted mainly of aquatic resources and, to a large extent, wild game, immediately eaten food-plants and seeds of grasses, e.g. sorghum. Moreover, the introduction of domestic animals (mainly cattle, sheep and goats) indicates that access to meat from the wild fauna was becoming scarce and hence, a new stable source of food was added to the already exploited terrestrial and aquatic resources.

Taking into consideration the other (closely interrelated) questions the archaeological evidence indicate that between c. 6000-5000 years B.P. animal husbandry had started to play a major role in the economy, supplemented by cultivation of wild cereals, e.g. sorghum and pearl millet, gathering of wild fruit-plants, fishing and hunting. Since animal husbandry was becoming a more important part of the economy, this implied nomadism and a change to a more mobile form of settlement were stimulated. There was, then, an incipient movement into areas "marginal" as regards the natural habitat of wild sorghum but optimal as regards the availability of other terrestrial and aquatic resources and pasture for livestock. Therefore, I will argue that the disappearance of archaeological sites in the Central Sudan after c. 5000 B.P. is a result of a changing settlement pattern due to an increased reliance on livestock and the stimulation of migratory movements and not a result of depopulation of the area. In other words, with increased reliance on livestock, mainly cattle, sometime after 5000 B.P. a mobile mode of life was developed in the savanna[19] environment prevailing in the area. The search for favourable conditions for the livestock, e.g. ample of pasture and water and avoidance of unfavourable conditions such as disease, seasonal shortage of pasture, etc. (cf. fig. no.56) dictated mobility, perhaps round the year. These included movements to areas outside the natural habitat of the wild cereals (cf. fig. no.56). To maintain a diet that consisted of cereals, people had to bring with them heads of these cereals and cultivate them during the rainy season in such favourable places as naturally flooded areas and rain-water trap (fans and head-waters of streams, along and within courses of seasonal streams and low lying areas and the flood plains of rivers). The regular cultivation of these cereals in these marginal areas led eventually (but unintentionally) to their domestication.Support for this argument

[19] More information on the migratory pattern of the savanna pastoralists is presented by Haaland (Haaland, G., 1968: 58-73).

Fig. no.56 Optimal conditions for animal keeping during the dry and wet seasons in the prehistoric Central Sudan.

Dry season	- A - B - C	+ A + B - C
Rainy season	+ A + B - C	+ A + B + C
Locality	away from the river-systems (hinterlands)	near the river-systems

Legend:

A	pasture
B	water
C	disease
+	present
-	absent

consists both of ethnographic studies and archaeoethnobotanical evidence, and to some extent of palaeoclimatic evidence.

The key role of livestock in the economy of prehistoric and the early historic Central Sudan can be inferred by considering the present Sudan pastoral communities who are aware of the fact that,

> '.... under pastoralism .... investment and growth take place automatically through the process of production .... [and] that considerable growth in pastoralism can take place with modest increase in labour .... [that] consumption needs can be satisfied not only by livestock products like milk ...., but also by direct consumption of animals .... [and that] security in times of adversity is greater with pastoralism' (McCown, R.L., et al, 1979: 298-299).

Archaeologically, an analysis of the faunal remains from the sites we excavated in the Central Sudan (Kadero, Um Direwia, Zakiab, Nofalab, Islang, Shaheinab, Rabak and the Jebel Tomat sites) clearly reflects the increased importance of livestock from 6000 years B.P. and onward (El-Mahi, A.T., 1982; Haaland, R. 1984: 39-51: Krzyzaniak, L., 1978: 159-172). We also have evidence of wild sorghum and pearl millet in the period between 6000 to 5000 years B.P. which we think it was cultivated. Data that furnish climatic indications, mainly floral evidence (Haaland, R., 1981: 195-196; Krzyzaniak, L., 1978: 159-172) and faunal evidence (El-Mahi, A.T., 1982), show that the summer rainfall was not less than 500 mm. Both the average rainfall and the presence of wild sorghum and pearl millet confirm that these species were indigenous to the Central Sudan (deWet, J.M.J., and J.P. Huckabay, 1967: 800). In addition, the material culture, the size of the sites, and the presence of numerous grinding tools and of pottery during the same period indicates permanent or base type settlements and small dry-season sites between 6000-5000 B.P. which again indicates seasonal movement during this period with the basis of the economy on livestock. But cultivation of wild sorghum and exploitation of aquatic and terrestrial resources (fishing, hunting and gathering) were also practised.

By c. 5000 years B.P. the course of events had markedly changed as the archaeological record indicates. There is no evidence after that period of settlements in the area of the Khartoum sites and in the area of the Rabak site; but the area of Jebel Tomat site seems to have been newly occupied. The most likely suggestion is that owing to its heavy reliance on livestock in a savanna environment (with a short rainy season and a long dry one) the population of the Central Sudan had a mobile way of life. Since livestock became increasingly important in the economy, people had to start a nomadic life due to insufficient pasture in the area of their permanent and semi-permanent settlements as compared with the growing numbers in their herds. In this connection ethnographic and ethnobotanic observations show that,

> 'the native flora, ... , was quickly altered by the introduction of cattle and sheep. Hamblin (1881: pp.87-89) observed that by 1882, only 15 years after the Mormons entered Utah, grazing had so reduced native seeds that the Indians were starving' (Steward, J.H., 1938: p.14). '[in addition] Indian testimony indicates that many areas now occupied especially by rabbit brush had much grass before the introduction of cattle and sheep' (ibid: 17).

It is most likely that 'overgrazing by domestic herds [during prehistoric times] near waterside settlement' (Phillipson, D.W., 1977: 64-65) also happened in some areas in the Central Sudan, e.g. in the Khartoum area. With regard to this, it has been observed that archaeological sites dated to after 5000 B.P. in the Central Sudan are more frequent in the hinterland (Fattovich, R., and M. Piperno, 1981: 26-30; Marks, A.E., et al, 1982: 47-50) than along the permanent watersides (mainly, the White, Blue and Main Niles and the Atbara river), indicating that due to the increasing importance of livestock and perhaps an increasing number of herds, there was a gradual process of nomadization that included the expansion of the range of movement, e.g. to the hinterlands in the east. This is further supported by the evidence recovered and geographical accounts from the area of ancient Meroe (cf. fig. no.1) which show that cattle-rearing - and hence nomadism - played an important role in life of the Meroitic population (Adams, W.Y., 1977: 301; Shinnie, P., 1967: 158-159; Strabo[20], 1949: 143). In addition, travellers' accounts from the northern part of the Central Sudan dated to A.D. 1813, show that the inhabitants of Berber who kept large numbers of cattle used to move with their herds in search of pasture during the winter, after the rains, as far east as the Bisharyyin area in the Red Sea mountains (Burckhardt[21], J.L., 1822: 213). It was also mentioned that these cattle- keepers used to live during their seasonal movements, like Bedouins, in huts and tents (ibid).

Taking into consideration evidence of palaeoenvironmental and palaeoclimatic conditions, studies on vegetation history and climate in general (with its rainfall parameter) indicate continuously deteriorating conditions in the Central Sudan since 5000 years ago (cf. fig. no.57). The present arid climate with its 160 mm of annual rainfall in the Khartoum area has its beginning 5000 years ago (Wickens, G.E., 1975 : 43-65;

---

20 Strabo (wrote a Geography about year 7 B.C. on the basis of notes he collected during his travels in Egypt during the years 25 to 19 B.C.) also stated that the people of Meroe lived also on sorghum which they cultivated along the seasonal streams and water courses. It is worth noting here that the areas described were (at the time 5000 B.P.) already outside the natural habitat of wild sorghum.

21 The same source states that sorghum was the principle produce and the basic food for both man and livestock (ibid).

Fig.no.57 **Climate shifts in Sudan from 9000 to present.**

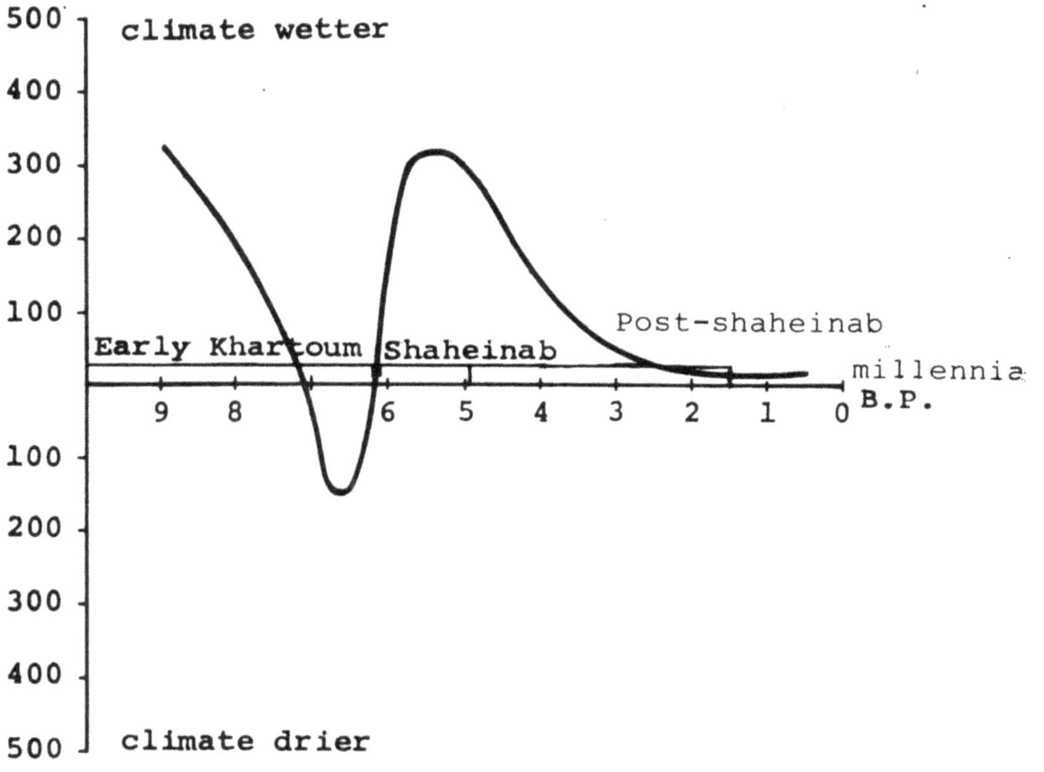

Note: the shape of the curve for each climatic period is not accurately known.
the curve is dependent on the point in time to which a particular piece of evidence refers.

After Wickens G.,E.,1975: 51 with some modifications;

years before present (B.P.)

1982: 23-30). The implication of such climatic deterioration is that there would have been insufficient pasture for the increasing numbers of livestock and that wild food-plants would have been becoming scarce and insufficient to depend upon. Furthermore the annual rainfall (160 mm) was less than required (500-1500 mm) for the natural growth of wild sorghum (deWet, J.M.J., and J.P. Huckabay, 1967: 800) and its cultivation have become more of a gamble due to the limited and unreliable rainfall in the area of Khartoum.

In the light of the foregoing, it seems reasonable to assume that with the development of migratory pastoralism, a new and more dynamic element was introduced into the economy of the Central Sudan allowing the population to exploit seasonally the rich savanna grazing resources. Initially, this pastoral niche must have been highly productive and led to a growth of human and animal population. Over time, this, unavoidably, must have led to pressure on grazing resources and possibly to environmental deterioration.

With regard to the development of events in the area of Rabak, conditions seem to have been different to those in the area of Khartoum. The general climatic deterioration and the marked decrease in the annual rainfall that affected the area further north, e.g. Khartoum, since early Holocene did not seem to have significantly affected the area of Rabak. The climatic reconstructions and vegetation history of the Central Sudan (cf. fig. no.8) place this area within the 500-700 mm rain isohyets during the middle Holocene and it has c. 400 mm of annual rainfall at the present time (cf. fig. no.10). Thus, by 4000 years B.P. this area was still within the natural range of wild growing sorghum. As a matter of fact some grain-impressions of wild sorghum were found in *in situ* potsherds at the site of Rabak and dated to 4490±100 B.P. (cf. appendix no.1; table no.15). Since the present rainfall in this area is only 20 mm short of the minimum requirement of wild growing sorghum (which should be 500 mm), this rainfall was also most likely adequate 4,000 years ago for the wild species to grow. Hence the domestication of sorghum would then, have been impossible due to cross-pollination, hybridization, backcrossing, etc. Therefore one reason for the apparent abandonment of this site may have been the evolving nomadic mode of life due to an increased reliance on livestock[22], mainly cattle.

Finally as regards the site of Jebel Tomat, the earliest evidence - so far - for the occupation of this site is radiocarbon dated to 4540±200 B.P. (Clark, J.D., and Stemler, A.B.L., 1975: 588-591; appendix no.1); some 500 (±200) years after the presumed disappearance of the sites of Shaheinab type from the area of Khartoum. Evidence for occupation of this site is lacking shortly after 1700-1600 years ago. The area of the Jebel Tomat receives at present c. 300 mm of annual rainfall (Wickens, G.E., 1982: 40). It

[22] For more information on the faunal remains and animal husbandry at the site of Rabak, see El-Mahi, A.T., and R. Haaland, 1984: 28-32; and Haaland, R., 1984: 39-51.

was probably receiving c. 450 mm when it was first occupied some c. 4500 years B.P., middle Holocene, which is rather low for the natural growth of wild sorghum but sufficient for its cultivation within and along naturally flooded and rain-water trapping areas (cf. fig. no.16). Thus, it seems to have been an ideal area for the mobile "pastoral" groups from the north and south to choose as an optimum area for the rainfed cultivation of sorghum and for pasture for their livestock. Biological evidence recovered from the site confirms the presence of domesticated cattle and domesticated sorghum. Therefore and in the light of the available archaeoethnobotanical, faunal, climatic and chronological evidence, the area of the Jebel Tomat seems to have been situated just outside the natural habitat of wild sorghum and to have been one to which pastoral groups were coming and practising regular cultivation of wild sorghum until it was finally (yet indeliberately) domesticated shortly before 1700 years B.P. Evidence for both domesticated sorghum *S. bicolor* (L.) Moench (Clark, J.D., and A.B.L. Stemler, 1975: 588-591) and wild species was recovered from this site.

9.6 Implications of the Alternative Explanation.

Agriculture, farming, mixed farming, etc. are terms often used by archaeologists (such as Reed,C.A.,1969: 361-364) to denote the cultivation of domesticated plants and animal husbandry and the different activities connected with these. The conventional use of the term may be attributed to the interdependencies between the processes of the domestication of plants and animals. Both seem to have started very nearly at the same time. Once they are established, the two usually go hand in hand although emphasis is often put on one more than the other. Thus there are uncertainties as to which actually came first, domestication of plants or of animals. Despite these uncertainties, there are frequently direct statements or implicit suggestions in archaeological interpretations[23] of the beginning of the food-producing economy in different parts of the world and in the Sudan in particular that cultivation and domestication of plants predated the domestication of animals. Among the published archaeological studies with such tendencies of interpretation are those of C.S. Chard (1975: 186-187); M.N. Cohen (1977: 282); R. Haaland (1987a: 207); C.A. Reed (1969: 261-264); A.B. Smith (1980: 489-501); etc. As regards the Central Sudan, it has been suggested that cultivation of plants was a pre-condition for the emergence of animal husbandry (Haaland, R., 1987a: 207).

The results of the analysis of our data (plant remains, fauna[24] and remains of material culture) strongly suggest the opposite development as regards the domestication

---

[23] Some of these interpretations are untestable as regards archaeological evidence.

[24] See El-Mahi, A.T., 1982; El-Mahi, A.T., and R. Haaland, 1984: 28-32.

of sorghum. In the light of these results, I propose that animal husbandry pre-dated the domestication of sorghum in the Central Sudan. If it had not been for the introduction of domesticated animals, their increased importance and finally the profound reliance on them in the economy of the Central Sudan; the occupants of this region would not have adopted pastoralism and mobility in search of pasture for their livestock. During these movements, migratory herders would have brought with them heads of sorghum to areas outside their natural habitat and cultivated them. In this manner, the "regularly" cultivated gene pool would unintentionally have been genetically isolated (either completely or to a large extent) and prevented from interbreeding with wild close relatives and hence the process of domestication would have been initiated. On the basis of this sequence of events, I will argue that the introduction of domestic animals, e.g. cattle, and the beginning of pastoralism in the Central Sudan were vital pre-conditions for the domestication of sorghum. This argument is also supported by the fact that wherever we recovered evidence of domesticated animals (from the sites in the Central Sudan), there was nearly always evidence of wild grasses (mainly sorghum and sometimes pearl millet) which seemed to have been exploited as a staple vegetal food. On the other hand, no evidence of wild or domesticated cereal (e.g. sorghum) was found prior to the earliest evidence of domesticated animals in this region. As a matter of fact our evidence shows the presence of the earliest domesticated animals and wild grasses in the same archaeological context. Even today it is most of the time the case that a successful traditional farmer tends to invest surplus in livestock and that a pastoralist resorts to mere farming only at times of crisis (Haaland, G., 1968: 58-73; McCown, R.L., et al, 1979: 298-299).

According to Phillipson 'pastoralism and agriculture are not necessarily interrelated' (Phillipson, D.W., 1977: 56). But our evidence from the Central Sudan clearly indicates the interdependency of agriculture and pastoralism. The origins of animal domestication in the Central Sudan are beyond the scope of this paper. It is, however, suggested that agriculture came to the Sudan from neighbouring regions and from what is now the southern Sahara (Hassan, F.A., 1986: 25-26; Phillipson, D.W., 1977: 78, 84). But from the sequence of events presented in this paper, I believe that the inception of agriculture in the Sudan and in the Central part in particular, was initiated locally. The long evolutionary history of gathering of food-plants, utilization of indigenous wild grasses and subsequently of their cultivation and domestication (as I discussed earlier) stands witness for an indigenous origin of agriculture in the Sudan.

APPENDIX I

Radiocarbon Dates

LABORATORIET FOR RADIOLOGISK DATERING

Adr.: Fysikkseksjonen, 7034 Trondheim - NTH

D A T E R I N G S R A P P O R T

Oppdragsgiver: Randi Håland
Historisk museum
Bergen

Lab. ref.	Oppdragsgivers ref.	Mate-riale	For-be-handl	Datert del	^{14}C alder før nåtid	Kalibrert alder (MASCA)	$\delta^{13}C$ ‰
	DF 1469						
T-5726	Rabak 107x/100y lag 3, Prøve 1 Sudan	Skjell	VI	85%	5860± 80	4750±150 BC	-4,1
T-5727	Abu Darbein 101x/90y lag 4, Prøve 3 Sudan	"	"	90%	7860±190		-5,3
T-5728	Abu Darbein 101x/90y lag 1, Prøve 4	"	"	90%	7700±140		-7,1

Dato: 4. oktober 1985

Laboratoriet for Radiologisk Datering

Steinar Gulliksen

Reidar Nydal

LABORATORIET FOR RADIOLOGISK DATERING

Adr.: Fysikkseksjonen, 7034 Trondheim - NTH

D A T E R I N G S R A P P O R T

Oppdragsgiver: Randi Håland
Historisk museum
Bergen

Lab. ref.	Oppdragsgivers ref.	Materiale	For-be-handl	Datert del	^{14}C alder før nåtid	Kalibrert alder (MASCA)	$\delta^{13}C$ ‰
	DF 1469						
T-6381	Abu Darbein 108x/96y Sudan	Skjell	VI	80%	8500±100		-9,2
T-6382	Aneïbis 100x/98y 1.1 area I prøve 3 Sudan	"	"	90%	7310±110		-8,9
T-6383	Aneibis 100x/98y 1.2 I prøve 4	"	"	90%	7290±150		-8,9
T-6384	Aneibis 61x/98y 1. I Area II prøve 5	"	"	90%	7450±100		-7,5
T-6385	Aneibis 43x/147y 1. I area IV prøve 6	"	"	90%	7730±110		-10,0
	*) Ikke målt						

Dato: 30. september 1986 Laboratoriet for Radiologisk Datering

Steinar Gulliksen Reidar Nydal

LABORATORIET FOR RADIOLOGISK DATERING

Adr.: Fysikkseksjonen, 7034 Trondheim - NTH

D A T E R I N G S R A P P O R T

Oppdragsgiver: Randi Håland
Historisk museum
Bergen

Lab. ref.	Oppdragsgivers ref.	Materiale	For-be-handl	Datert del	^{14}C alder før nåtid	Kalibrert alder XXXXXXX (MASCA) Stuiver & Pearson	$\delta^{13}C$ ‰
	DF 1469						
T-7480	Aneibis 103x/96y 1.II Area 1, Sudan	Skjell	VI	90%	7460±130	BC6440-6130	-6,0*
T-7481	Aneibis 56x/93y 1.1 Area 2	"	"	90%	6760±170	BC5760-5490	"
T-7482	Aneibis 64x/107y 1.3 Area 3	"	"	90%	7850±140	BC7030-6490	"
T-7483	Aneibis 49x/147y 1.5 Area 4	"	"	90%	7720±110	BC6680-6440	"
T-7484	El Damer 57x/140y og 1.II-III	"	"	90%	7690±100	BC6610-6430	"
T-7485	El Damer 105x/187y 1.I	"	"	70%	8390± 50		"
	*) Ikke målt						

Dato: 1/2-1988

Laboratoriet for Radiologisk Datering

Steinar Gulliksen   Reidar Nydal

Radiocarbon dates for Shaqadud

Provenience		*Years BP*	*Lab. number*
Cave	— level 16	3615 ± 88	SMU 1133
	— level 38	4123 ± 86	SMU 1128
	— level 54	4059 ± 65	SMU 1127
	— level 71	4046 ± 101	SMU 1208

After Marks,A.E.,1984:32

LABORATORIET FOR RADIOLOGISK DATERING

Adr.: Fysikkseksjonen, 7034 Trondheim - NTH

DATERINGSRAPPORT

Oppdragsgiver: Randi Håland, Historisk Museum, Bergen

Lab. ref.	Oppdragsgivers ref.	Mate-riale	For-be-handl	Datert del	^{14}C alder for nåtid	Kalibrert alder (MASCA)	[illegible] ‰
	Skjell fra Sudan DF 993						
T-3697	Rute 104x/88y, Um Direiwal	skjell	VI	90%	6010± 90	4920± 130BC	÷ 5.6
T-3698	Prøvestikk I, " "II	"	ingen	0.7g	5000± 300	3660± 390BC	"
T-3699	G-61 I, Esh Shaheinab	"	VI	90%	5550± 90	4440± 80BC	÷ 6.0
T-3700	Rute 4, lag II, Nofalab	"	VI	90%	5290± 100	4175± 175BC	÷ 6.4
T-3701	Rute 4, lag V, "	"	VI	90%	5520± 130	4420± 90BC	÷ 5.6
T-3880	Rute 18-G-1708, lag II Islang, Sudan	"	VI	90%	5870±110	4750±170BC	÷5.6

Dato: 20. januar 1981

Laboratoriet for Radiologisk Datering

Steinar Gulliksen

Reidar Nydal

(After Magid, A. 1982: 110)

LABORATORIET FOR RADIOLOGISK DATERING

Adr.: Fysikkseksjonen, 7034 Trondheim - NTH

D A T E R I N G S R A P P O R T

Oppdragsgiver: Randi Håland
Historisk Museum
Bergen

Lab. ref.	Oppdragsgivers ref.	Mate-riale	For-be-handl	Datert del	^{14}C alder før nåtid	Kalibrert alder (MASCA)	$\delta^{13}C$ ‰
	DF 1469						
T-5132	Rabak 107x/100y level II prøve 1 Rabak, Sudan	Skjell	VI	90%	4490±100	3270±110 BC	-1,9
T-5133	Rabak 107x/100y level 6 prøve 2	"	"	90%	6050±100	4990±110 BC	-2,3*
T-5134	Rabak 107x/100y level 15 prøve 3	"	"	90%	6020±130	4890±200 BC	-5,0
T-5135	Jebel Mir 100x/100y level prøve 4 Jebel Mir, Sudan	"	"	90%	170± 90	Yngre enn AD 1530	-2,6
T-5136	Jebel Tomat site 2b level I prøve 5 Jebel Tomat, Sudan	"	"	90%	1770± 80	AD195±85	+0,3
T-5137	Jebel Tomat site 2b level III prøve 6	"	"	90%	4140± 90	2795±175 BC	+0,6
	*) Ikke målt						

Dato: 8. februar 1984

Laboratoriet for Radiologisk Datering

Steinar Gulliksen

Reidar Nydal

A P P E N D I X II

Preparation of samples for SEM examination

Preparation of samples for SEM examination.

HOW TO COAT SAMPLES THAT DO NOT RESIST TEMPERATURES HIGHER THAN 40-45°C

1 connect the black power lead for the thermocouple module to the socket on the rear panel.Now, the stage temperture will stay at about 5°C during the coating.(the surface temperature of the sample may reach c. 40°C)

2 place the samples inside the coating chamber(max.6 stubs).close the chamber properly.

3 open the outlet valve of the Argon-bottle.(the little aluminium knob) Do not touch the regulation-valve(the black one) or the main valve.

4 turn"Operation" switch on the front panel to "Pump" position and wait until the vacum gauge indicates about 0.1 torr(the first blue mark).

5a open leak valve for about 30 sec. and flush system with Argon.close leak valve and allow system to pump down to 0.02 torr(the second blue mark) if possible.

5b if it is hard to reach 0.02 torr,you can adjust the leak valve and try to reach 0.02 torr or better.

6 you should flush the system as described in(5a) at least two times and end up with a vacum better than 0.02 torr.

7 set the timer to an adequate value,i.e 2.5 to 3.0 min. for Gold/Palladium target or 3.0 to4.0 min. for Gold target is normally sufficient.

8 turn"Operation"switch to "Set HT" position and turn voltage control to 2.5 kV. the mA-meter will now indicate 5 to 10 mA or less.

9 NB. THIS OPERATION IS VERY CRITICAL.
gradually open the leak valve until the mA-meter reads 19 to 21 mA (blue mark).when the needle starts moving,turn the knob slowly very little and wait and see.never exceed 25 mA.

10 when the needle of the mA-meter has reached about 20 mA , press timer button and switch to"Timer" position.

11 when the specimens are coated, turn down the voltage control knob and switch off the "Operation" switch.Ventilate the system with Argon.After 1.0 min. open the chamber.do not forget to close the Argon leak valve.

12 take out your samples.when all finished,close the Argon outlet valve on the Argon bottle.

.....(c o n t.)

HOW TO COAT SAMPLES THAT CAN RESIST A SURFACE TEMPERATURE OF 70 - 75°C.

1 disconnect the black power lead for the thermocouple module from the rear of the panel.the stage temperature will stay at c.25°C during the coating but the surface temperature may reach 70° to 75°C.

2 place your samples inside the coating chamber(max. 6 stubs).close the chamber properly.

3 open the outlet valve of the Argon bottle(the little aluminium knob). Do not touch the regulation-valve(the black one) or the main valve.

4 turn "Operation"switch on the front panel to "pump" position and wait until the vacum gauge indicates about 0.1 torr(the first blue mark).

5a open leak valve for about 30 sec. and flush system with Argon.close leak valve and allow system to pump down to 0.02 torr(second blue mark) if possible.

5b if it is hard to reach 0.02 torr,you can adjust the leak valve so that the pressure rises to between 0.1 to 0.2 torr.leave it like this for a few minutes ,then close the valve and try to reach 0.02 torr or better.

6 flush the system as described in no. (5a) at least two times and end up with a vacum better than 0.02 torr.

7 set the timer to an adequate value, i.e 2.5 to 3 min. for Gold/Palladium target or 3 to 4 min. for Gold target is normally sufficient.

8 turn "Operation" switch to"Set HT" position and turn voltage control knob to 2.5 kV. the mA-meter will now indicate 5 to 10 mA or less.

9 NB. THIS OPERATION IS VERY CRITICAL:
gradually open the leak valve until the mA-meter reads 19 to 21 mA (blue mark).when the needle sarts moving,turn the knob slowly and very little and wait and see. never exceed 25 mA.

10 when the needle of the mA has reached about 20 mA,press timer button. switch to"Timer" position.

11 when the specimens are coated , turn down the voltage control knob, then close the Argon leak valve properly,but do not overtighten, and finally switch off the "Operation" switch.

12 ventilate the system by pulling the knob on the top of the chamber.take out your samples.when all is finished, close the Argon outlet valve of the Argon-bottle.

A P P E N D I X III

Experiments on carbonization of seeds and their impressions in pottery

Table no. 1 Sorghum vulgare L.: Carbonization
(measurements in mm.)

STATE	CATALOG.-NO.	1 L.	1 B.	1 T.	2 L.	2 B.	2 T.	3 L.	3 B.	3 T.	4 L.	4 B.	4 T.	5 L.	5 B.	5 T.	6 L.	6 B.	6 T.
FRESH	A	4.60	3.70	1.50	5.30	4.90	4.00	4.70	4.30	3.00	4.20	4.00	3.00	5.10	5.00	3.40	4.60	3.60	2.50
CARBONIZED	A	4.30	4.90	1.90	5.10	5.50	4.10	4.50	5.00	3.00	4.00	4.30	3.20	5.00	5.30	3.60	5.00	5.10	4.00
FRESH	B	4.90	5.20	3.10	4.70	5.00	3.30	4.80	4.70	3.50	5.00	5.00	3.50	4.30	4.00	2.50	5.00	4.00	2.90
CARBONIZED	B	5.00	5.10	4.00	4.90	5.00	4.20	4.90	5.10	3.90	4.80	5.10	3.90	4.20	4.30	3.00	5.00	4.80	3.30
FRESH	C	4.50	4.90	3.00	5.40	4.00	3.10	4.70	5.20	3.20	5.80	4.50	3.40	4.30	3.90	2.90	5.00	5.10	3.00
CARBONIZED	C	4.50	4.70	3.50	5.00	4.40	3.50	4.70	5.20	3.40	5.50	4.60	3.50	4.10	4.40	3.00	5.10	4.90	3.30
FRESH	D	4.60	5.00	2.90	5.10	4.70	3.50	5.40	4.80	3.20	4.50	4.30	3.00	4.40	4.20	2.50	4.20	3.80	2.00
CARBONIZED	D	5.00	4.80	3.80	4.70	5.10	3.90	5.40	5.00	3.80	4.50	4.50	3.50	4.20	4.30	3.00	4.00	4.20	2.80
FRESH	E	4.80	4.40	3.00															
CARBONIZED	E	4.60	4.80	3.50															
FRESH	F	5.10	4.70	3.60	4.90	4.20	3.10	4.80	4.80	3.20	5.60	4.40	3.00	5.00	5.30	3.70	5.50	4.50	2.50
CARBONIZED	F	5.00	5.00	3.80	4.20	4.40	3.30	4.60	4.90	3.50	5.50	4.80	3.30	5.20	5.10	4.00	5.50	4.50	2.90
FRESH	G	5.20	4.50	3.20	4.20	4.30	3.00	5.00	4.50	2.80	4.90	4.80	3.70	5.00	4.00	3.10	3.70	3.60	2.90
CARBONIZED	G	4.80	5.10	3.50	4.50	4.50	3.20	4.80	4.50	3.10	4.90	5.20	4.00	4.70	4.20	3.20	3.70	3.70	3.10
FRESH	H	5.20	4.80	3.70	4.90	3.80	3.00	5.10	4.10	3.30	5.10	4.20	3.20	5.30	4.50	3.20	4.80	4.20	3.40
CARBONIZED	H	4.90	5.00	3.90	4.10	4.00	3.20	4.80	4.20	3.50	4.50	4.40	3.60	5.00	4.60	3.50	4.50	4.50	3.60
FRESH	I	4.90	4.50	3.40	3.70	3.70	2.60	4.70	4.80	3.40	5.10	4.70	3.20	5.30	4.40	3.00	5.00	4.50	3.10
CARBONIZED	I	4.60	4.60	3.60	3.30	4.40	2.80	4.80	4.70	3.60	4.80	5.00	3.50	4.90	4.60	3.30	5.00	4.70	3.50
FRESH	J	4.70	4.60	2.90															
CARBONIZED	J	4.60	4.90	3.10															

Table no. 2 Pennisetum americanum (L.)K.Schum.: Carbonization (measurements in mm.)

STATE	CATALOG.-NO.	1 L.	1 B.	1 T.	2 L.	2 B.	2 T.	3 L.	3 B.	3 T.	4 L.	4 B.	4 T.	5 L.	5 B.	5 T.	6 L.	6 B.	6 T.
FRESH	A	2.20	1.70	1.50	2.20	1.70	1.50	2.40	1.70	1.50	2.50	2.00	1.70	2.50	1.90	1.60	2.10	1.80	1.60
CARBONIZED	A	2.10	1.70	1.60	2.10	1.90	1.60	2.10	1.80	1.60	2.40	2.10	1.70	2.40	2.00	1.70	2.10	1.80	1.60
FRESH	B	2.50	2.00	1.70	2.00	1.60	1.40	3.00	2.10	1.60	2.50	2.00	1.50	2.40	1.80	1.60	1.80	1.50	1.10
CARBONIZED	B	2.20	2.20	1.80	1.80	1.70	1.50	2.70	2.30	2.00	2.30	2.10	1.70	2.20	2.10	1.70	1.70	1.60	1.20
FRESH	C	2.30	1.80	1.60	2.70	1.50	1.40	2.30	1.80	1.50	2.90	2.40	1.70	2.00	1.50	1.20	2.10	1.70	1.50
CARBONIZED	C	2.20	1.90	1.60	2.30	1.60	1.50	2.20	1.90	1.70	2.85	2.50	1.80	1.80	1.60	1.30	2.00	1.80	1.60
FRESH	D	1.60	1.20	1.00	2.00	1.80	1.60	2.50	1.80	1.80	2.30	1.60	1.10	2.00	2.00	1.60	2.50	1.70	1.50
CARBONIZED	D	1.50	1.30	1.10	1.90	1.80	1.60	2.30	1.90	1.90	2.10	1.80	1.10	1.95	2.00	1.65	2.30	1.80	1.60
FRESH	E	2.50	2.00	1.60															
CARBONIZED	E	2.40	2.00	1.70															
FRESH	F	2.20	1.70	1.50	2.00	1.60	1.20	2.50	2.00	1.50	2.00	1.50	1.30	2.00	1.50	1.30	2.20	1.90	1.40
CARBONIZED	F	2.00	2.20	1.60	1.90	1.70	1.30	2.40	2.20	1.50	1.80	1.70	1.40	1.90	1.70	1.40	2.00	2.00	1.50
FRESH	G	2.50	1.80	1.50	2.20	1.80	1.40	2.40	1.90	1.40	2.50	1.90	1.50	2.10	1.80	1.10	2.40	1.80	1.40
CARBONIZED	G	2.40	2.00	1.50	2.10	1.90	1.50	2.20	2.00	1.40	2.20	2.20	1.90	2.10	2.00	1.40	2.10	1.90	1.60
FRESH	H	2.20	1.70	1.50	2.20	1.90	1.60	2.30	1.70	1.50	2.20	1.70	1.30	2.20	1.50	1.10	2.10	1.60	1.40
CARBONIZED	H	2.10	2.00	1.50	2.10	2.20	1.60	2.20	1.80	1.60	2.00	2.00	1.50	1.90	1.70	1.30	1.90	1.70	1.50
FRESH	I	1.80	1.40	1.20	2.30	2.10	1.70	1.70	1.30	1.00	2.30	1.80	1.40	2.50	1.70	1.60	1.70	1.50	1.30
CARBONIZED	I	1.70	1.50	1.40	2.30	2.30	1.60	1.40	1.50	1.20	2.20	1.90	1.50	2.30	2.00	1.70	1.60	1.60	1.40
FRESH	J	2.30	1.70	1.40															
CARBONIZED	J	2.20	2.00	1.50															

Table no. 3 Sorghum arundinaceum (Desv.)Stapf.: Carbonization (measurements in mm.)

STATE	CATALOG. NO.	1 L.	1 B.	1 T.	2 L.	2 B.	2 T.	3 L.	3 B.	3 T.	4 L.	4 B.	4 T.	5 L.	5 B.	5 T.	6 L.	6 B.	6 T.
FRESH	A	7.10	1.80	0.70	7.00	1.80	0.60	7.00	1.50	1.00	8.90	1.80	0.90	6.40	1.50	1.00	6.50	1.90	0.90
CARBONIZED	A	6.90	1.70	1.00	6.20	1.20	1.10	6.60	1.40	1.10	8.00	1.40	1.00	6.00	1.20	1.00	6.00	1.50	1.00
FRESH	B	6.70	1.10	0.50	8.20	2.00	0.90	8.30	1.90	1.10	6.50	1.80	1.80	8.20	2.20	1.20	8.10	1.60	1.00
CARBONIZED	B	5.70	0.90	0.60	7.40	1.50	1.00	7.00	1.40	0,90	5.80	1.50	1.90	7.10	1.60	1.30	7.00	1.30	1.00
FRESH	C	7.40	2.00	1.00	8.60	2.00	0.90	7.80	2.00	1.00	8.20	1.60	1.00	6.00	1.00	0.40	6.50	1.00	0.30
CARBONIZED	C	7.00	1.50	1.10	7.50	1.50	0.90	7.00	1.50	1.00	7.40	1.20	0.70	5.40	0.90	0.50	6.00	0.90	0.50
FRESH	D	6.50	1.80	0.90	6.00	1.60	0.80	6.20	1.80	1.00	8.50	2.00	0.70	8.60	2.00	1.00	6.70	1.70	1.30
CARBONIZED	D	5.70	1.50	1.10	5.50	1.30	1.10	5.70	1.40	1.00	8.00	1.60	1.00	7.70	1.50	1.10	6.00	1.30	1.20
FRESH	E	8.00	1.80	1.00	6.50	2.10	0.90	8.50	2.00	1.20	6.00	1.80	0.90	6.70	1.50	0.90	8.00	1.50	1.30
CARBONIZED	E	7.00	1.30	0.90	6.00	1.60	1.10	7.50	1.50	1.30	6.00	1.50	1.10	5.80	1.10	1.00	7.00	1.20	0.90

Table no. 4 Echinochloa pyramidis (L.)Hitchc.: Carbonization (measurements in mm.)

STATE	CATALOG. NO.	1 L.	1 B.	1 T.	2 L.	2 B.	2 T.	3 L.	3 B.	3 T.	4 L.	4 B.	4 T.	5 L.	5 B.	5 T.	6 L.	6 B.	6 T.
FRESH	A	4.30	1.70	1.35	2.70	1.25	1.20	2.75	1.30	1.15	3.40	1.50	1.00	3.35	1.50	1.25	2.60	1.35	1.25
CARBONIZED	A	4.00	1.80	1.45	2.50	1.35	1.35	2.50	1.45	1.30	3.10	1.70	1.15	3.15	1.65	1.35	2.35	1.35	1.35
FRESH	B	2.65	1.30	1.00	2.70	1.15	1.05	3.65	1.25	1.20	3.00	1.45	1.10	3.50	1.15	1.25	2.65	1.35	1.25
CARBONIZED	B	2.50	1.35	1.15	2.25	1.25	1.25	2.80	1.40	1.30	2.80	1.55	1.25	3.40	1.30	1.30	2.60	1.45	1.35
FRESH	C	4.65	1.80	1.35	4.00	1.75	1.25	3.50	1.25	1.15	2.75	1.30	1.35	3.65	1.45	1.45	3.65	1.60	1.45
CARBONIZED	C	4.20	1.95	1.45	3.55	1.85	1.30	3.25	1.30	1.25	2.65	1.35	1.45	3.00	1.60	1.55	3.35	1.75	1.60
FRESH	D	2.60	1.05	1.25	2.55	1.10	1.30	2.55	1.05	1.25	3.00	1.50	1.85	3.35	1.65	0.95	3.75	1.60	1.05
CARBONIZED	D	2.50	1.15	1.35	2.55	1.10	1.30	2.50	1.20	1.40	3.00	1.60	1.90	3.25	1.80	1.05	3.60	1.80	1.15
FRESH	E	3.75	1.45	1.35	2.80	1.25	1.10	3.65	1.50	1.25	3.55	1.25	1.10	3.65	1.35	1.00	3.35	1.30	1.45
CARBONIZED	E	3.50	1.55	1.45	2.25	1.40	1.20	3.30	1.70	1.30	3.25	1.35	1.15	3.00	1.45	1.10	3.15	1.45	1.50

Table no. 5 Setaria verticillata (L.)Beauv.: Carbonization (measurements in mm.)

STATE	CATALOG.-NO.	1 L.	1 B.	1 T.	2 L.	2 B.	2 T.	3 L.	3 B.	3 T.	4 L.	4 B.	4 T.	5 L.	5 B.	5 T.	6 L.	6 B.	6 T.
FRESH	A	2.65	1.30	1.10	2.60	1.20	1.00	2.45	1.50	1.10	3.30	1.25	1.15	2.80	1.25	1.10	3.00	1.30	1.10
CARBONIZED	A	2.25	1.35	1.15	2.60	1.30	1.00	2.25	1.60	1.15	3.05	1.25	1.30	2.65	1.35	1.20	2.80	1.25	1.15
FRESH	B	2.80	1.15	1.15	3.50	1.00	1.25	2.85	1.20	1.20	2.80	1.30	1.15	2.80	1.25	1.05	2.90	1.25	1.15
CARBONIZED	B	2.55	1.25	1.25	3.25	1.05	1.30	2.70	1.30	1.20	2.60	1.35	1.25	2.25	1.35	1.10	2.70	1.30	1.25
FRESH	C	2.90	1.15	1.20	2.75	1.25	1.20	2.85	1.25	1.25	2.90	1.05	1.00	2.90	1.25	1.15	3.05	1.30	1.10
CARBONIZED	C	2.55	1.25	1.30	2.25	1.25	1.20	2.60	1.35	1.35	2.75	1.20	1.15	2.80	1.35	1.25	2.85	1.45	1.25
FRESH	D	2.75	1.50	1.00	2.85	1.25	1.00	2.80	1.35	1.10	3.65	1.40	1.20	2.90	1.25	1.25	2.95	1.25	1.25
CARBONIZED	D	2.65	1.60	1.15	2.70	1.35	1.15	2.65	1.45	1.20	2.95	1.45	1.30	2.75	1.35	1.30	2.55	1.30	1.30
FRESH	E	3.00	1.25	1.10	2.75	1.35	1.20	2.25	1.40	1.25	2.75	1.25	1.00	2.80	1.30	1.25	2.90	1.25	1.25
CARBONIZED	E	2.80	1.35	1.20	2.65	1.40	1.25	2.45	1.50	1.25	2.70	1.35	1.15	2.70	1.35	1.30	2.80	1.35	1.30

Table no. 6 Triticum aestivum L.: Carbonization (measurements in mm.)

STATE	CATALOG.-NO.	1			2			3			4			5			6		
		L.	B.	T.	L.	B.	T.	L.	B.	T.	L.	B.	T.	L.	B.	T.	L.	B.	T.
FRESH	A	7.00	3.80	3.50	6.90	3.70	3.20	6.50	3.60	2.80	6.50	3.50	2.30	6.30	3.30	2.80	6.50	3.50	2.50
CARBONIZED		6.70	4.50	4.00	6.40	4.50	3.50	6.00	4.00	3.50	6.00	4.40	2.90	5.80	3.80	3.00	6.00	4.00	2.80
FRESH	B	7.10	3.90	3.50	6.90	3.60	3.40	7.00	4.00	3.00	6.30	3.50	2.90	6.50	3.20	3.00	6.30	3.60	3.20
CARBONIZED		6.80	4.60	3.50	6.40	4.50	3.60	6.60	4.80	3.30	6.00	3.90	3.00	6.10	4.00	3.20	5.80	4.40	3.50
FRESH	C	6.70	3.60	2.70	6.50	3.50	3.40	6.60	3.60	3.50	6.80	3.60	3.00	6.50	3.80	3.00	6.40	3.50	2.90
CARBONIZED		6.50	4.40	3.40	6.00	4.00	3.50	6.30	4.20	3.50	6.50	4.30	3.40	5.90	4.40	3.30	6.00	4.50	3.20
FRESH	D	6.30	3.30	2.00	7.00	3.70	3.50	6.60	3.60	2.80	6.10	3.30	2.20	6.10	3.60	3.00	6.30	3.00	3.20
CARBONIZED		5.90	3.50	2.60	6.30	4.50	3.50	6.10	3.90	3.50	5.70	3.70	3.00	5.80	4.10	3.20	5.60	4.00	3.80
FRESH	E	6.50	3.80	2.70															
CARBONIZED		6.00	4.90	3.60															
FRESH	F	6.50	4.00	3.00	7.20	4.00	2.90	7.00	4.20	2.80	6.70	3.60	2.70	7.10	4.50	3.20	6.50	4.20	3.40
CARBONIZED		6.00	4.90	3.70	6.80	4.60	3.10	6.40	4.70	3.20	6.30	4.10	3.20	6.40	5.00	3.80	6.20	5.00	3.80
FRESH	G	6.50	3.10	2.50	7.00	3.80	3.30	6.40	3.20	2.40	7.00	3.60	2.50	6.70	3.50	2.70	6.40	3.40	3.00
CARBONIZED		5.90	4.10	3.30	6.40	5.00	4.10	5.70	4.00	3.00	6.80	4.20	3.20	6.50	4.40	3.20	5.90	5.30	3.60
FRESH	H	6.70	3.90	3.40	6.50	3.40	3.10	6.80	4.10	3.20	6.60	3.80	2.60	6.30	3.30	2.50	6.80	3.60	2.80
CARBONIZED		6.30	5.00	4.00	5.90	4.30	3.60	6.40	4.30	3.80	6.00	4.20	3.20	5.80	4.20	2.90	6.50	4.60	3.30
FRESH	I	6.60	3.20	2.70	7.00	3.80	3.50	6.30	3.40	2.40	6.10	3.30	2.40	6.40	3.50	2.80	6.80	3.80	2.80
CARBONIZED		6.20	4.40	3.30	6.70	4.80	4.20	6.00	4.20	3.00	5.90	4.00	2.90	6.00	4.20	3.10	6.40	4.80	3.40
FRESH	J	6.30	3.40	2.70															
CARBONIZED		6.00	4.10	3.00															

Table no. 7 Hordeum vulgare L.: Carbonization (measurements in mm.)

STATE	CATALOG.-NO.	1 L.	1 B.	1 T.	2 L.	2 B.	2 T.	3 L.	3 B.	3 T.	4 L.	4 B.	4 T.	5 L.	5 B.	5 T.	6 L.	6 B.	6 T.
FRESH	A	8.20	3.60	3.00	7.50	3.00	2.50	8.50	3.40	2.50	7.50	3.50	2.60	7.70	3.70	3.00	8.70	3.50	2.50
CARBONIZED	A	7.90	4.20	3.30	7.00	3.20	2.80	7.30	4.20	3.10	7.10	4.40	3.50	7.70	4.50	3.30	8.40	3.80	3.30
FRESH	B	9.00	3.80	3.00	8.30	3.10	2.20	8.20	3.30	2.50	8.50	3.50	2.60	7.70	3.00	2.40	7.40	3.00	2.30
CARBONIZED	B	9.00	4.40	3.50	7.20	3.20	2.70	7.70	3.60	3.40	8.00	4.00	3.50	7.40	3.90	3.00	7.20	3.70	2.90
FRESH	C	8.20	3.40	2.50	8.00	3.60	3.10	7.20	2.60	2.20	8.90	3.50	2.80	8.00	3.80	3.00	7.40	2.60	2.20
CARBONIZED	C	7.80	3.90	3.20	7.80	4.00	3.50	7.00	2.90	2.70	8.20	4.00	3.30	7.80	4.30	3.60	7.10	3.50	2.80
FRESH	D	8.50	3.00	2.50	7.00	3.20	2.20	8.90	3.00	2.30	7.50	3.60	2.60	8.00	3.10	2.60	8.90	3.00	2.50
CARBONIZED	D	8.20	3.20	3.10	7.00	4.20	2.50	8.50	4.00	3.20	7.50	4.20	3.00	7.70	4.00	3.30	8.70	3.70	3.50
FRESH	E	8.60	3.60	3.20															
CARBONIZED	E	8.60	4.30	3.50															
FRESH	F	8.00	3.50	2.50	7.70	3.50	3.00	7.80	3.00	2.10	8.50	2.80	2.20	8.40	2.90	2.30	8.30	3.50	2.50
CARBONIZED	F	7.90	4.00	3.40	7.50	4.10	3.30	7.70	4.00	2.50	8.30	3.50	2.50	8.20	3.60	2.70	8.00	4.00	3.00
FRESH	G	9.50	3.80	3.00	7.90	3.40	2.50	8.20	2.90	2.00	7.60	3.20	2.20	8.00	3.30	2.50	8.00	3.00	2.30
CARBONIZED	G	8.70	4.00	3.60	7.70	3.70	3.30	7.80	3.70	2.90	7.30	3.50	2.80	7.90	3.80	2.80	7.70	3.30	2.50
FRESH	H	8.50	3.50	2.90	7.00	3.00	2.10	8.20	3.60	3.00	8.20	3.60	3.00	8.20	3.50	2.50	7.90	2.50	2.40
CARBONIZED	H	8.50	4.20	3.10	6.80	3.70	2.80	8.00	4.00	3.50	8.00	3.90	3.30	8.00	4.00	3.30	7.50	2.90	2.90
FRESH	I	7.90	3.70	3.00	8.50	3.60	3.00	8.40	3.60	2.90	9.00	3.50	3.10	8.00	3.50	2.60	9.60	2.50	2.20
CARBONIZED	I	7.50	4.00	3.40	8.20	4.10	3.50	8.00	4.00	3.40	8.50	4.00	3.50	7.70	3.90	3.30	9.20	3.00	2.80
FRESH	J	6.60	2.80	2.10															
CARBONIZED	J	6.50	3.00	2.50															

Table no. 8

Sorghum vulgare L.: Impressions
(measurements in mm.)

STATE	CATALOG.-NO.	1 L.	1 B.	1 T.	2 L.	2 B.	2 T.	3 L.	3 B.	3 T.	4 L.	4 B.	4 T.	5 L.	5 B.	5 T.	6 L.	6 B.	6 T.	7 L.	7 B.	7 T.	8 L.	8 B.	8 T.	9 L.	9 B.	9 T.	10 L.	10 B.	10 T.
FRESH	A	4.60	5.40		4.50	4.50		4.60	4.60		3.40	2.60		4.50	4.20		7.00	4.90		4.80	4.60		4.20	4.60		4.50	4.80		4.80	4.80	
SOAKED	A	5.00	6.00		4.80	5.20		4.60	5.00		3.50	3.00		4.50	4.80		7.60	5.40		5.20	5.00		4.40	5.20		4.90	5.50		5.30	5.50	
IMPRESSION	A	4.40	5.00		3.90	4.00		4.00	4.30		2.90	2.50		4.10	4.00		6.10	4.50		4.00	3.90		3.90	4.40		4.00	4.30		4.10	4.30	
FRESH	B	4.50	4.50		4.60	5.00		4.40	4.30		4.20	5.30		3.40	2.30		4.70	4.70		4.20	4.60		4.30	4.50		3.30	2.60		4.50	4.70	
SOAKED	B	5.00	5.20		5.20	5.50		4.70	4.90		5.00	6.00		3.50	2.70		5.00	5.20		4.50	5.00		4.90	5.00		3.50	3.00		5.00	5.30	
IMPRESSION	B	4.20	4.00		4.40	4.60		3.80	4.00		4.00	4.90		2.80	2.00		4.30	4.50		4.00	4.10		4.00	4.10		3.00	2.20		4.00	4.00	
FRESH	C	4.00	4.50		5.20	5.30		4.20	4.30		4.90	5.20		4.30	4.20		4.20	4.40		4.60	5.50		4.30	4.50		6.00	4.50		4.80	4.80	
SOAKED	C	4.70	4.70		5.30	5.50		4.50	4.70		5.20	5.80		5.00	5.00		4.40	4.80		5.00	6.10		4.60	5.00		6.30	5.10		5.20	5.20	
IMPRESSION	C	3.70	4.20		4.80	4.90		4.00	3.80		4.20	4.60		4.00	4.00		4.00	4.20		4.30	4.70		4.00	4.00		5.60	4.20		4.50	4.40	
FRESH	D			3.00			3.00			3.50			4.10			3.10			2.50			3.00			3.40			1.70			2.50
SOAKED	D			4.00			4.00			3.80			4.70			3.60			3.20			3.50			4.30			2.40			3.00
IMPRESSION	D			2.80			2.80			3.00			3.60			2.80			2.50			2.90			3.20			1.60			2.40
FRESH	E			3.00			3.00			2.50			2.60			2.50			1.70			3.10			2.50			3.20			2.70
SOAKED	E			3.90			3.70			3.20			3.30			3.40			2.20			3.10			2.90			4.00			3.00
IMPRESSION	E			3.00			2.90			2.40			2.40			2.10			1.50			2.90			2.30			3.00			2.40
FRESH	F			3.30			3.60			2.70			2.70			2.00			1.80			2.50			4.10			3.00			2.50
SOAKED	F			4.00			3.90			3.20			3.60			2.60			2.40			3.00			4.90			3.50			3.10
IMPRESSION	F			3.00			3.10			2.60			2.70			2.00			1.50			2.00			3.80			2.80			2.30

Table no. 9

Pennisetum americanum,(L.)K. Schum.: Impressions
(measurements in mm.)

STATE	CATALOG.-NO.	1			2			3			4			5			6			7			8			9			10		
		L.	B.	T.	L.	B.	T.	L.	B.	T.	L.	B.	T.	L.	B.	T.	L.	B.	T.	L.	B.	T.	L.	B.	T.	L.	B.	T.	L.	B.	T.
FRESH	A	2.50	2.00		2.50	2.00		2.00	1.90		1.90	1.80		2.30	1.80		2.40	1.90		2.40	1.70		2.30	1.80		2.40	1.80		2.20	1.80	
IMPRESSION		2.40	1.85		2.40	1.90		2.00	1.90		1.90	1.70		2.10	1.70		2.40	1.75		2.30	1.70		2.20	1.70		2.40	1.85		2.20	1.75	
FRESH	B	1.50	1.30		1.80	1.50		2.30	1.80		2.40	1.60		2.20	1.70		2.10	1.70		2.10	1.90		2.20	2.00		2.30	1.70		2.30	1.60	
IMPRESSION		1.50	1.20		1.75	1.40		2.35	1.65		2.30	1.65		2.20	1.70		2.00	1.65		2.05	1.80		2.00	1.95		2.20	1.65		2.20	1.55	
FRESH	C	1.80	1.30		1.70	1.40		2.30	1.90		2.20	1.80		2.30	1.90		2.00	1.90		2.00	1.90		2.40	1.90		1.90	1.10		2.40	2.10	
IMPRESSION		1.70	1.30		1.60	1.30		2.00	1.70		2.20	1.80		2.25	1.95		1.90	1.85		1.85	1.85		2.35	1.80		1.75	1.10		2.75	1.95	
FRESH	D			1.70			1.50			1.50			2.00			1.80			1.70			1.90			1.90			1.80			1.80
IMPRESSION				1.60			1.50			1.50			1.95			1.75			1.65			1.85			1.80			1.75			1.75
FRESH	E			1.00			1.40			1.30			1.50			1.10			1.70			1.10			1.50			1.50			1.50
IMPRESSION				0.95			1.35			1.30			1.45			1.05			1.50			1.10			1.55			1.25			1.50
FRESH	F			1.60			1.40			1.80			1.50			1.20			1.50			1.90			1.50			1.80			1.00
IMPRESSION				1.50			1.50			1.75			1.40			1.15			1.50			1.75			1.35			1.75			0.95

Table no. 10

Sorgum arundinaceum (Desv.)Stapf.: Impressions
(measurements in mm.)

STATE	CATALOG. NO.	1 L.	1 B.	1 T.	2 L.	2 B.	2 T.	3 L.	3 B.	3 T.	4 L.	4 B.	4 T.	5 L.	5 B.	5 T.
FRESH	A	6.90	1.60		7.00	1.70		7.50	1.20		8.10	1.90		7.00	1.70	
IMPRESSION	A	5.00	1.40		6.10	1.50		5.40	1.00		6.30	1.90		5.00	1.50	
FRESH	B	7.70	2.00		6.10	1.50		8.00	1.40		8.00	1.70		8.30	1.30	
IMPRESSION	B	6.10	1.70		5.50	1.40		7.20	1.20		6.50	1.40		7.40	1.00	
FRESH	C	7.50	1.60		6.00	1.60		7.00	1.70		6.70	1.80		6.00	1.60	
IMPRESSION	C	6.60	1.50		6.00	1.50		6.40	1.50		6.50	1,50		5.80	1.60	

STATE	CATALOG. NO.	6 L.	6 B.	6 T.	7 L.	7 B.	7 T.	8 L.	8 B.	8 T.	9 L.	9 B.	9 T.	10 L.	10 B.	10 T.
FRESH	A	9.00	1.80		8.50	1.60		6.20	1.20		9.00	1.90		8.90	2.70	
IMPRESSION	A	6.10	1.50		6.40	1.50		5.10	1.00		6.50	1.70		8.00	2.40	
FRESH	B	7.20	1.70		11.5	2.20		7.70	2.40		9.00	1.70		8.40	1.80	
IMPRESSION	B	6.80	1.40		10.1	1.90		7.00	2.00		6.70	1.50		8.00	1.60	
FRESH	C	7.50	1.60		7.30	1.80		8.50	1.90		6.80	1.80		7.20	1.80	
IMPRESSION	C	6.90	1.50		6.50	1.50		7.20	1.60		6.20	1.60		6.70	1.40	

Table no. 11 Echinochloa pyramidis (L.)Hitchc.: Impressions
(measurements in mm.)

STATE	CATALOG.-NO.	1 L.	1 B.	1 T.	2 L.	2 B.	2 T.	3 L.	3 B.	3 T.	4 L.	4 B.	4 T.	5 L.	5 B.	5 T.	6 L.	6 B.	6 T.
FRESH	A	3.00	1.25		3.30	1.45		3.75	1.55		3.55	1.45		3.15	1.35		3.85	1.35	
IMPRESSION		2.90	1.25		3.15	1.35		3.60	1.50		3.50	1.50		3.10	1.20		3.60	1.25	
FRESH	B	3.15	1.25		3.45	1.30		3.50	1.35		3.00	1.4		3.10	1.25		3.50	1.30	
IMPRESSION		3.15	1.25		3.30	1.25		3.35	1.25		2.90	1.40		3.25	1.25		3.25	1.20	
FRESH	C	3.00	1.25		3.10	1.30		3.35	1.25		3.10	1.50		3.50	1.45		3.10	1.10	
IMPRESSION		3.00	1.30		3.00	1.25		3.10	1.15		3.05	1.50		3.35	1.35		3.00	1.05	
FRESH	D	3.25	1.10		4.00	1.30		3.25	1.25		2.80	1.15		3.50	1.15		3.35	1.20	
IMPRESSION		3.15	1.00		3.90	1.30		3.25	1.30		2.80	1.00		3.35	1.05		3.20	1.25	
FRESH	E	4.00	1.45		3.10	1.25		3.50	1.25		3.55	1.30		3.25	1.20		3.00	1.00	
IMPRESSION		3.80	1.35		3.00	1.20		3.35	1.15		3.40	1.20		3.00	1.10		2.65	0.85	

Table no. 12 Setaria verticillata (L)Beaquv.: Impressions (measurements in mm.)

STATE	CATALOG.-NO.	1 L.	1 B.	1 T.	2 L.	2 B.	2 T.	3 L.	3 B.	3 T.	4 L.	4 B.	4 T.	5 L.	5 B.	5 T.
FRESH	A	2.20	1.30		3.20	1.20		3.10	1.30		3.00	1.20		2.80	1.20	
IMPRESSION		2.00	1.25		2.50	1.25		2.40	1.20		2.75	1.15		2.25	1.20	
FRESH	B	2.10	1.10		2.20	1.30		3.50	1.30		3.10	1.30		2.30	1.20	
IMPRESSION		2.00	1.20		2.20	1.30		3.40	1.25		2.35	1.20		2.25	1.25	
FRESH	C	3.00	1.20		2.60	1.20		2.70	1.20		2.70	1.20		3.50	1.20	
IMPRESSION		2.45	1.10		2.40	1.20		2.50	1.10		2.50	1.15		3.10	1.10	
FRESH	D	3.00	1.30		3.10	1.10		2.90	1.20		3.30	1.20		2.50	1.20	
IMPRESSION		2.60	1.25		3.00	1.20		2.75	1.20		3.20	1.15		2.30	1.10	
FRESH	E	2.60	1.20		3.00	1.30		3.00	1.30		2.60	1.30		2.90	1.30	
IMPRESSION		2.25	1.15		2.85	1.25		2.70	1.20		2.50	1.25		2.75	1.25	
FRESH	F	2.80	1.30		3.00	1.30		2.70	1.20		3.80	1.30		3.00	1.20	
IMPRESSION		2.50	1.30		2.85	1.25		2.50	1.25		3.15	1.20		2.80	1.15	
FRESH	G			1.00			0.90			1.20			1.00			1.00
IMPRESSION				1.00			1.00			1.10			1.00			0.95
FRESH	H			1.30			1.10			1.20			1.30			1.00
IMPRESSION				1.20			1.05			1.20			1.20			1.05
FRESH	I			1.20			1.10			1.00			1.00			1.10
IMPRESSION				1.20			1.05			0.95			1.05			1.05
FRESH	J			1.30			1.20			1.20			1.30			1.20
IMPRESSION				1.20			1.20			1.15			1.20			1.15
FRESH	K			1.10			1.10			1.10			0.90			1.20
IMPRESSION				1.10			1.05			1.10			0.85			1.15
FRESH	L			1.10			1.10			1.40			1.10			1.20
IMPRESSION				1.05			1.15			1.20			1.05			1.10

Table no. 13 Pennisetum pedicellatum Trin.: Impressions (measurements in mm.)

STATE	CATALOG.-NO.	1			2			3			4			5			6		
		L.	B.	T.	L.	B.	T.	L.	B.	T.	L.	B.	T.	L.	B.	T.	L.	B.	T.
FRESH	A	12.0	2.50		10.0	3.00		10.0	2.50		9.00	1.80		8.50	2.00		8.00	2.00	
IMPRESSION	A	10.0	2.30		9.20	2.90		9.30	2.10		9.00	1.80		8.40	2.00		7.80	1.90	
FRESH	B	8.00	2.10		6.50	1.30		8.00	2.20		9.00	2.30		6.50	2.30		7.50	2.10	
IMPRESSION	B	7.40	1.70		6.50	1.30		7.60	2.00		8.80	2.10		6.50	2.20		6.90	2.00	
FRESH	C	8.50	2.70		8.00	2.30		7.50	2.70										
IMPRESSION	C	8.10	2.50		8.00	2.20		7.10	2.50										

Table no. 14 Sorghum verticilliflorum (Steud.)Stapf.: Impressions (measurements in mm.)

STATE	CATALOG.-NO.	1 L.	1 B.	1 T.	2 L.	2 B.	2 T.	3 L.	3 B.	3 T.	4 L.	4 B.	4 T.	5 L.	5 B.	5 T.
FRESH	A	5.40	1.80		5.70	1.80		6.50	2.00		7.30	1.90		9.20	2.30	
IMPRESSION		5.10	1.60		5.50	1.60		6.50	2.00		6.90	1.70		8.50	2.00	

Table no. 15 Sorghum aethiopicum (Hack.)Rupr.: Impressions (measurements in mm.)

STATE	CATALOG.-NO.	1 L.	1 B.	1 T.	2 L.	2 B.	2 T.	3 L.	3 B.	3 T.	4 L.	4 B.	4 T.	5 L.	5 B.	5 T.
FRESH	A	7.50	3.30		8.50	2.80		12.0	2.70		-	-		-	-	
IMPRESSION		7.30	3.20		8.10	3.00		11.5	2.40		-	-		-	-	

A P P E N D I X IV

Distribution,Habit,and Habitat of the Macro-
fossil plant species recovered from the C.Sudan

Distribution,habit,and habitat of the macrofossil plant species recovered from C.Sudan.

Plant species	Family	Distribution	Habit	Habitat
Boerhavia spp.	Nyctaginaceae	Widespread	Prostrate, ascending or spreading herb	Lowland plains , arable land and waste-places
Capparis spp.	Capparidaceae	Widespread	Climbing shrub or small trees	Wide-range of habitats
Celtis integrifolia	Ulmaceae	Central and S.Sudan	Deciduous tree	Khor and Nile bank
Crotalaria spp.	Papilionaceae (Leguminaceae)	Widespread	Annual or perennial herb or shrub	Wide-range of habitats
Grewia tenax	Liliaceae	C.& S.Sudan	Shrubs or small trees	Lowland plains
Panicum turgidum	Poaceae(Graminaceae)	N.& C.Sudan	Herbs	Coastal plains desert or stony places
Pennisetum spp.	Poaceae(Graminaceae)	Widespread	Annual or perennial herb	Lowland plains, khor-banks, moist places,upland grass land, arable and fallow lands
Sesamum spp.	Pedaliaceae	Widespread	Herbs	Lowland plains, alluvial soils & arable lands
Setaria spp.	Poaceae(Graminaceae	Widespread	Annual or Perennial herb	Basalt clay depressions,lowland plains shady places and flood-plains
Sida alba	Malvaceae	Widespread	Herbs	Foothills,lowland plains and arable lands
Solanum dubium	Solanaceae	N.& C.Sudan	Woody herb	Lowland plains
Sorghum spp.	Poaceae(Graminaceae)	Widespread	Annual or perennial herb	Khor and Nile banks and marshy places
Ziziphus spina-christi	Rhamnaceae	N.& C.Sudan	Evergreen shrubs or herbs	Nile banks and lowland plains

(cont.)

Plant species	Family	Distribution	Habit	Habitat
-	Chenopodiaceae	Widespread	Herbs or undershrubs	Lowland plains & as arable weed
-	Cruciferae	Widespread	Annual or perennial herb	Lowland plains, arable weed & aquatic
-	Cucurbitaceae	Widespread	Climbing or prostrate herbs	Wide-range of habitats
-	Leguminaceae	Widespread	Herbs, shrubs or trees	Wide-range of habitats
-	Molluginaceae	Widespread	Herbs	Mainly shallow clay or silt-depressions

A P P E N D I X V

Results of measurements of one hundred gouges

Results of measurements of one hundred (randomly sampled) gouges .from the sites in the Khartoum area.(measurements in cms.)

Index no.	Museum-shelf no.	Specimen label no.	Site	Maximum length	Maximum Breadth			Maximum Thickness			Type of Gouge and othe comments
					Distal	Middle	Proximal	Distal	Middle	Proximal	
1	Bx-3	I-59- S	Shaheinab	7.5	3.4	3.2	0.7	1.5	1.2	0.2	Flaked on both faces. Distal end truncated by a single blow.
2	Bx-3	no label	Shaheinab	6.4	4.0	3.8	1.6	0.9	1.9	1.5	Flaked on both faces. Distal end is flaked. Proximal end is broad and thick
3	Bx-3	Q-76-S	Shaheinab	8.1	3.8	3.8	0.8	1.9	2	0.3	Flaked on both faces. Distal end truncated by a single blow.
4	Bx-3	no label	Shaheinab	6.8	3.9	3.6	1.0	1.8	1.8	0.7	Flaked on both faces. Distal end truncated by single blow creating an oblique surface.
5	no. label	52h	Nofalat	6.5	4.0	3.6	0.8	2	1.8	0.6	Flaked on both faces. Distal end is truncated by a single blow.
6	no label	100e	Nofalab	5.1	3.5	3.3	0.5	1.4	1.4	0.3	Flaked on both faces. Truncation by a single blow created flat distal end

Index no.	Museum-shelf no.	Specimen label no.	Site	Maximum length	Maximum Breadth			Maximum Thickness			Type of Gouge and othe comments
					Distal	Middle	Proximal	Distal	Middle	Proximal	
7	no label	K-C-42	Kadero	5.1	3.8	0.8	1.8	1.6	1.7	0.4	Flaked on both faces. Similar in other details to no. 6
8	no label	K-C-41	Kadero	6.1	4.7	4.0	1.6	1.8	1.6	0.3	Flaked on both faces. Distal end is flaked.
9	no. label	K-C-104	Kadero	4	3.3	2.9	2.3	1.6	1.7	1.1	Flaked on both faces. Distal end is flaked.
10	no. label	100x/100y	Umm Direiwa	5.5	3.4	3.0	0.9	1.5	1.1	0.4	Flaked on both faces. Distal end is truncated by a single blow.
11	no. label	100x/100y	Umm Direiwa	2.2	3.3	2.7	0.5	1.3	1.2	0.5	Flaked on both faces. Distal end is truncated by a single blow. Proximal end broken and retouched. and used as scraper.
12	no. label	100x/102y	Umm Direiwa	5.3	2.8	2.7	2.4	1.5	1.4	1.3	Flaked on both faces. Distal end is flaked.
13	Bx-3	D-81	Shaheinab	6.2	5.6	5.5	5.3	2.8	2.7	2.7	Polished on both faces. Distal end is flaked Proximal end is broad.

Index no.	Museum-shelf no.	Specimen label no.	Site	Maximum length	Maximum Breadth			Maximum Thickness			Type of Gouge and othe comments
					Distal	Middle	Proximal	Distal	Middle	Proximal	
14	Bx-3	N-88 (20-40cm)	Shaheinab	5.8	3.8	3.3	2.3	0.9	1.1	1.0	Polished on both faces. Distal end is flaked. Proximal end is broken
15	Bx-3	no label	Shaheinab	4.9	3.7	3.4	2.3	1.0	1.0	0.8	Polished on both faces. Similar to no. 14 in other details.
16	Bx-3	100x/100y	Umm -Direiwa	5.7	3.4	3.1	1.3	1.7	1.5	0.4	Polished on both faces. Distal end is flaked.
17	Bx-3	100x/100y	Umm-Direiwa	5.6	3.4	3.1	2.8	1.7	1.6	1.2	Polished on both faces. Distal end is flaked.
18	no label	94x/100y	Umm Direiwa	4.4	3.0	2.8	2.6	1.1	1.2	1.1	Polished on both faces. Distal end is flaked. Proximal end is broken. Lower face is better polished than the upper one.
19	no label	96x/100y	Umm Direiwa	5.4	3.6	3.8	3.3	1.7	1.6	1.5	Polished on both faces. Distal end is flaked.
20	Bx-24	N - 82 (20-40cm)	Shaheinab	6.0	4.4	4.1	1.2	1.4	1.5	1.2	Flaked on the upper face and polished on the Lower one. Distal end is truncated by a single blow creat-

Index no.	Museum-shelf no.	Specimen label no.	Site	Maximum length	Maximum Breadth			Maximum Thickness			Type of Gouge and othe comments
					Distal	Middle	Proximal	Distal	Middle	Proximal	
											ing oblique surface.
21	Bx-24	no label	Shaheinab	5.8	3.8	3.4	1.3	1.4	1.4	0.8	Flaked on the upper face and polished on the lower one. Truncation of the distal end made it emerge into the upper face.
22	Bx-24	no label	Shaheinab	7.4	4.1	3.8	0.7	1.6	1.7	1.0	Flaked on the upper face and polished on the lower one Distal end is flaked. Proximal end is broken.
23	Bx-24	N-85 (0 -20cm)	Shaheinab	5.8	3.8	3.0	0.6	1.2	1.5	0.4	Flaked on the upper face and polished on the lower one. Distal end is broken.
24	Bx-24	M-79 (5-20cm)	Shaheinab	6.0	3.2	3.0	0.6	1.2	1.2	0.3	Flaked on the upper face and polished on the lower one. Distal end is flaked.
25	Bx-24	M-92	Shaheinab	6.1	3.8	3.3	0.8	1.6	1.4	0.4	Flaked on the upper face and polished on the lower one.

Index no.	Museum-shelf no.	Specimen label no.	Site	Maximum length	Maximum Breadth			Maximum Thickness			Type of Gouge and othe comments
					Distal	Middle	Proximal	Distal	Middle	Proximal	
											Distal end is flaked.
26	Bx-24	5047	Shaheinab	7.2	3.9	3.5	1.0	1.8	1.5	0.4	Flaked on the upper face and polished on the lower one. Distal end is truncated by a single blow creating a flat surface.
27	Bx-24	L-66	Shaheinab	6.3	3.4	3.2	0.8	1.4	1.3	0.4	Flaked on the upper face and polished on the lower one. Distal end is flaked.
28	Bx-24	N-82 (0-20cm.)	Shaheinab	5.0	4.1	3.8	2.5	1.2	1.2	0.4	Flaked on the upper face and polished on the lower one. Distal end is truncated by a single blow creating an oblique surface.
29	no label	52h	Nofalab	6.7	3.2	3.1	2.7	1.6	1.7	0.6	Flaked on the upper face and polished on the lower one. Distal end is flaked
											Flaked on the upper

Index no.	Museum-shelf no.	Specimen label no.	Site	Maximum length	Maximum Breadth			Maximum Thickness			Type of Gouge and othe comments
					Distal	Middle	Proximal	Distal	Middle	Proximal	
30	no label	S	Zakiab	6.7	4.0	4.0	2.3	1.6	1.6	1.4	face and polished on the lower one. Polishing Is carefully done towards the distal end. Distal end is flaked.
31	no label	S	Zakiab	5.7	3.3	2.9	1.6	1.2	1.2	0.6	Flaked on the upper face and polished on the lower one. Distal end is truncated by a single blow creat-ing an oblique surface.
32	no label	S	Zakiab	5.7	4.0	3.7	1.3	1.4	1.4	0.4	Flaked on the upper face and polished on the lower one. Distal end is flaked.
33	no label	100x/102y	Umm Direiwa	4.9	3.6	3.4	3.0	1.3	1.5	1.1	Flaked on the upper face and polished on the lower one. Distal end is flaked.
34	no label	100x/100y	Umm Direiwa	4.5	3.9	3.8	1.7	1.6	1.6	1.0	Flaked on the upper face and polished on the lower one. Distal end is flaked.

Index no.	Museum-shelf no.	Specimen label no.	Site	Maximum length	Maximum Breadth			Maximum Thickness			Type of Gouge and othe comments
					Distal	Middle	Proximal	Distal	Middle	Proximal	
35	no label	96x/100y	Umm Direiwa	3.0	3.7	3.7	3.7	1.2	-	0.4	Flaked on the upper face and polished on the lower one. Distal end is flaked. Proximal end is broken.
36	no label	100x/100y	Umm Direiwa	4.0	3.0	2.7	0.9	1.0	0.9	0.4	Flaked on the upper face and polished on the lower one Distal end is broken.
37	no label	104x/72y	Umm Direiwa	3.2	2.7	2.4	0.9	1.2	1.2	0.5	Flaked on the upper face and polished on the lower one. Distal end is flaked.
38	no label	94x/100y	Umm Direiwa	2.8	3.1	3.0	2.3	1.6	1.5	1.5	Flaked on the upper face and polished on the lower one. Distal end is flaked.
39	no label	K-B-102	Kadero	5.5	3.6	3.3	1.5	1.1	1.1	0.2	Flaked on the upper face and polished on the lower one. Distal end is flaked. Proximal end is thin and sharp.

Index no.	Museum-shelf no.	Specimen label no.	Site	Maximum length	Maximum Breadth			Maximum Thickness			Type of Gouge and othe comments
					Distal	Middle	Proximal	Distal	Middle	Proximal	
40	no label	K-C-40	Kadero	4.8	3.8	3.8	2.8	1.3	1.3	0.5	Flaked on the upper face and polished on the lower one Distal end is flaked.
41	no label	K-B-100	Kadero	5.3	3.4	3.0	1.1	1.2	1.2	0.6	Flaked on the upper face and highly polished on the lower one. Distal end is flaked.
42	B-24	K-88-S	Shaheinab	6.4	3.4	3.2	0.7	1.5	1.2	0.3	Flaked on the upper face and partly flaked partly polished on the lower one. Distal end is flaked.
43	B-23	no label	Shaheinab	5.6	3.4	2.7	1.1	1.5	1.6	0.4	Flaked on the upper face and partly flaked partly polished on the lower one. Distal end is truncated by a single blow creating a flat surface.
44	Bx-24	M-81 (0-20cm)	Shaheinab	6.1	3.4	3.0	0.5	0.9	1.0	0.3	Flaked on the upper face-partly flaked partly polished on the lower, one

Index no.	Museum - shelf no.	Specimen label no.	Site	Maximum Length	Maximum Breadth			Maximum Thickness			Type of Gouge and other comments
					Distal	Middle	Proximal	Distal	Middle	Proximal	
45	Bx-24	5047	Shaheinab	6.3	3.2	3.4	0.4	1.6	1.7	0.2	Flaked on the upper face and partly flaked partly polished on the lower one. Distal end is flaked.
46	Bx-24	no label	Shaheinab	6.0	3.7	3.4	0.4	1.8	1.4	0.3	Flaked on the upper face, partly flaked partly polished on the lower one. Distal end is flaked.
47	Bx-24	L-59-S	Shaheinab	5.4	3.9	3.5	1.3	1.9	1.5	0.5	Flaked on the upper face partly flaked partly polished on the lower one Distal end is flaked
48	Bx-25	N-88 (20-40cm)	Shaheinab	5.7	3.4	3.2	2.8	1.2	1.4	1.2	Flaked on the upper face, partly flaked and partly polished on the lower one. Distal end is truncated by a single blow creating oblique surface.
49	Bx-24	N-80 (40-60cm)	Shaheinab	6.0	3.6	3.3	0.5	1.4	1.6	0.4	Flaked on the upper face partly flaked partly polished on the lower one Distal end is polished

Index no.	Museum - shelf no.	Specimen label no.	Site	Maximum Length	Maximum Breadth			Maximum Thickness			Type of Gouge and other comments
					Distal	Middle	Proximal	Distal	Middle	Proximal	
50	Bx-24	K-80- S	Shaheinab	5.8	4.0	3.6	1.2	1.6	1.5	0.4	Flaked on upper face, partly flaked partly polished on the lower one Distal end is flaked.
51	Bx-24	N-80 (0-20cm)	Shaheinab	5.3	3.8	3.8	3.5	1.9	1.9	1.8	Flaked on upper face partly flaked, partly polished on the lower one. Distal end is flaked
52	Bx-24	N-81 (0-20cm)	Shaheinab	7.2	4.3	4.3	0.9	1.4	1.5	0.4	Flaked on upper face partly flaked partly polished on the other. Distal end is flaked.
53	Bx-24	J-60 (50-60cm)	Shaheinab	5.6	3.2	3.9	0.8	1.4	1.5	0.3	Flaked on the upper face, partly flaked partly polished on the lower one Distal end is flaked
54	Bx-24	5047	Shaheinab	5.1	5.1	4.6	2.6	1.6	1.4	0.6	Flaked on the upper face, partly flaked partly polished on the lower one Distal end is flaked. Proximal end is thin and sharp.

Index no.	Museum - shelf no.	Specimen label no.	Site	Maximum Length	Maximum Breadth			Maximum Thickness			Type of Gouge and other comments
					Distal	Middle	Proximal	Distal	Middle	Proximal	
55	Bx-24	R-80,S	Shaheinab	7.0	3.3	3.1	0.6	1.4	1.5	0.5	Flaked on the upper face, partly flaked partly polished on the lower one. Large flakes are removed from the upper face. Distal end is flaked
56	Bx-24	H-67,S	Shaheinab	7.0	3.9	3.8	0.6	1.5	1.5	0.6	Flaked on the upper face, partly flaked partly polished on the lower one. Part of the coftex surface is still retained. Distal end is flaked.
57	Bx-24	N-84,S	Shaheinab	7.8	2.8	2.8	0.4	1.0	1.3	0.3	Flaked on the upper face, partly flaked partly polished on the lower one Distal end is flaked
58	Bx-24	0-80 (0-20cm)	Shaheinab	5.7	3.6	3.4	0.8	1.5	1.4	0.4	Flaked on the upper face, partly flaked partly polished on the lower one Distal end is flaked

Index no.	Museum - shelf no.	Specimen label no.	Site	Maximum Length	Maximum Breadth			Maximum Thickness			Type of Gouge and other comments
					Distal	Middle	Proximal	Distal	Middle	Proximal	
59	Bx-24	no label	Shaheinab	6.0	3.6	3.4	0.7	1.5	1.4	0.4	Flaked on the upper face, partly flaked partly polished on the lower one. Proximal end is thin and sharp. Distal end is flaked.
60	Bx-24	M-66 (40cm)	Shaheinab	5.6	3.4	3.3	0.6	1.5	1.3	0.4	Flaked on the upper face, partly flaked partly polished on the lower one. Distal end is flaked
61	Bx-24	S	Shaheinab	6.1	3.6	3.5	1.4	1.5	1.4	0.4	Flaked on the upper face, partly flaked partly polished on the lower one. Distal end is flaked.
62	Bx-24	S	Shaheinab	5.5	3.5	3.2	1.4	1.5	1.5	0.5	Flaked on the upper face, partly flaked partly polished on the lower one. Distal end is flaked Proximal end is polished

Index no.	Museum - shelf no.	Specimen label no.	Site	Maximum Length	Maximum Breadth			Maximum Thickness			Type of Gouge and other comments
					Distal	Middle	Proximal	Distal	Middle	Proximal	
63	Bx-24	5047	Shaheinab	6.3	3.1	3.0	1.5	1.5	1.4	0.5	Flaked on the upper face, partly flaked partly polished on the lower one Distal end is truncated by a single blow creating an oblique surface.
64	Bx-24	no label	Shaheinab	6.7	3.4	3.2	1.0	1.4	1.7	0.3	Flaked on the upper face, partly flaked partly polished on the lower one. Distal end is flaked.
65	Bx-24	J-59 (0-20cm)	Shaheinab	6.1	4.1	3.7	1.3	1.7	1.5	0.4	Flaked on the upper face, partly flaked partly polished on the lower one. Distal end is flaked.
66	Bx-24	L-59 (50-70cm)	Shaheinab	5.7	3.7	3.6	0.8	1.5	1.4	0.5	Flaked on the upper face, partly flaked partly polished on the lower one Distal end is flaked.
67	Bx-24	no label	Shaheinab	5.2	3.8	3.4	1.6	1.9	1.4	0.4	Flaked on the upper face, partly flaked

Index no.	Museum - shelf no.	Specimen label no.	Site	Maximum Length	Maximum Breadth			Maximum Thickness			Type of Gouge and other comments
					Distal	Middle	Proximal	Distal	Middle	Proximal	
68	Bx-24	P-84,S	Shaheinab	5.9	2.8	2.8	1.3	1.3	1.2	0.4	Flaked on the upper face, partly flaked; partly polished on the lower one. Distal end is truncated by a single blow creating an oblique surface.
69	Bx-24	no label	Shaheinab	5.1	3.3	3.0	1.9	1.5	1.5	0.6	Flaked on the upper face, partly flaked partly polished on the lower one. Similar to no. 68 in other details.
70	Bx-24	R-80,S	Shaheinab	5.2	3.1	2.9	1.8	1.5	1.5	0.3	Flaked on the upper face, partly flaked partly polished on the lower one. Similar to nos. 68, 69 in other details.
71	Bx-24	N-66 (40-60cm)	Shaheinab	5.8	3.5	3.0	1.5	1.5	1.5	0.4	Flaked on the upper face, partly flaked, partly polished on the lower one. Distal end is flaked

Index no.	Museum - shelf no.	Specimen label no.	Site	Maximum Length	Maximum Breadth			Maximum Thickness			Type of Gouge and other comments
					Distal	Middle	Proximal	Distal	Middle	Proximal	
72	Bx-24	N-77	Shaheinab	5.3	3.2	3.0	0.7	1.9	1.9	0.4	Flaked on the upper face, partly flaked partly polished on the lower one. Distal end is flaked.
73	Bx-24	M-66	Shaheinab	7.2	3.2	3.0	0.8	1.7	1.8	0.4	Flaked on the upper face, partly flaked partly polished on the lower one Similiar to no. 67 in other details.
74	Bx-24	N-84 (50cm+)	Shaheinab	5.6	3.2	3.0	1.1	1.4	1.3	0.3	Flaked on the upper face, partly flaked partly polished on the lower one. Distal end is flaked
75	Bx-24	no label	Shaheinab	6.3	3.3	3.1	0.6	1.5	1.4	0.3	Flaked on the upper face, partly flaked partly polished on the lower one. Distal end is truncated by a single blow creating an oblique surface. Proximal end is thin.

Index no.	Museum - shelf no.	Specimen label no.	Site	Maximum Length	Maximum Breadth			Maximum Thickness			Type of Gouge and other comments
					Distal	Middle	Proximal	Distal	Middle	Proximal	
76	Bx-24	0-81 (20-40cm)	Shaheinab	5.7	3.0	3.0	1.6	1.5	1.5	0.5	Flaked on the upper face, partly flaked partly polished on the lower one. Similar to no. 51 in other details.
77	Bx-24	no label	Shaheinab	5.2	3.3	3.1	1.0	1.3	1.5	0.4	Flaked on the upper face, partly flaked partly polished on the lower one. Both flaking and polishing are carefully done. Distal end is flaked.
78	Bx-24	no label	Shaheinab	5.0	3.5	3.2	1.8	1.6	1.5	1.0	Flaked on the upper face, partly flaked partly polished on the lower one Distal end is flaked
79	Bx-24	M-83 (50+)	Shaheinab	4.0	4.0	3.5	1.1	1.4	1.6	0.4	Flaked on the upper face, partly flaked partly polished on the lower one Distal end is flaked

Index no.	Museum - shelf no.	Specimen label no.	Site	Maximum Length	Maximum Breadth			Maximum Thickness			Type of Gouge and other comments
					Distal	Middle	Proximal	Distal	Middle	Proximal	
80	no label	no label	Zakiab	5.9	3.6	3.3	0.9	1.2	1.2	0.8	Flaked on the upper face, partly flaked partly polished on the lower one. Flaking and polishing are carefully done. Distal end is flaked.
81	no label	no label	Zakiab	5.1	3.6	3.4	2.1	1.6	1.4	0.4	Flaked on the upper face, partly flaked partly polished on the lower one. Similar to no. 80 in other details
82	no label	no label	Zakiab	6.2	3.5	3.2	1.0	1.4	1.4	0.4	Flaked on the upper face, partly flaked partly polished on the lower one. Similar to no. 79 in other details.
83	no label	90x/100y	Umm Direiwa	5.7	3.1	3.1	1.3	1.3	1.2	0.3	Flaked on the upper face, partly flaked partly polished on the lower one. Similar to nos. 79 and 82 in other details.

Index no.	Museum - shelf no.	Specimen label no.	Site	Maximum Length	Maximum Breadth			Maximum Thickness			Type of Gouge and other comments
					Distal	Middle	Proximal	Distal	Middle	Proximal	
84	no label	65x/56y	Umm Direiwa	5.0	4.1	3.8	1.7	1.8	1.6	0.4	Flaked on the upper face, partly flaked partly polished on the lower one. Distal end is flaked
85	no label	96x/100y	Umm Direiwa	3.8	3.3	3.1	1.3	1.3	1.3	0.4	Flaked on the upper face, partly flaked partly polished on the lower one. Distal end is flaked. It seems like it has been resharpened several times.
86	no label	110x/100y	Umm Direiwa	3.2	2.9	2.6	0.8	1.1	1.1	0.3	Flaked on the upper face, partly flaked partly polished on the lower one Similar to no.79 in other details.
87	no label	98x/110y	Umm Direiwa	4.2	3.2	3.0	1.2	1.3	1.3	0.4	Flaked on the upper face, partly flaked partly polished on the lower one. Similar to no. 86 in other details.

Index no.	Museum - shelf no.	Specimen label no.	Site	Maximum Length	Maximum Breadth			Maximum Thickness			Type of Gouge and other comments
					Distal	Middle	Proximal	Distal	Middle	Proximal	
88	no label	94x/100y	Umm Direiwa.	3.8	3.9	2.4	0.8	1.3	1.3	0.4	Flaked on the upper face, partly flaked partly polished on the lower one Distal end is flaked.
89	no label	104x/72y	Umm Direiwa	2.7	2.8	2.6	0.8	1.9	1.0	0.4	Flaked on the upper face, partly flaked partly polished on the lower one. Distal end is flaked. The tool seem to have been reshaped several times, e.g. as indicated by redution of size.
90	no label	94x/84y	Umm Direiwa	4.0	3.0	2.8	2.6	1.2	1.3	1.1	Flaked on the upper face, partly flaked partly polished on the lower one. Distal end is flaked.
91	no label	130x/104y	Umm Direiwa	4.6	3.5	3.5	2.1	1.1	1.1	1.0	Flaked on the upper face partly flaked partly polished on the lower one Distal end is flaked.

Index no.	Museum - shelf no.	Specimen label no.	Site	Maximum Length	Maximum Breadth			Maximum Thickness			Type of Gouge and other comments
					Distal	Middle	Proximal	Distal	Middle	Proximal	
92	no label	100x/102y	Umm Direiwa	3.2	3.5	3.5	3.5	1.3	1.5	1.3	Flaked on the upper face, partly flaked partly polished on the lower one. Similar to specimens nos. 79 and 86, it seems as it has been re-sharpened several Times.
93	no label	122x/100y	Umm Direiwa	3.9	3.3	3.2	3.0	1.6	1.6	1.6	Flaked on the upper face, partly flaked partly polished on the lower one Similar to no. 92 it seems as it has been resharpened several times.
94	no label	100x/102y	Umm Direiwa	6.0	3.6	3.3	1.3	1.3	1.5	0.4	Flaked on the upper face, partly flaked partly polished on the lower one. Flaking on the partly polished face is carefully done. Distal end is flaked
95											Flaked on the upper face, partly flaked

Index no.	Museum - shelf no.	Specimen label no.	Site	Maximum Length	Maximum Breadth			Maximum Thickness			Type of Gouge and other comments
					Distal	Middle	Proximal	Distal	Middle	Proximal	
95	no label	K-B-103	Kadero	3.5	3.4	3.2	1.8	1.1	1.1	0.6	partly polished on the lower one Distal end is flaked.
96	no label	K-B-101	Kadero	5.1	3.6	3.3	1.6	1.5	1.5	0.3	Flaked on the upper face, partly flaked partly polished on the lower one. Distal end is flaked
97	no label	K-B-102	Kadero	5.1	3.3	2.9	1.1	1.5	1.5	0.3	Flaked on the upper face, partly flaked partly polished on the lower one Distal end is flaked
98	no label	K-C-31	Kadero	5.4	3.5	3.3	1.3	1.3	1.3	0.3	Flaked on the upper face, partly flaked, partly polished on the lower one. Similar to no.94 in other details
99	Bx-23	K-85 (0-20cm)	Shaheinab	7.3	3.4	3.4	1.6	1.3	1.8	1.4	Partly flaked partly polished on both faces Distal end is flaked
100	Bx-24	K-44 (0-20cm)	Kadero	6.2	3.5	3.3	0.8	1.5	1.4	0.4	Similar to no.98 in all details.

APPENDIX VI

Pollination of *Pennisetum* spp. and *Sorghum* spp.

Pennisetum spp.

POLLINATION:

"Sampson (1936) gives a summary based on Indian accounts.Thespike emerges about 10 weeks after sowing.The styles begin to protrude 2-3 days later, first at the top of the inflorescence and moving downwards over a period of 24 hours.They take two days to reach their maximum length,remain fresh for about a day and then begin to dry.After the styles have started to dry up the anthers from the hermaphrodite flowers in the middle of the spike begin to emerge and proceed upwards and downwards.The first flush of anthesis is completed in about two days;a day later the anthers from the staminate flowers emerge and a second flush of anthesis continues for a further two days.The plant is thus markedly protogynous and cross-pollination normally occurs.Where there is overlapping of the female and male phases in the same inflorescence or in different inflorescences on the same plant some self-pollination can occur and seed may be set when the head is bagged.The interval between the female and male phases may be lengthened by lowering the temperature or by growing the plants under irrigation in the dry season.
Grain begins to develop about eight days after the emergence of the spike it is fully developed 20-30 days later,and it takes about 40days from the fertilization to ripening.

GERMINATION:
under favourable conditions the seeds germinate about five days after planting.A dormancy period of several weeks after harvesting have been reported."

Sorghum spp.

POLLINATION:

"The inflorescence begins to flower when the peduncle has completed its elongation.The first flowers to open are at the top of the uppermost panicle branch and opening continues downwards;flowers in the same horizontal plane in the panicle open about the same time.The pedicelled spikelets bloom 2-4 days after the sessile spikelet on the same branch.
Flowering is spread over a period of 6-15 days,but 6-9 days is more usual with the peak blooming about half way through the flowering period.The time of spikelet opening is between 8.0 p.m. and 8.30 a.m.; opening is rapid,caused by swelling of the lodicules;it is complete in about ten minutes, and the stigmas and anthers emerge.The flowers remain open for about two hours.Stigmas are receptive for 1-2 days before blooming, and may remain so for up to a week if not pollinated.Pollen is shed soon after the anthers become pendent and remains viable for 3-6 hours.Pollination may occur from stamens in the same flowers,from other flowers in the same inflorescence,or from other inflorescences;it is by wind.
The average amount of cross pollination is about 5%,but much higher figures are occasionally reported:The degree of cross-pollination is influenced by the wind and the panicle rype,open heads being more liable to crossing than compact ones.Cleistogamy has been observed in some cvs in Ghana.Pollen germinates immediately if falls on the receptive stigma and fertilization takes place about two hours later.

(cont.)

GERMINATION:

Some cvs show seed dormancy for the first month after harvesting.The seed will remain viable for considerable period,provided it is properly stored, but, all too often under poor storage conditions, it is severely attacked by insects which may make it difficult to keep seed between planting-seasons.A laboratory germination of 90 % may be expected to give about 50 % germination in the field. The radicle produced on germination is later replaced by adventitious roots arising from the lowest nodes of the stem. The seedling appear above ground in about 7 days."

(After Purseglove,J.W.,1985: 210-211,274-275)

Bibliography

Abu-Saleem, M.A.
1980 *The Saqia* . Institute of Afro-Asian Studies, University of Khartoum.

Adams, W.Y.
1977 *Nubia. Corridor to Africa.* Allen Lane, Penguin Books Ltd. London.

1982 The 1982 excavations at Qasr Ibrim, Egyptian Nubia. *Nyame Akuma* no.21,pp.20-30.

Adamson, D.A., R. Gillespie and M.A.J. Williams
1982 Palaeogeography of the Gezira and the lower Blue and White Nile Valleys. M.A.J. Williams and Adamson (eds.). *A Land Between Two Niles*. pp.165-220. A.A. Balkema/Rotterdam.

Addison, F.
1949 *Jebel Moya, Volume I & II. The Welcome excavations in the Sudan.* Oxford University Press.

Alexander, J.
1969 The indirect evidence for domestication. P.J. Ucko and G.W. Dimbleby (eds.): *The Domestication and Exploitation of Plants and Animals*. pp.123-129. The Garden City Press, London.

Allan, W.N., and R.J. Smith
1948 Irrigation in the Sudan. J.D. Tothill (ed.): *Agriculture in the Sudan.* pp.593-632. Oxford University Press, London.

Andrew, G.
1948 Geology of the Sudan. J.D. Tothill (ed.): *Agriculture in the Sudan.* pp.84-128. Oxford University Press, London.

Andrews, F.W.
1948 The vegetation of the Sudan. J.D. Tothill (ed.): *Agriculture in the Sudan*. pp.32-61. Oxford University Press, London.

Arkell, A.J.
1937 The Tigda or Reaping Knife in Darfur. *Sudan Notes and Records*, vol.20, part 2. pp.306-307.

1949 *Early Khartoum.* Oxford University Press, London.

1953 *Shaheinab.* Oxford University Press, London.

Arkell, A.J., and P.J. Ucko
1965 Review of Predynastic Development in the Nile Valley. *Current Anthropology*, vol.6, no.2. pp.145-165.

Arnold, C.D.
1947 *An Introduction to Palaeobotany*. New York, London.

Bacon, G.H.
1948 Crops of the Sudan. J.D. Tothill (ed.): *Agriculture in the Sudan.* pp.302-400. Oxford University Press, London.

Barbour, K.M.
1964 *The Republic of the Sudan. A Regional Geography.* Second impression. University of London Press Ltd.

Barich, B.R.
1978 *Neue Ausgrabungen im Acacus-Gebirge. Sahara.* pp.222-245. Gesamtherstellung Greven & Bechtold, Köln.

Bedigian, D.
1984 *Sesamum indicum L. Crop origin, diversity, chemistry, and ethnobotany.* (Unpublished Ph. D.), University of Illinois: Urbana.

Bender, B.
1975 *Farming in Prehistory: From Hunter-Gatherers to Food Producers.* John Baker, London.

Binford, L.R.
1971 Post-Pleistocene Adaptation. S. Struever (ed.): *Prehistoric . Agriculture* pp.22-49. N.H.P. Garden City, New York.

Blackmore, S., and E. Toothill
1984 *The Penguin Dictionary of Botany.* Market House Book Ltd., Aylesbury, England.

Bohrer, V.L., and K.R. Adams
1977 Ethnobotanical Techniques and Approaches at Salmon Ruins, New Mexico. *Eastern Mexico University Contributions in Anthropology Vol.8 no.1.*

Boserup, E.
1974 *The Conditions of Agricultural Growth.* Seventh Reprint, Chicago.

Braidwood, R.J.
1962 The earliest village communities of southwestern Asia reconsidered. *Atti del VI Congresso Internazionale della Scienza Preistoriche e Protostoriche*, vol. 1, pp.115-126. Collegio Romano, Rome.

Bray, W.
1977 From foraging to farming in early Mexico. J.V.S. Megaw (ed.): *Hunters, Gatherers and First Farmers Beyond Europe.* pp.225-250. Leicester University Press.

Burckhardt, J.L.
1822 *Travels in Nubia.* Second edition. London: John Murray, Alberle Street.

Burnett, J.R.
1948 Crop production. J.D. Tothill (ed.): *Agriculture in the Sudan.* pp.275-310. Oxford University Press, London.

Butzer, K.W.
1976 *Early Hydraulic Civilization in Egypt: A Study in Cultural Ecology.* The University of Chicago Press.

Caneva, I.
1983 Raciocarbon Dates from Saggai 1 (Sudan). I. Caneva (ed.): *Pottery Using Gatherers and Hunters at Saggai 1 (Sudan). Preconditions for Food-Production.* pp.149-154. Origini XII, Rome.

1983 Wavy Line Pottery from Saggai 1: An Essay of Classification. I. Caneva (ed.): *Pottery Using Gatherers and Hunters at Saggai (Sudan): Preconditions for Food-Production.* pp.155-190. Origini XII, Rome.

1983 Microlithism and Functionality in the Saggai 1 Industry. I. Caneva (ed.): *Pottery Using Gatherers and Hunters at Saggai (Sudan): Preconditions for Food-Production.* pp.209-233. Origini XII, Rome.

Caton-Thompson, G., and E.W. Gardener
1934 *The Desert Fayum.* The Royal Anthropological Institute of Great Britain and Ireland - Upper Bedford Place.

Chards, C.S.
1975 *Man in Prehistory.* McGraw-Hill, New York.

Clark, J.D.
1944 The use of the Bored Stone in Abyssinia. *Man*, no.25, pp.31-32.

1973 Recent archaeological and geomorphological field studies in the Sudan: Some preliminary results. *Nyame Akuma*, no.3. pp.54-64.

1976 Prehistoric population and pressures favouring plant domestication. J.R. Harlan, J.M.J. deWet and A.B.L. Stemler (eds.): *Origins of African Plant Domestication.* pp.67-105. Mouton Publishers, The Hague, Paris.

1984a Prehistoric cultural continuity and change in the Central Sudan in the early Holocene period. J.D. Clark and S.A. Brandt (eds.): *From Hunters to Farmers.* pp.113-126. University of California Press. Berkely, London.

1984b The domestication process in Northeast Africa: ecological change and adaptive strategies. L. Krzyzaniak and M. Kobusiewicz (eds.): *Origins and Early Developments of Food-Producing Cultures in North-eastern Africa.* pp.25-42. Polish Academy of sciences. Poznan, Poland.

Clark, J.D., J.L. Phillips, and P.S. Staley
1975 *Interpretation of Prehistoric Technology from Ancient Egyptian Bows and Arrows and their Relevance for African Prehistory.*

Clark, J.D., and A.B.L. Stemler
1975 Early domesticated sorghum from Central Sudan. *Nature*, volume 254, no.5501, pp.588-591.

Cohen, M.N.
1977 *The food crisis in prehistory: Overpopulation and the origins of agriculture.* Yale University Press.

Costantini, L.
1983 Bioarchaeological Analysis of soil samples and Potsherds from the site of Saggai 1. I. Caneva (ed.): *Pottery Using Gatherers and Hunters at Saggai (Sudan): Preconditions for Food-Production*. pp.47-49. Origini XII, Rome.

Costantini, L., R. Fattovich, E. Pardini, and M. Piperno
1982 Preliminary report of archaeological investigations at the site of Mahal Teglinos (Kassala). *Nyame Akuma*, no.21, pp.30-33.

Crawford, O.G.S., and F. Addison
1951 *Abu Geili. The Welcome excavations in the Sudan, volume III*. Oxford University Press. London. New York.

de Schlippe, P.
1956 *Shifting Cultivation in Africa: The Zande System of Agriculture*. London.

Dimbleby, G.W.
1967 *Plants and Archaeology*. London.

Doggett, H.
1965 The development of cultivated sorghums. J. Hutchinson (ed.): *Essays on Crop Plant Evolution*. pp.50-69. London.

deWet, J.M.J., and J.P. Huckabay
1967 The origin of Sorghum bicolor. II. Distribution and domestication. *Evolution*, 21, pp.787-802.

deWet, J.M.J., J.R. Harlan, and E.G. Price
1970 Origin and variability in the spontanea complex of Sorghum bicolor. *American Journal of Botany*, 57(6), pp.704-707.

deWet, J.M.J., J.R. Harlan, and E.G. Price
1976 Variability in Sorghum bicolor. J.R. Harlan, J.M.J. deWet and A.B.L. Stemler (eds.): *Origins of African Plant Domestication*. pp.453-464. Mouton Publishers. The Hague, Paris.

Duham, D.,
1955 *Royal Cemeteries of Kush, volume II. Nuri*. Published by the Museum of Fine Arts, Boston, Massachusetts.

1957 *Royal Cemetaries of Kush, volume IV. Royal tombs at Meroe and Barkal*. Published by the Museum of Fine Arts, Boston, Massachusetts.

1963 *Royal Cemeteries of Kush, volume V. The West and South Cemeteries at Meroe*. Published by the Museum of Fine Arts, Boston, Massachusetts.

1970 *The Barkal Temples*. Museum of Fine Arts, Boston, Massachusetts.

El-Mahi, A.T.
1982 *Fauna, Ecology and Socio-Economic Conditions in the Khartoum Nile Environment. Case Study from four Neolithic Sites in Central Sudan.* (Unpublished Ph.D.). University of Bergen.

El-Mahi, A.T. and R. Haaland
1984 Archaeological research in the area of Rabak and Atbara, Sudan. *Nyame Akuma*, no.24/25, pp.28-32.

El-Tayeb, G.
1972 *Forestry and Landuse in the Sudan.* Khartoum University Press.

Fadlalla, B.
1987 *The Principle Vegetation Types in Sudan and their Ecology.* (Manuscript).

Fattovich, R., and M. Piperno
1981 Survey of the Gash Delta: November 1980. *Nyame Akuma*, no.19, pp.26-30.

Flannery, K.V.
1965 The ecology of early food production in Mesopotamia. *Science*, 147, pp.1247-1256.

1969 Origins and ecological effects of early domestication in Iran and the Middle East. P.J. Ucko, and G.W. Dimbleby (eds.): *The Domestication and Exploitation of Plants and Animals.* pp.73-100. The Garden City Press, London.

Fægri, K. and J. Iversen
1975 *Textbook of Pollen Analysis.* Copenhagen, Denmark.

Fægri, K. and L. Van Der Pijl
1979 *The Principles of Pollination Ecology.* 3rd. revised edition. London.

Gould, R.A.
1980 *Living Archaeology.* Cambridge University Press.

Greene, H.
1948 Soils of the Anglo-Egyptian Sudan. J.D. Tothill (ed.): *Agriculture in the Sudan.* pp.144-175. Oxford University Press. London.

Haaland, G.
1968 Economic Determinants in Ethnic Processes. F. Barth (ed.): *Ethnic Groups and Boundaries.* pp.58-73. Universitetsforlaget, Norway.

1982 Problems in Savannah Development. G. Haaland (ed.): *Problems of avannah Development, African Savannah Studies no.19.* pp.1-38. University of Bergen.

Haaland, R.
1981 *Migratory Herdsmen and Cultivating Women. The Structure of Neolithic Seasonal Adaptation in the Khartoum Nile Environment.* University of Bergen.

1984 Continuity and Discontinuity. How to Account for Two Thousand Years Gap in the History of the Khartoum Nile Environment. *Norwegian Archaeological Review*, vol. 17, no. 1. pp.39-51.

1987a Socio-economic Differentiation in the Neolithic Sudan. *Cambridge Monographs in African Archaeology 20. BAR International Series 350.*

1987b Problems in the Mesolithic and Neolithic Culture History in the Central Nile Valley, Sudan. T. Hagg (ed.): *Nubian Culture Past and Present.* pp.47-74. Almqvist & Wiksell International, Stockholm, Sweden.

Harlan, J.R.
1977 The origins of cereal agriculture in the old world. A.C. Reed (ed.): *Origins of Agriculture.* pp.357-384. Mouton Publishers. The Hague, Paris.

Harlan, J.R., and J. Pasquereau
1969 Décure Agriculture in Mali. *Economic Botany*, 23. pp.70-74.

Harlan, J.R., and J.M.J. deWet
1971 Toward a rational classification of cultivated plants. *Taxon.* vol.20, pp.509-517.

Harlan, J.R., and J.M.J. deWet
1972 A simplified classification of cultivated Sorghum. *Crop science*, vol.12 (March/April) pp.172-176.

Harlan, J.R., J.M.J. deWet, and E.G. Price
1973 Comparative evolution of cereals. *Evolution* 27, pp.311-325.

Harlan, J.R., J.M.J. deWet, and A.B.L. Stemler
1976 Plant domestication and indigenous African agriculture. J.R. Harlan, J.M.J. deWet, and A.B.L. Stemler (eds.): *Origins of African plant domestication.* pp.3-19. Mouton Publishers. The Hague, Paris.

Harlan, J.R., J.M.J. deWet, and A.B.L. Stemler
1976 A foot note. J.R. Harlan, J.M.J. deWet, and A.B.L. Stemler (eds.): *Origins of African Plant Domestication.* p.434. Mouton Publishers. The Hague, Paris.

Harlan, J.R., and A.B.L. Stemler
1976 The Races of Sorghum in Africa. J.R. Harlan, J.M.J. deWet, and A.B.L. Stemler (eds.): *Origins of African Plant Domestication.* pp.465-478. Mouton Publishers. The Hague, Paris.

Harlan, J.R. and D. Zohary
1966 Distribution of wild wheat and barley. *Science*, 153, pp.1075-1080.

Harris, D.R.
1976 Traditional Systems of Food Production and the Origins of Agriculture in West Africa. J.R. Harlan, J.M.J. deWet, and A.B.L. Stemler (eds.): *Origins of African Plant Domestication*. pp.311-356. Mouton Publishers. The Hague, Paris.

Hassan, F.A.
1981 *Demographic Archaeology*. Academic Press. New York, London, Paris.

1986 Ancient Nubian Culture: Classification and Interpretation. T. Hagg (ed.): *Nubian Culture Past and Present*. pp.1-36. Almqvist & Wiksell International. Stockholm, Sweden.

Hawkes, J.G.
1969 The ecological background of plant domestication. P.J. Ucko, and G.W. Dimbleby (eds.): *The Domestication and Exploitation of Plants and Animals*. pp.17-30. The Garden City Press. London.

Helbaek, H.
1955B The Botany of the Valhagar Iron Age Field. M. Stenberger (ed.): *Valhagar, A Migration Period Site in Gotland, Sweden*. Stockholm.

1959 Domestication of food plants in the Old World. *Science*, vol. 130, no.3372 (July-December). pp.365-374. Washington 5, D.C.

1969 Plant collecting, dry farming and irrigation agriculture in prehistoric Deh Luran. F. Hole, K. Flannery, and J.A. Neely (eds.): *Prehistory and Human Ecology of the Deh Luran Plains*. pp.383-424. University of Michigan.

Herzog, R.
1957 *Die Nubier*. Akademie-Verlag-Berlin.

Heslop-Harrison, J.
1973 *New concepts in flowering-plant taxonomy*. (reprint) Heinemann (H.E.B.). London.

Higgs, E.
1976 Archaeology and Domestication. J.R. Harlan, J.M.J. deWet, and A.B.L. Stemler (eds.): *Origins of African Plant Domestication*. pp.29-39. Mouton Publishers. The Hague, Paris.

Hornby, A.S., E.V. Gatenby and H. Wakefield
1971 *Advanced Dictionary of Current English*. Second edition. Oxford University Press.

Ibrahim, F.N.
1984 *Ecological Imbalance in the Republic of the Sudan - with reference to Desertification in Darfur*. Volume 6. Bayreuth.

Ireland, A.W.
1948 The Climate of the Sudan. J. Tothill (ed.): *Agriculture in the Sudan*. pp.62-83.

Jarman, M.N., A.J. Legge
and J.A. Charles
1972 Retrieval of Plant Remains from Archaeological Sites by Froth Cambridge: The University Press.

Khabir, A.M.
1987 New radiocarbon dates for Sarurab 2 and the age of the Early Khartoum tradition. *Current Anthropology*. Volume 28, number 3. pp.377-380.

Klichowska, M.
1978 Preliminary results of palaeoethnobotanical studies on plant impressions on potsherds from the Neolithic settlement at Kadero. *Nyame Akuma*, no.12. pp.42-43.

Krzywinski, K. and
E.C. Soltvedt
1988 A Medieval Brewery (1200-1450) at Bryggen,Bregen.*The Bryggen Papers Supplementary Series No.3*. .pp.1-68.Norwegian University Press

Krzyzaniak, L.
1978 New light on early food production in the central Sudan. *Journal of African History* XIX, 2, pp.159-172.

1979 Polish excavations at Kadero. *Nyame Akuma*, no.15. pp.67-69.

Lang-Anderson, R.
1908 *Text Book of Egyptian Agriculture*. Cairo.

Lebon, J.H.G.
1965 Land Use in Sudan. Sir D. Stamp (ed.): *The World Land Use Survey*. Monograph No.4. London.

Lee, R.B.
1968 What hunters do for a living, or how to make out on scarce resources. R.B. Lee and I. deVore (eds.): *Man the Hunter*. pp.30-48. Aldine Publishing Company, Chicago.

1969 !Kung Bushmen Subsistence-an input-output analysis.A.P.Vayda (ed.):*Environment and Cultural Behaviour*. pp.47-79.Natural History Press, New-York.

MacFall, R.P., and
J.C. Wollin
1983 *Fossils for Amateurs*. Van Nostrand Reinhold Company, Second revised edition. New York, London.

Magid, A.
1981 Archaeological excavations on the west bank of the river Nile in the Khartoum area. *Nyame Akuma*, no.18. pp.42-45.

1982 *The Khartoum neolithic in the light of archaeoethnobotany. A case study from the Nofalab and the Islang sites*. Unpublished M.A., University of Khartoum, Sudan.

1984 Macrobotanical remains from Shaqadud: Interim Note. *Nyame Akuma*, no.24/25. pp.27-28

in press The exploitation of plants in the Eastern Sahel (Sudan): c.5000-2000 B.C. in the *Proceeding of the 1984 conference on Food-Production in North east Africa*, held in Poznan, Poland.

Magid,A., and K.,Krzywinski
1988 Extraction of Pollen from Pottery: Techniue and objectives S.Indrelid, S.,Kaland andB.Solberg(eds.):*Arkeologiske Skrifter No.4* .pp.196 - 204. Historisk Museum,Universitetet i Bergen.

Manger, L.O.
1982 Cultivation Systems and the Struggle for Household Viability under Conditions of Desert Encroachment. G. Haaland (ed.): *Problems of Savannah Development, The Sudan Case*. Bergen Occasional Papers in Social Anthropology, No.19. pp.133-166. Bergen.

1983 *The Sand Swallows Our Land*. Bergen Occasional Papers in Social Anthropology, No.24. Bergen.

Marks, A.E.
1984 Butana project 1983-84. *Nyame Akuma*, no.24/25. pp.32-33.

Marks, A.E., A. Mohammed-Ali, T.R. Hays, and Y.M. Elamin
1980 Survey of Northern Butana. *Nyame Akuma*, no.16. pp.30-35.

Marks, A.E., A.Mohammed-Ali, T.R. Hays, and Y.M. Elamin
1982 Butana Archaeological Project: 1981 Field Work. *Nyame Akuma*, no.20. pp.47-50.

Marks, A.E., A. Mohammed-Ali, J. Peters, and R. Robertson
1985 The Prehistory of the Central Nile Valley as Seen from Its Eastern *Archaeology*, vol.12. pp.261-277.

Marshall, L.
1976 *The ! Kung of Nyae Nyae*. Harvard University Press,Cambridge, Massachusetts and London.

Martin, A.C., and W.D. Barkley
· 1973 *Seed Identification Manual*. University of California Press. Berkeley, Los Angeles & London.

McCown, R.L., G. Haaland, and C. de Haan
1979 The Interaction Between Cultivation and Livestock Production in Semi-Arid Africa. Reprint from A.E. Hall, H. Cannell & H.W. Lawton eds.): *Ecological Studies, Volume 34, Agriculture in Semi-Arid Environments*. pp.297-332. Berlin, New York.

Meyers, J.T.
1971 The Origins of Agriculture: An evaluation of three hypotheses. S. Struever (ed.): *Prehistoric Agriculture*. pp.101-121. Garden City, New York.

Mohammed-Ali, A.S.
1982 The Neolithic Period in the Sudan c. 6000-2500 B.C. *Cambridge Monographs in African Archaeology 6, BAR International Series 139.* Oxford OX2, England.

1984 Early Food-production in Northeast Africa. Krzyzaniak, L., and M. Kobusiewicz (eds.): *Origins and Early Development of Food-producing Cultures in North-eastern Africa.* pp.65-72. Polish Academy of Sciences, Poznan, Poland.

Mubarak, M.O., E.A. Bari, G.E. Wickens, and M.A.J. Williams
1982 The vegetation of the central Sudan. M.A.J. Williams, and D.A. Adamson (eds.): *A Land Between Two Niles.* pp.143-164. A.A. Balkema/Rotterdam.

Nelson, H.D.
1982 Country Profile. H.D. Nelson (ed.): *Sudan: a country study.* pp. xiii-xxi. Washington D.C.

Netting, R. McC.
1968 *Hill Farmers of Nigeria, Cultural Ecology of the Kofyar of the Jos Plateau.* University of Washington Press. Seattle and London.

Newman, J.L.
1970 *The Ecological Basis for Subsistence Among the Sandawe of Tanzania. Natural Academy of Science.* Washington.

Nicolaisen, J.
1963 *Ecology and Culture of the Pastoral Tuareg.* The National Museum of Copenhagen.

Obeid, M., and A. Mahmoud
1969 The vegetation of Khartoum Province. *Sudan Notes and Records,* vol.50. pp.134-158. Khartoum.

O'Keefe, P., and B. Wisner
1977 Landuse and Development. *African Environment Special Report* 5. IAI. London.

Pfeiffer, J.E.
· 1976 A Note on the Problem of Basic Causes. J.R. Harlan, J.W.J. deWet, and A.B.L. Stemler (eds.): *Origins of African Plant Domestication.* pp.23-28. Mouton Publishers. The Hague, Paris.

Phillipson, D.W.
1977 *The prehistory of eastern and southern Africa.* Heinemann Educational Books Ltd. London WIX 8AH.

Portères, R.
1976 African Cereals: Eleusine, Fonio, Black Fonio, Teff;Brachiaria, and A.B.L. Stemler (eds.): *Origins of African Plant Domestication.* pp.405-452. Mouton Publishers. The Hague, Paris.

Purseglove, J.W.
1985 *Tropical Crops: Monocotyledons.* Longman Group Ltd. Harlow, Essex, England.

Reed, C.A.
1969 The patterns of animal domestication in the prehistoric Near East. P.J. Ucko, and G.W. Dimbleby (eds.): *The Domestication and Exploitation of Plants and Animals*. pp.361-380. The Garden City Press. London WC2.

Renfrew, J.M.
1973 *Palaeoethnobotany. The Prehistoric Food Plants of the Near East and Europe*. Methuen and Co. Ltd.

Rinehart, R.
1982 Historical Setting. H.D. Nelson (ed.): *Sudan: a country study*. p.88. Washington D.C.

Rowley-Conwy, P.
1986 Archaeological evidence for crop-plants in Lower Nubia, c. 700 B.C. - A.D. 1800: Some preliminary results. R. Dehlin and T. Hägg (eds.): *Sixth International Conference for Nubian Studies, Abstracts of Communications*. pp.110-111. Klassisk Institutt, Universitetet i Bergen.

Sahlins, M.
1972 *Stone age economics*. Aldine-Atherton. Chicago.

Save-Söderbergh, T.
1982 *SJE Volume 6. The Late Nubian Cemeteries*. Berlings, Arlov - Sweden.

Schery, R.W.
1954 *Plants for man*. London W.C.1. Great Britain.

Sebbins, G.L.
1971 *Chromosomal evolution of higher-plants*. Edward Arnold (publishers) Ltd. London.

Semenov, S.A.
1970 *Prehistoric Technology. An experimental study of the oldest tools and artefacts from traces of manufacture and wear*. Second English impression. Bath: Adams & Dart. London.

Shaw, T.
1977 Hunters, gatherers and first farmers in West Africa. J.V.C. Megaw (ed.): *Hunters, Gatherers and First Farmers Beyond Europe*. pp.69-125. Leicester University Press.

1978 *Nigeria*. London: Thames and Hudson.

Shinnie, P.L.
1967 *Meroe: A Civilization of Sudan*. T. and H. London.

1971 The Sudan. P.L. Shinnie (ed.): *The African Iron Age*. pp.89-107. Clarendon Press. Oxford, London.

1985 Iron Working at Meroe. R. Haaland and P. Shinnie (eds.): *African Iron Working - Ancient and Traditional*. pp.28-36. Norwegian University Press. Norway.

Shinnie, P.L., and
R.J. Bradley
1980 *The capital of Kush. I. Meroe excavations 1965-1872.* Akademie - Verlag - Berlin.

Smith, A.B.
1980 Domesticated cattle in the Sahara and their introduction into west Africa. *Quaternary Environments and Prehistoric Occupation in Northern Africa.* pp.489-501. A.A. Balkema/Rotterdam.

Snowden, J.D.
1936 *The cultivated races of Sorghum.* London: Allard and son.

Stemler, A.B.L.
1980 Origins of plant domestication in the Sahara and the Nile Valley. M.A.J. Williams and H. Faure (eds.): *The Sahara and the Nile: Quaternary Environment and Prehistoric Occupation in Northern Africa.* pp.503-526. A.A. Balkema/Rotterdam.

1984 The transition from food collecting to food production in northern Africa. J.D. Clark and S.A. Brandt (eds.): *From Hunters to Farmers.* pp.127-131. University of California Press. Berkeley.

Stemler, A.B.L., and
R.H. Falk
1981 SEM of Archaeological Plant Specimens. *Scanning Electron Microscopy, III.* pp.191-196. SEM Inc.

Steward, H.J.
1938 Basin-plateau aboriginal socio-political groups. *The Smithsonian Institution Bureau of American Ethnology, Bulletin 120.*

Sutton, J.E.G.
1974 The aquatic civilization of middle Africa. *Journal of African History* 15. pp.527-546.

Swartz, D.
1971 *Collegiate Dictionary of Botany.* The Ronald Press Company. New York.

Tothill, J.D.
1948 A note on the origins of the soils of the Sudan from the point of view of the man in the field. J.D. Tothill (ed.): *Agriculture in the Sudan.* pp.129-143. Oxford University Press. London.

Tracey, C.B., and
J.W. Hewison
1948 Northern Province. J.D. Tothill (ed.): *Agriculture in the Sudan.* pp.736-761. Oxford University Press. London.

Trigger, B.G.
1965 *History and Settlement in Lower Nubia.* Yale University.

Tubiana, M.-J., and
J. Tubiana
1977 *The Zaghawa from an Ecological Perspective.* A.A. Balkema/ Rotterdam.

Vail, J.R.
1982 Geology of the Central Sudan. M.A.J. Williams & D.A. Adamson (eds.): *A Land Between Two Niles*. pp.51-62. A.A. Balkema/ Rotterdam.

Washburn, S.L., and C.S. Lancaster
1975 The Evolution of Hunting. I. de Vore and R.B. Lee (eds.): *Man the Hunter*. pp.293-303. Aldine Publishing Company. Chicago. (Fifth printing. Origin al 1968).

Wendorf, F.
1968 *The Prehistory of Nubia*, vol. I & II. Southern Methodist University. Dallas.

Wendorf, F., and R. Schild
1976 Ground Grain Use in the Late Palaeolithic of the Lower Nile Valley. J.R. Harlan, J.M.J. deWet and A.B.L. Stemler (eds.): *Origins of African Plant Domestication*. pp.269-288. The Hague, Paris.

Whitaker, D.P.
1983 The economy. H.D. Nelson (ed.): *Sudan: a country study*. pp.139-196. Third edition (first printing). The American University. Washington D.C.

Whiteman, A.J.
1971 *The Geology of the Sudan Republic*. Oxford University Press. London.

Wickens, G.E.
1975 Changes in the Climate and Vegetation in the Sudan 20000 B.P. *Boissiera 24*, pp.43-65.

Wickens, G.E.
1982 Palaeobotanical Speculations and Quaternary Environments in the Sudan. M.A.J. Williams & D.A. Adamson (eds.): *A Land Between Two Niles*. pp.23-50. A.A. Balkema/Rotterdam.

Williams, M.A.J., D.A. Adamson, and H.H. Abdulla
1982 Landforms and Soils of the Gezira: A. Quaternary Legacy of the Blue and White Nile rivers. M.A.J. Williams and D.A. Adamson (eds.): *A Land Between Two Niles*. pp.111-142. A.A. Balkema/Rotterdam.

Woodburn, J.
1970 *Hunters and gatherers. The material culture of the nomadic Hadza*. London. The British Museum.

Yarnell, R.A.
1971 Early Woodland Plant Remains and the Question of Cultivation. S. Struever (ed.): *Prehistoric Agriculture*. pp.550-556. N.H.P. Garden City. New York.

Zarattini, A.
1983 Ground Stone Implements from Saggai 1. I. Caneva (ed.): *Pottery Using Gatherers and Hunters at Saggai (Sudan): Preconditions for Food-Production*. pp.234-242. Origini XII. Rome.

CAMBRIDGE MONOGRAPHS IN AFRICAN ARCHAEOLOGY

General Editor

John Alexander, M.A., Ph.D., F.S.A.

St. John's College
Cambridge CB2 1TP, England

Manuscripts should be submitted to
Dr. Alexander in the first instance.

Titles already published:

1. 'The Niger Delta: Aspects of its Prehistoric Economy and Culture' by Nwanna Nzewunwa. BAR-S75.

2. 'Prehistoric Investigations in the Region of Jenne, Mali' by Susan Keech McIntosh and Roderick J. McIntosh. BAR-S89.

3. 'Off-Site Archaeology and Human Adaptation in Eastern Africa: An Analysis of Regional Artefact Density in the Amboseli, Southern Kenya' by Robert Foley. BAR-S97.

4. 'Later Pleistocene Cultural Adaptations in Sudanese Nubia' by Yousif Mukhtar el Amin. BAR-S114.

5. 'Settlement Patterns in the Iron Age of Zululand: An Ecological Interpretation' by Martin Hall. BAR-S119.

6. 'The Neolithic Period in the Sudan, c. 6000 - 2500 B.C.' by Abbas S. Mohammed-Ali. BAR-S139.

7. 'History and Ethnoarchaeology in Eastern Nigeria: A Study of Igbo-Igala relations with special reference to the Anambra Valley' by Philip Adigwe Oguagha and Alex Ikechukwu Okpoko. BAR-S195.

8. 'Meroitic Settlement in the Central Sudan: An Analysis of Sites in the Nile Valley and the Western Butana' by Khidir Abdelkarim Ahmed. BAR-S197.

9. 'Economy and Technology in the Late Stone Age of Southern Natal' by Charles Cable. BAR-S201.

10. 'Frontiers: Southern African Archaeology Today' ed. M. Hall, G. Avery, D.M. Avery, M.L. Wilson and A.J.B. Humphreys. BAR-S207.

11. 'Archaeology and History in Southern Nigeria: The ancient linear earthworks of Benin and Ishan' by P.J. Darling. BAR-S215.

12. 'The Latest Stone Age of Southernmost Africa' by Janette Deacon. BAR-S213.

13. 'Fisher-Hunters and Neolithic Pastoralists in East Turkana, Kenya' by John Webster Barthelme. BAR-S254.

14. 'The Archaeology of Central Darfur (Sudan) in the 1st Millennium A.D.' by Ibrahim Musa Mohammed. BAR-S285.

15. 'Stable Carbon Isotopes and Prehistoric Diets in the South-Western Cape Province, South Africa' by Judith Sealy. BAR-S293.

16. 'L'art rupestre préhistorique des massifs centraux sahariens' by Alfred Muzzolini. BAR-S318.

17. 'Spheroids and Battered Stones in the African Early and Middle Stone Age' by Pamela R. Willoughby. BAR-S321.

18. 'The Royal Crowns of Kush' by L. Török. BAR-S338.

19. 'The Later Stone Age of the Drakensberg Range and its foothills' by H. Opperman. BAR-S339.

20. 'Socio-Economic Differentiation in the Neolithic Sudan' by Randi Haaland. BAR-S350.

21. 'Later Stone Age Settlement Patterns in the Sandveld of the South-Western Cape Province, South Africa' by Anthony Manhire. BAR-S351.

22. 'L'art rupestre du Fezzan septentrional (Libye) by Jean-Loïc le Quellec. BAR-S365.

23. 'Archaeology and Environment in the Libyan Sahara' edited by Barbara E. Barich. BAR-S368.

24. 'The Early Farmers of Transkei, Southern Africa' by J.M. Freely. BAR-S378.

25. 'Later Stone Age Hunters and Gatherers of the Southern Transvaal' by Lyn Wadley. BAR-S380.

26. 'Prehistoric Cultures and Environments in the Late Quaternary of Africa' edited by John Bower and David Lubell. BAR-S405.

27. 'Zooarchaeology in the Middle Nile Valley: A Study of Four Neolithic Sites near Khartoum' by Ali Tigani ElMahi. BAR-S418.

28. 'L'Ancienne Métallurgie du Fer à Madagascar' by Chantal Radimilahy. BAR-S422.

29. 'El Geili: The History of a Middle Nile Environment, 7000 B.C. - A.D. 1500' edited by I. Caneva. BAR-S424.

30. 'The Ethnoarchaeology of the Zaghawa of Darfur (Sudan)' by Natalie Tobert. BAR-S445.

31. 'Shellfish in Prehistoric Diet' by W.F. Buchanan. BAR-S455.

32. 'Houlouf I: Archéologie des sociétés protohistoriques du Nord-Cameroun' by Augustin Holl. BAR-S456.

33. 'The Predynastic Lithic Industries of Upper Egypt' by Diana L. Holmes. BAR-S469.

www.ingramcontent.com/pod-product-compliance
Ingram Content Group UK Ltd.
Pitfield, Milton Keynes, MK11 3LW, UK
UKHW061844041026
15591UKWH00015B/904

9 780860 546641